T0100703

Pervasive Computing in Healthcare

Pervasive Computing in Healthcare

Edited by
Jakob E. Bardram
Alex Mihailidis
Dadong Wan

CRC Press
Taylor & Francis Group
Boca Raton London New York

CRC Press is an imprint of the
Taylor & Francis Group, an **informa** business

CRC Press
Taylor & Francis Group
6000 Broken Sound Parkway NW, Suite 300
Boca Raton, FL 33487-2742

First issued in paperback 2019

ISBN-13: 978-0-8493-3621-8 (hbk)
ISBN-13: 978-0-367-38988-8 (pbk)

Library of Congress Cataloging-in-Publication Data

Pervasive computing in healthcare / edited by Jakob E. Bardram, Alex Mihailidis, and Dadong Wan.
    p. ; cm.
Includes bibliographical references and index.
ISBN-13: 978-0-8493-3621-8 (hardcover : alk. paper)
ISBN-10: 0-8493-3621-X (hardcover : alk. paper)
    1. Medical informatics. 2. Ubiquitous computing. 3. Medical telematics. 4. Patient self-monitoring. 5. Biosensors.   I. Bardram, Jakob E. II. Mihailidis, Alex. III. Wan, Dadong.
    [DNLM: 1. Biomedical Technology--instrumentation. 2. Computers. 3. Biosensing Techniques--instrumentation. 4. Delivery of Health Care--organization & administration. 5. Information Systems--instrumentation. 6. Telemedicine--instrumentation.   W 26.55.C7 P471 2007]

R859.7.U27P47 2007
610.28--dc22   2006020620

Foreword

Pervasive computing has become a very fashionable area for computer science, and many researchers are making much progress in the field. The size of the healthcare sector in the developed world in particular means that there are potentially vast markets for new technologies. It is almost inevitable that researchers, developers, and manufacturers will investigate the potential of pervasive computing in healthcare. There are, however, some dangers when computer scientists and engineers blindly jump on this particular bandwagon. The major danger—which has sadly been seen all too often—is that of technology being the driving force with little or no consideration of the healthcare sector's real needs and wants. Technological push can work in some sectors (personal computers and mobile telephones are two examples), but the healthcare sector has substantially different characteristics from more traditional business and domestic marketplaces.

For example, users of healthcare technology fall into two groups: patients or clients, and healthcare professionals. The former are very likely to be ill, disabled, or old, whereas professionals may have to use equipment in busy situations, in addition to being under significant stress. Both groups may have to operate equipment in hostile or unusual environments such as operating theaters, intensive care wards, and car accident sites. Even domestic environments, which have been modified to cope with an ill or disabled person, provide nonstandard situations. Similarly, a greater range of clothing (hospital gowns, nightwear, and protective clothing) may be worn than in more traditional situations. Further challenges include the need to protect equipment from a range of bodily fluids and chemicals while not compromising the requirement for a sterile environment.

Thus, in healthcare, neither the users nor the environments will be ordinary. As part of my research into design for computer systems to support older and disabled people, I have suggested the concept of "ordinary and extraordinary human computer interaction." This draws parallels between the computing challenges of ordinary (able-bodied) people operating in an extraordinary (high work load, stress, environmentally extreme) situation and extraordinary (disabled) people operating in an ordinary (office) environment.[1] The healthcare situation, however, also includes extraordinary people operating in an extraordinary environment. It thus provides very substantial challenges for designers and engineers, but these challenges are well worth tackling because the potential rewards of relieving suffering are very great.

If pervasive computing is going to be successfully applied to healthcare, it is vital that design and development teams fully understand the users and the environments in which their systems are to be deployed. Both of these are likely to be alien to design teams. Even if designers have been temporarily disabled or hospitalized at some stage in their lives, they are unlikely to have concentrated on the pervasive computing design challenges at the time! It is thus absolutely vital that they become aware of not only the potential of pervasive computing technology but also the sensory, motor, and cognitive characteristics of the users for whom they are designing and the environments in which users are likely to use the equipment. It is only after designers have absorbed this background information that they are in a position to investigate the real needs and wants of the healthcare sector.

The design and development process itself needs an understanding of human factor issues, but it is also important for the research team to contain healthcare professionals and, where possible, patients and clients. There are a number of ethical issues in the use of patients and clients, however, particularly if they are minors or cannot give informed consent. My group suggests the concept of "mutual inspiration," whereby both computing professionals and potential users work together to produce a solution, and we also address how to facilitate interaction with users by using a range of techniques including theater work.[2]

There can be substantial ethical issues in research in this field, and it will often be necessary for the research and development process to have the approval of an ethics committee. In addition there will be ethical issues of privacy and consent with the introduction of pervasive technology and these need to be thought through at an early stage of the design process.

The final hurdle for the healthcare sector is providing appropriate evidence that one's equipment is effective, but there are significant challenges for showing the effectiveness of pervasive computing. Because of the dominance of pharmaceutical interventions, the norm for clinical evaluations is to have large "n" randomized double-blind studies that include a control group. These are often impractical or impossible for systems involving pervasive computing. (How do you fool a user into not using a PDA or conceal this from an observer?) Developers of technology not only need to consider appropriate ways to evaluate their systems, but they also must have a robust defense of their methodology.

I have outlined the various aspects of changes presented to developers of pervasive computing for healthcare. A team that develops pervasive computing systems for healthcare needs a wide range of background knowledge. This book provides an excellent grounding in the various aspects of the process that need to be considered; each chapter is written by authors who have long and distinguished careers in this field. I recommend this book to all those venturing down this exciting and rewarding path.

—Alan F. Newell, Ph.D., MBE, FRSE

References

1. Newell, A.F. and Gregor, P. Human computer interfaces for people with disabilities. In Helander, M., Landauer, T.K., and Prabhu, P., editors, *Handbook of Human-Computer Interaction*. Amsterdam: Elsevier, 1997, pp. 813–824.
2. Newell, A.F. Older people as a focus for inclusive design. *Gerontechnology* 4:4 (March 2006), 190–199.

Acknowledgments

We would like to first and foremost acknowledge the tireless efforts of each of our contributors and authors, without whom this book would not exist. The chapters reflect the wide range of skill, expertise, and knowledge of each of our authors, and they were selected to be a part of this volume because they represent the leading edge in the fields of computer science, healthcare, and pervasive computing.

Furthermore, we would like to thank all of our reviewers who spent many hours providing us with thoughtful and insightful comments and feedback.

A special thank you goes to Jennifer Boger at the University of Toronto, who dedicated much time to assisting the editors in reviewing all of the chapters and compiling the final volume. We are indebted to her service.

Finally, we would like to thank those at CRC Press who gave us the opportunity to pursue this project. Without their confidence, patience, and support we would have not been able to achieve our goal.

—**Jakob, Alex, and Dadong**

Introduction

Why pervasive healthcare?

Most countries will face the same set of healthcare challenges in upcoming years: skyrocketing costs needed for caring for an increased number of elders; a rapid increase in lifestyle-related and chronic diseases; a demand for new medical treatments and technologies; and a shortage in the number of available clinicians, nurses, and other caregivers. The statistics and numbers are overwhelming. To name but one example, medical care for people with chronic diseases such as arthritis, asthma, cancer, diabetes, and heart disease accounts for more than 75 percent of healthcare expenditures in the United States.[1] These challenges are evident in most industrialized countries, whether a country has a publicly funded welfare and healthcare system as in Canada and Europe or whether it is based on insurance as in the United States. Furthermore, these challenges will affect all parties involved in the healthcare system, including patients and their families, employers, governments, clinicians, caregivers, and administrators. The current healthcare model can no longer meet these challenges simply because it is poorly suited to serving an increasing number of chronically ill and elderly people. The current Western healthcare model, which is concentrated around highly specialized and centralized experts located in large hospitals focusing on acute care, can be compared to a centralized data processing model evident in a mainframe computer.[2] This mainframe healthcare model needs to be transformed into a more distributed and highly responsive "healthcare" processing model, where locally available and distributed systems can help empower patients to manage their own health in the form of wellness management, preventive care, and proactive intervention. A distributed healthcare model that pervades the everyday lives of citizens is much more suited to managing the kind of lifestyle and chronic disease challenges underlying the current healthcare system.

The goal of *pervasive healthcare* is to *enable* this distributed and pervasive care model for health and wellness management through the use of information and communication technology. In this respect and for the purpose of this book, we define pervasive healthcare in two ways. First, it is the application of pervasive computing (or ubiquitous computing, proactive computing, ambient intelligence) technologies for healthcare, health, and wellness management. Second, it is about making healthcare available every-where, anytime, and to anyone. In essence, pervasive healthcare addresses a set of related technologies and concepts that help integrate healthcare more seamlessly to our everyday lives, regardless of space and time.[3] Mark Weiser, the father of ubiquitous computing, stated that "the most profound technologies are those that disappear."[4] In that sense, pervasive computing may be con-sidered as the opposite to virtual reality. While in virtual reality the user enters the world created by computers, in pervasive computing it is the computing that enters the physical world and bridges the gap between the virtual and physical worlds. This bridging is perhaps best described by three important enabling technologies: ubiquitous computing, ubiquitous commu-nication, and intelligent user-friendly interfaces.

Ubiquitous computing refers to the integration of computing power (microprocessors) and sensing (sensors) into anything, including not only traditional computers, personal digital assistants, and printers, but also everyday objects and environments. Ubiquitous communication means enabling anytime and anywhere the communication of anything with any-thing else, not only between people but also between the objects of which the computing is part. Important ubiquitous communication technologies include ad hoc networking and wireless communication, such as low-power, short-range networks. Intelligent user-friendly interfaces enable natural interaction and control of the environment by the users, or inhabitants of the ambient environment. The interfaces support natural communication (e.g., speech, gestures) and multimodal interactions, which take into account user preferences, personalities, and usage context. The envisioned pervasive computing infrastructure provides a seamless environment of computing, networking, and user interfaces. It is context aware in that it has senses and the required intelligence to interpret sensory information and make reason-able decisions when taking actions.

One of the most important applications for pervasive computing technologies is healthcare, including wellness and disease management, and support for independent and assisted living. For example, developments in sensors, and more generally measurement technology, make it possible to obtain physiological data from wearable or embedded sensors. Ubiquitous communication based on mobile phone networks, WiFi, and other wireless technologies makes it possible to deliver and access data including measure-ments, person-to-person communications, and health information anywhere and anytime. Mobile devices provide ubiquitous user interfaces for users ranging from healthcare professionals to average citizens. We have just begun to unleash the vast potential that this technology has to offer for healthcare delivery. In addition to health monitoring, pervasive healthcare also has great potential in social computing. For example, these technologies can be used by relatives, family members, and peers of chronically ill persons to remain in contact no matter where people are physically located. Furthermore, pervasive healthcare also helps patients to manage their own diseases better and helps healthcare professionals to communicate and collaborate.

A historical perspective

Technology has always played a central role in medical diagnosis, treatment, follow-up, monitoring, and prevention. *Medical engineering* has been central to medical progress, and most medical diagnoses and treatments are only possible through the use of quite advanced technology. These technologies range from the simple stethoscope to advanced scanning techniques, such as x-rays and magnetic resonance imaging (MRI). In parallel, *medical informatics* covers the discipline of using computers for medical purposes. Medical informatics is, however, primarily concerned with *informatics*, which is the use of information technology for storing, managing, and accessing medical data and information. This research is traditionally built on more general information systems (IS) research that studies the use of computers for information processing in large organizations. Telemedicine, which can be viewed as a subdiscipline of medical informatics, uses communication technology for establishing medical consultations and conferences over distances. Common to medical informatics and telemedicine is that these research approaches have focused on the use of computers by medical professionals for electronic record keeping and for conducting *indirect* medical treatment over distances. As such, there has been little focus on the use of computer technology for direct patient treatment, where patients play active roles in their healthcare management.

Pervasive healthcare takes a slightly different approach by focusing much more explicitly on the use of new pervasive computing technology for patient self-treatment and self-care (i.e., empowering the patient to take a more active role in managing and treating disease). In particular, pervasive healthcare pays special attention to the tools and services that put the patient at the center of the healthcare process. This includes support for patient self-management, self-care, preventive efforts, cooperation between the patient and the healthcare institutions, cooperation between home and hospital, self-monitoring, remote monitoring, remote consultation, and assistive technologies.

More recently, there has been growing interest in the research and development of healthcare-related technologies from the more traditional application-oriented disciplines of computer science such as human–computer interaction (HCI), computer-supported cooperative work (CSCW), and the emerging research community in pervasive and ubiquitous computing, including wireless sensor networks for healthcare (e.g., body sensor networks). A particularly interesting observation is that large computing technology companies such as Intel, IBM, Microsoft, Nokia, Cisco, Samsung, and Sony are currently investing a huge amount of resources to researching and developing technologies and applications for healthcare. While these companies traditionally supply basic computer and communication hardware and software, they now see healthcare, especially patient-centered healthcare, as one of the biggest application areas for computer technology. We believe that pervasive computing holds great potential for transforming the business of healthcare by helping improve the quality of care while reducing overall healthcare costs. This new model of care bears significant economic consequences and will become especially relevant to the chronically ill and the elderly, who are the heaviest users of healthcare services.

Pervasive healthcare—A multidisciplinary research agenda

Given this historically diverse background for pervasive healthcare, and the multidisciplinary nature of the research, it is not surprising that the field draws on research from a wide range of different professions. On the surface, "pervasive healthcare" is defined as the application of pervasive computing in healthcare. However, "pervasive computing" is itself not a well-defined computer science area but a multidisciplinary research agenda involving technological-oriented research on topics such as hardware, communications, embedded hardware and software, software infrastructures, sensor technology, distributed computing, CSCW, HCI, and sociological studies of the use of technology. Hence, pervasive computing is not something one simply buys and installs in a healthcare setting. Rather, it is a new kind of technology that has not yet been defined and that must be shaped according to the changing needs and challenges of healthcare. The creation of pervasive healthcare technologies and services involves a wide range of professions, including doctors with various specialties, nurses, caregivers, therapists, engineers, computer scientists, human factor professionals, industrial designers, and patients and citizens. As pointed out by a number of authors in this book, innovation in pervasive healthcare requires a community of dedicated researchers and practitioners to cooperate closely in the design, development, and evaluation of the technology while focusing closely on the users. Furthermore, pervasive healthcare applications represent only some of the dramatic changes currently taking place in the healthcare industry. To be effective, these innovations must go hand in hand with other system-level changes, including government policies, business processes, and reimbursement models.

As a result of its multidisciplinary nature, pervasive healthcare lies at the intersection of different research paradigms. There are fundamental differences between research methodologies employed in medicine, engineering, computer science, design, and ethnography. These disciplines have their roots in noncompatible philosophical traditions. Therefore, we cannot claim that pervasive healthcare subsumes to one particular scientific approach or methodology. This is most apparent in what we call the "fundamental methodological challenge" of pervasive healthcare.[3] Typical research in pervasive computing uses experimental computer science methods,[4] where researchers design, develop, program, and evaluate prototypes of a new technology. The original ubiquitous computing technologies created at Xerox PARC are an excellent example of this approach. The "proof of concept" is a term often used to denote a prototype, which illustrates and implements the important aspects of a computer system that one wants to demonstrate. Such an experimental approach becomes highly problematic when dealing with health-related research. Modern evidence-based medicine is rooted in statistical significance, in which one has to demonstrate with significant confidence that a treatment or cure works and that it has minimal side effects. This proof is generated through clinical trials, which often involve great numbers of human subjects with experimental and control groups. To set up such a clinical trial running over several months or years clearly takes much more than a proof-of-concept prototype. One must have the resources to design, develop, implement, and maintain a full-fledged computer system to be used by thousands of real-world users. Conducting such large trials presents a huge financial challenge, especially for researchers who are used to a more traditional computer science research. With this in mind, perhaps one methodological contribution of pervasive healthcare is developing alternative ways of bringing such healthcare technologies to market without the full rigor of traditional clinical trials. This is especially significant in light of the increasing acceptance of consumer-grade devices and the future importance of consumer-centric healthcare.

The book

This book provides an introduction and an overview of the new emerging field of pervasive computing in healthcare or, simply, pervasive healthcare. Drawing from the contributions of leading researchers in this field, we provide readers with in-depth discussions of relevant topics. This book offers the first known comprehensive resource on the application of pervasive computing to healthcare. It covers such broad topics as the current healthcare system and its challenges, core computer science approaches underlying pervasive healthcare, and leading research on specific software architectures and systems for pervasive healthcare, including how they can be applied within hospitals, homes, and public spaces. It also discusses a wide range of issues related to development and research methods in pervasive healthcare systems, including human factors, clinical trials, evidence-based medicine, and potential business models.

Chapter 1 provides an overview of healthcare, diseases, and disabilities. It begins with a discussion of shifting demographics and their impact on the healthcare system. Next, it describes in detail the current healthcare system in the United States. Specifically, it covers home- and community-based healthcare, outpatient care, hospitals, and assisted-care institutions. For each, the authors describe major challenges and how technologies have been and could be used to address these challenges. The chapter concludes with a call for large-scale, clinical trial-like studies of pervasive healthcare technologies to demonstrate their clinical efficacy and to accelerate their adoption by the healthcare system.

Chapter 2 introduces a number of key computer science principles underlying pervasive computing and its applications described in subsequent chapters. Pollack and Peintner provide a technical baseline to ensure that a wide range of readers can benefit from this book, including first-timers who have never been exposed to this area and those who are from computer science backgrounds. The chapter focuses on three core computer science topics within pervasive healthcare: (1) pervasive computing technologies, protocols, and devices; (2) essentials for creating intelligent applications, including machine learning and artificial intelligence techniques; and (3) privacy and security approaches and technologies for keeping sensitive patient health data private and secure. The chapter provides sufficient background to enable a reader to understand the remaining chapters in this book. Special emphasis is given to key computational trade-offs that typically arise in the design of pervasive healthcare systems. For example, these trade-offs include those between making a system easy to use and ensuring a high level of security; between collecting maximal sensor information and minimizing power use so that battery replacement is infrequent; and between having frequent updates of a highly detailed, fine-grained model of patient activity and having efficient, real-time computation.

Chapter 3 describes some key challenges in using computers in large, modern hospitals and how current software and hardware technology is evolving to meet these challenges. The chapter starts by discussing the core challenges in deploying contemporary computer technology designed for office use to a hospital setting. It then describes the current state of computer technology in hospitals, including electronic patient records (EPRs); picture, archiving, and communication systems (PACS); and intensive care unit (ICU) monitoring systems. Next, the chapter looks at several examples of present research in deploying pervasive computing technology in hospitals, including mobile computing, location- and context-aware computing, wireless communication, software infrastructures, support for cooperation and social awareness, and multimodal interaction with computers during a surgical operation. The chapter ends by discussing a core set of considerations in the design, development, and deployment of pervasive computer technology in hospitals.

Chapter 4 describes new pervasive technologies that can be used in the home and community for people with cognitive disabilities. It describes two possible avenues for assistance. The first is using an outdoor activity recognition system based through a global positioning system (GPS) to help people who make occasional cognitive errors recover safely. The second is an indoor activity recognition system based on a wearable computing platform and radio frequency identification (RFID) tags designed to monitor common activities in a home. In the outdoor case, the chapter demonstrates that such a system could successfully be built now. It also shows that, with a simple, intuitive user interface, such a system could serve as a valuable aid to those cognitively impaired. In the indoor case, the authors demonstrate that a single technology can subsume many previous activity recognition techniques in a way that is robust, easily deployable, and accurate at a fine level of granularity.

Chapter 5 identifies the main usage models and applications for mobile and personal health and wellness management and monitoring systems, often called mHealth applications. Advancements in sensor technology, wireless communications, and information technology in general give rise to new ways of providing healthcare and wellness or disease management that supports extended independent living at home and improvement of quality of life for individuals. The chapter examines personal disease or wellness management enabled by pervasive personal digital devices and their ubiquitous communication capabilities. These devices allow ubiquitous access to health information, flexible and time- and place-independent access to communications with health professionals or personal trainers, and support tools for personal health or wellness monitoring. The chapter focuses on the out-of-hospital use of mobile devices such as PDAs and mobile phones for disease or health management.

Chapter 6 describes sensors and wearable technologies for pervasive healthcare. It begins with a discussion of key challenges in healthcare and thus establishes the need for pervasive healthcare. It then describes the principal modules of a patient-centric pervasive healthcare system that addresses the typical user's requirements. Next, it discusses the role of sensors and wearable sensor systems for biomedical monitoring of patients with chronic ailments such as hypertension, cardiac disease, and diabetes. The authors establish the need for integrated multiparameter sensing and use the Smart Shirt as an example to illustrate the development of a fabric-based sensor network for pervasive healthcare. The chapter concludes with descriptions of the challenges and opportunities for research and development in pervasive healthcare.

Chapter 7 presents current research in assistive technologies. People with disabilities have long been early adopters of pervasive computing technologies. Smart homes and mobile computing devices help people compensate for physical, sensory, or cognitive limitations. This compensation provides increased educational, vocational, and recreational opportunities and improved quality of life. The chapter describes technologies for assisting people with different disabilities, including physical, sensory, communication, and cognitive impairments. The applications of pervasive computing, including mobile computing and smart environments, are described. The chapter looks at current state-of-the-art, ongoing research and the application of emerging technologies. It also examines the implications of pervasive computing technologies to the design of new assistive technologies. The chapter concludes with a description of a number of specific technologies for augmentative and alternative communication (AAC), including wheelchairs and walkers, wayfinding, reminder systems for taking medicine, and smart environments.

Chapter 8 addresses the challenges associated with human factors and the usability of healthcare systems. The chapter focuses on how pervasive computing can be integrated into healthcare practices so that users can maximize the benefits of new computing capabilities. Topics discussed include requirements engineering for new applications, the use of workflow studies in designing unobtrusive and effective system–user interactions, design and testing of pervasive applications in geriatrics, the application of usability engineering methods, and issues of user testing. A range of applications are used to illustrate human-centered methodological approaches in the design and evaluation of pervasive healthcare systems. Examples include a computerized patient record system that is integrated with a range of new pervasive applications, and the relation of pervasive computing to telemedicine applications, location awareness, the need for integrating standards, and usability considerations in mobile and Internet-based medical applications. The chapter also discusses the importance of an improved understanding and consideration of human cognition and communication modalities, including speech, handwriting, head-mounted displays, and combined approaches.

Chapter 9 presents methods for technology innovation in pervasive healthcare. This chapter introduces the requirements, recommendations, and routes for commercializing medical devices and gives pervasive healthcare researchers a basic understanding of how medical applications are assessed and regulated. First, the author defines the concept of health technology assessment, which serves as the framework within which the effectiveness and the value of a device are evaluated. Second, the regulations pertaining to medical devices are explained in some detail, including definitions, classifications, and processes for obtaining ethical and clinical approval, with examples from Europe and North America. Some newly proposed concepts in clinical trials are introduced that may, if adopted, be particularly suited to medical devices. Also covered in this chapter are deployment and data issues. The latter arises from the link between the predominantly independent worlds of devices and computer networks that are now coming together in pervasive healthcare. Third, the product design process for medical devices is described, concentrating on quality systems, standards, and recommendations for good practice in validation and human factors. Some of these recommendations are already finding their way into updated regulations in this rapidly changing field. The conclusion section reiterates the multidisciplinary role of the pervasive healthcare engineer and points to sources of additional information and advice.

Chapter 10 discusses user evaluations in pervasive healthcare with a special focus on real-world deployment and assessment of the technology. Central to the success of pervasive healthcare technologies is the end users' acceptance of the systems. Hence, methods for proper user evaluations and redesign during the development and design of a piece of technology are fundamental to pervasive healthcare. This applies both for concrete medico-technical equipment, like a new blood pressure monitor, as well as more overall systems and infrastructures, such as home monitoring. This chapter presents state-of-the-art methods and discusses some of the special requirements concerning medical technologies. The authors especially discuss methods for putting an evaluation in context in order to establish how the technology can work in a complex clinical or community-based environment.

Finally, Chapter 11 discusses the business aspects of pervasive healthcare. It describes potential business opportunities and implications of pervasive healthcare technologies, including remote patient monitoring, to key stakeholders involved. First, the authors provide an overview of the key enabling technologies, including consumer/sensing devices, wireless networks, analytic engines, decision support, and collaboration solutions. Second, the chapter presents the economic and business case for pervasive healthcare. Specifically, it describes how such technologies can be used to address the root causes of the current healthcare crisis by redefining care management to meet the critical needs of the chronically ill and the elderly population and by helping reduce emergency room visits and hospital stays. Third, the authors highlight a number of successful commercial pervasive healthcare applications in the marketplace and describe how these applications have demonstrated compelling business values to key healthcare stakeholders. The chapter concludes with a discussion about major challenges and barriers ahead in adopting pervasive healthcare technologies, including reimbursements, resistance among physicians and patients, standards, data security, and usability.

Pervasive healthcare is an exciting, emerging research area that is bound to play an important role in an increasingly aging society. By providing the first known book fully devoted to this field, we attempt to give our readers a holistic view by covering a variety of topics, including basics of healthcare and computing; examples of applications of pervasive and mobile healthcare in settings such as hospitals, communities, and homes; assistive technologies; human factors; evaluation methods; and business models. We also try hard to make this book appealing to readers with computer science or healthcare backgrounds. Given the broad scope of this field, the rapid rate of change in both technologies and healthcare, and the length constraint of this book, there are inevitably interesting and relevant topics that are missing here. Nevertheless, we believe that this book provides a solid foundation on which current and future researchers and practitioners can build and use to further their endeavors that will eventually make these technologies truly pervasive, especially to those who need them the most.

—Jakob E. Bardram, Alex Mihailidis, and Dadong Wan

References

1. Centers for Disease Control and Prevention. Chronic disease prevention. See www.cdc.gov/nccdphp/overview.htm.
2. Dishman, Eric. Inventing wellness systems for aging in place, *IEEE Computer* 37:5 (May 2004), 34–41.

3. Korhonen, Ilkka, and Bardram, Jakob E. Guest editorial introduction to the special section on pervasive healthcare, *IEEE Transactions on Information Technology in Biomedicine* 8:3 (September 2004), 229–234.
4. Weiser, Mark. The computer for the 21st century, *Scientific American* 265:3 (September 1991), 94–104.

Editors

Dr. Jakob E. Bardram is an associate professor in the Department of Computer Science of the University of Aarhus, Denmark, and is the manager of the Centre for Pervasive Healthcare in Denmark. Dr. Bardram's research interests include software architecture, pervasive computing, human–computer interaction, software engineering, and computer-supported cooperative work. He has focused specifically on the design and development of pervasive computing systems in healthcare, for both patients at home and staff working in hospitals. Dr. Bardram has more than fifty international publications and his work is done in close cooperation with industry and healthcare organizations.

Dr. Alex Mihailidis is an assistant professor in the Department of Occupational Science and Occupational Therapy at the University of Toronto with cross appointments in Biomedical Engineering and Computer Science. He has conducted research in the field of pervasive computing and intelligent systems in healthcare for the past eight years, and he has published or submitted over thirty publications. He has focused specifically on intelligent systems for elder care and wellness. He holds several major research grants from internationally recognized funding agencies and industrial partners to support this work, including from the Canadian and American Alzheimer Associations, Intel Corporation, Natural Sciences and Engineering Research Council of Canada (NSERC), and Canadian Institutes of Health Research (CIHR).

Dr. Dadong Wan is a senior researcher with Accenture Technology Labs, the research and development organization for Accenture. For the past ten years, Dr. Wan has investigated how emerging technologies, specifically ubiquitous computing, can be used to create new kinds of consumer experiences and business opportunities. He is the inventor of the Magic Medicine Cabinet, the world's first smart medicine cabinet that integrates face recognition, RFID, and health-monitoring devices to provide consumers with compliance support, vital sign monitoring, and personalized health information. Currently, his research focuses on patient-centered, connected healthcare using sensors, wireless networks, and service-oriented architecture. Dr. Wan's work is widely covered in the media, including by the *Wall Street Journal*, *Financial Times*, *Wired*, BBC, CNN, ABC News, and TechTV.

Contributors

Tim Adlam
Bath Institute of Medical
  Engineering
Bath, United Kingdom

Heribert Baldus, Ph.D.
Philips Research Laboratories
Aachen, Germany

Jakob E. Bardram, Ph.D.
University of Aarhus
Aarhus, Denmark

Elizabeth Borycki, Ph.D.
University of Victoria
Victoria, British Columbia, Canada

Michael P. Craven, Ph.D.
University of Nottingham
Nottingham, United Kingdom

Teresa Dunn
Bath Institute of Medical
  Engineering
Bath, United Kingdom

Jesus Favela, Ph.D.
Centro de Investigación Científica y
  de Educación Superior de
  Ensenada
Ensenada, Mexico

Dieter Fox, Ph.D.
University of Washington
Seattle, Washington, United States

Sundaresan Jayaraman, Ph.D.
Georgia Institute of Technology
Atlanta, Georgia, United States

Henry A. Kautz, Ph.D.
University of Washington
Seattle, Washington, United States

Jeffrey Kaye, M.D.
Oregon Health & Science University
  and Portland Veterans Affairs
  Medical Center
Portland, Oregon, United States

Ilkka Korhonen, Ph.D.
Technical Research Center of
  Finland
Tampere, Finland

Andre Kushniruk, Ph.D.
University of Victoria
Victoria, British Columbia,
  Canada

Lin Liao
Google Inc.
Kirkland, Washington, United
  States

Edmund LoPresti, Ph.D.
AT Sciences, LLC
Pittsburgh, Pennsylvania,
  United States

Elina Mattila, M.Sc.
Technical Research Center of
  Finland
Tampere, Finland

Alex Mihailidis, Ph.D., P.Eng.
University of Toronto
Toronto, Canada

Alan Newell, Ph.D.
University of Dundee
Dundee, Scotland

Roger Orpwood
Bath Institute of Medical
  Engineering
Bath, United Kingdom

Sungmee Park, M.Sc., M.F.A.
Georgia Institute of Technology
Atlanta, Georgia, United States

Donald J. Patterson, Ph.D.
University of California
Irvine, California, United States

Bart Peintner, Ph.D.
Artificial Intelligence Center,
  SRI International
Menlo Park, California,
  United States

Martha E. Pollack, Ph.D.
University of Michigan
Ann Arbor, Michigan,
  United States

Niilo Saranummi, Ph.D.
Technical Research Center of
  Finland
Tampere, Finland

Luis E. Taveras, Ph.D.
Accenture
Florham Park, New Jersey,
  United States

Dadong Wan, Ph.D.
Accenture Technology Labs
Chicago, Illinois, United States

Tracy Zitzelberger, M.P.H
Oregon Health & Science
  University
Portland, Oregon, United States

Contents

Section I: Introduction and overview to pervasive healthcare

Chapter 1   Overview of healthcare, disease, and disability 3
Jeffrey Kaye and Tracy Zitzelberger

Chapter 2   Computer science tools and techniques 21
Martha E. Pollack and Bart Peintner

**Section II: Architectures, systems, and technologies
for pervasive healthcare**

Chapter 3   Pervasive computing in hospitals 49
Jakob E. Bardram, Heribert Baldus, and Jesus Favela

Chapter 4   Pervasive computing in the home and community 79
Donald J. Patterson, Henry A. Kautz, Dieter Fox, and Lin Liao

Chapter 5   Mobile and personal health and
             wellness management systems 105
Elina Mattila, Ilkka Korhonen, and Niilo Saranummi

Chapter 6   Sensors and wearable technologies
             for pervasive healthcare 135
Sungmee Park and Sundaresan Jayaraman

Chapter 7   Assistive technologies 161
Edmund LoPresti

Section III: Design and development of pervasive healthcare technologies

Chapter 8   Human factors and usability of
        healthcare systems 191
Andre Kushniruk and Elizabeth Borycki

Chapter 9   Routes and requirements for realizing
        pervasive medical devices 217
Michael P. Craven

Chapter 10   User evaluation in pervasive healthcare 243
Tim Adlam, Roger Orpwood, and Teresa Dunn

Chapter 11   The business of pervasive healthcare 275
Dadong Wan and Luis E. Taveras

Index 299

Introduction and overview to pervasive healthcare

chapter one

Overview of healthcare, disease, and disability

Jeffrey Kaye
Oregon Health & Science University and Portland Veterans Affairs Medical Center, Portland, Oregon
Tracy Zitzelberger
Oregon Health & Science University, Portland, Oregon

Contents

1.1 Introduction 3
1.2 Demographics, disease, and disability......................................4
    1.2.1 Consequences of the demographic shift 5
    1.2.2 Burden of disability and care needs 6
    1.2.3 Disparately burdened groups:
           Socioeconomic status and gender 7
1.3 The healthcare model: Transforming with technology 8
    1.3.1 Home- and community-based health...........................9
    1.3.2 Outpatient care 11
    1.3.3 Hospitals...13
    1.3.4 Assisted-care communities and institutions..............15
1.4 Conclusions...17
References 18

1.1 Introduction

This chapter summarizes the impact of current trends in disease and disability on the delivery of healthcare using pervasive technology. We review the remarkable demographic shifts of the past century that have been driven by increases in life expectancy as a result of advances in public health as well as in medical technology. The impact of these trends is further discussed with regard to the need to focus healthcare not only on its traditional core (providing point of care diagnosis and acute intervention) but also increasingly on managing growing numbers of people aging with chronic disease outside of traditional hospital or clinic settings. In this context we review how a broad range of technologies using pervasive computing may provide important new approaches to improve the provision of healthcare. How these technologies may affect diagnosis and treatment in specific settings such as the home or doctor's office or as a hospitalized patient or resident of a continuing care facility are considered in turn with special emphasis given in conclusion to the power of pervasive computing to effectively integrate and blur the boundaries of these conventionally isolated healthcare settings. Finally, we conclude with a consideration of what these new technologies will require to become widely adopted such as standardization and adequately powered clinical trials demonstrating efficacy and safety.

1.2   *Demographics, disease, and disability*

Healthcare is an endeavor that constantly looks forward while forever being judged by what has just passed. In part, this is the result of the perception that medicine and health sciences are driven by rationally designed breakthroughs rather than by changes in small increments, which are often the result of discovery in unsuspected corners. Although conceptually tied to cures and revolutions in care, the concept of "improvement" in health nevertheless operates on a playing field where improvement is measured not by an absolute sea of change, but in small waves of change relative to recent standards of care. With time, the weight of evidence, balanced by cost and social norms, leads to the new standard of the day. This transition has been most evident when one observes the unprecedented growth in health and well-being marked by the striking gain in life expectancy of almost forty years during the twentieth century.[1] This remarkable increase (in developed countries, at least) is often attributed to advances in medicine and healthcare technology, when in fact the major advances are likely to be more closely linked to advances in fields not typically considered "medical." Advances in areas such as sanitation, food sources, and workplace safety have probably played a much greater role in this growth in life expectancy than inventions of medical devices or pharmaceuticals, save for vaccines.

Such advances have led to a dramatic demographic shift of increased survival to increasingly advanced ages. The young are and remain healthier, resulting in more elderly in successive generations. Also, as populations have aged in most developed countries, birth rates have declined, leading to a decline in the relative number of young while the older group continues to grow (see Figure 1.1). Baby boomers are now transitioning from a middle-aged to an older-aged cohort. As the baby boomers represent large percentages of the U.S. and Canadian populations, the focus of healthcare

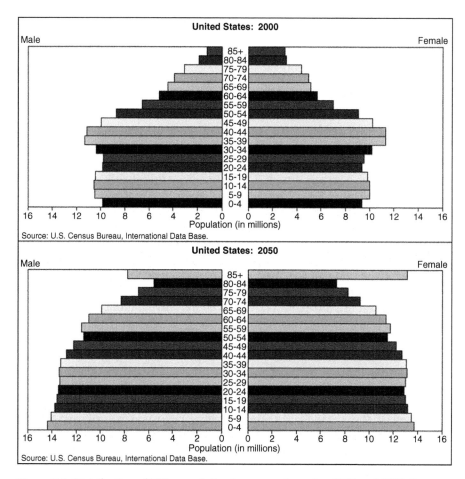

Figure 1.1 Distribution of U.S. population by age and gender: 2000 and 2050. Source: U.S. Census Bureau, Population Division, International Programs Center. International Data Base Population Pyramids at http://www.census.gov/ipc/www/idbpyr.html.

in North America is perceptibly shifting to accommodate this aging cohort's needs. Additionally, those in the "sandwich generation" are balancing the demands of caring for elderly parents at the same time as they are raising young children.

1.2.1 Consequences of the demographic shift

Along with this increasingly aged population, the profile of those at risk for disease and disability has accordingly shifted. The fastest-growing threats to public health are currently age related. People now survive the once-fatal illnesses and diseases of youth and live on to experience chronic and debilitating conditions of advanced age. In 1900, the leading causes of death in the United States were pneumonia, tuberculosis, and gastrointestinal diseases. Now, people succumb most often to heart disease, cancer, and cerebrovascular disease.[2] Accordingly, populations experiencing these demographic changes now face causes of mortality and morbidity that are strongly age associated. Age, rather than infectious agents, is the greatest risk factor for the current top causes of mortality.

1.2.2 Burden of disability and care needs

The impact of the demographic shift is apparent in causes of morbidity as well as mortality. Americans over sixty-five are more likely to incur medical expenses than younger Americans and their average cost per expense is higher: $6,140 as compared with $2,127 for those under sixty-five.[3] Chronic conditions, especially loss of mobility and cognitive function, not only fuel healthcare costs but also prevent the elderly from living independently in their homes. Changes leading to the loss of independence may be subtle and slow to develop, such as loss of mobility due to arthritis, a visual impairment, or a cognitive decline. According to Centers for Disease Control and Prevention statistics, 80 percent of adults over sixty-five report having at least one chronic illness, while 50 percent have at least two.[2] Predominant causes of disability and impairment in adults over sixty-five include arthritis, heart disease, diabetes, vision loss, and hearing loss. The prevalence of these conditions is presented by age group in Figure 1.2.[2] On the other hand, loss of independence may be precipitated by more acute events, such as strokes, accidental injuries like falls, medication errors, or acute systemic illness. These events frequently result in the need for alternate care situations, such as relocation to a nursing home or a rehabilitation facility.

Together, these acute and chronic conditions form a major target for the technological interventions and remediations that are discussed in detail in

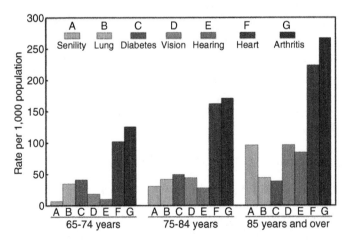

Figure 1.2 Chronic conditions causing limitation of activity, 2002–2003. Source: National Center for Health Statistics, United States, 2005.[2] Figure 20. Accessed via http://www.cdc.gov/nchs/hus.htm.

the following chapters. Technological advances in healthcare may yield a considerable return on investment, not just in terms of saving healthcare dollars, but more importantly and perhaps more certainly in longer independent living and improved quality of life. A recent economic report from the New Millennium Research Council provides just one impressive example from the communication technology field. It estimates that current use of broadband technology will yield up to $927 billion in savings and benefits for Americans over sixty-five or for those living with a disability, with an additional $532 billion to $847 billion saved with targeted dissemination of these services.[4]

All of these trends create unique needs and challenges for care that are specific to the age group affected. In general, the penetration of technology into the lives of the current elderly is quantitatively and qualitatively different from that of their children (the "boomers") as well as their grandchildren, who may be more familiar and comfortable with using such technology. These differences have practical implications. Emphasis in healthcare systems is appropriately placed on the elderly as the consumers with the greatest needs for healthcare. Technology designed for elders may need to be fundamentally altered to adjust to their needs. The concept of universal technical solutions should be inclusive in trying to accommodate seniors who may not use a cell phone or computer as well as teenagers who habitually use instant messaging and video games. Ironically, the technology culture of the younger generation may contribute in part to poorer health as they age. The current intense media culture that is highly embedded with messages designed to promote fast food and a sedentary lifestyle is largely responsible for the epidemic of obesity in the United States.[5]

1.2.3 Disparately burdened groups: Socioeconomic status and gender

Other challenges extend beyond the age-based continuum outlined above. Aside from age, the different strata of society and cultural backgrounds from which people originate or reside play an important role in the impact of technology on health. Those coming from lower socioeconomic groups are likely to have less technology experience or access while simultaneously having above average needs for healthcare. Among those over sixty-five, women and poorer Americans on average experience a heavier burden of chronic conditions. Research with health-related quality-of-life data from the Centers for Disease Control and Prevention found that the number of unhealthy days per month reported by adults over age sixty-five varied significantly with socioeconomic status and gender. Women consistently reported more unhealthy days per month than men of the same age, regardless of socioeconomic status. Similarly, those with lower socioeconomic status also reported more unhealthy days than those with higher status.[6]

Although this chapter refers primarily to U.S. and comparable populations, the goals and principles of pervasive computing technologies applied to healthcare are highly relevant throughout the world. Aging of populations, accidental injuries, war and violence, automobile crashes, chronic diseases, and many factors contribute to the rise in disabilities worldwide. There are approximately six hundred million people worldwide living with disabilities. Of those six hundred million, 80 percent live in poor countries where access to basic and rehabilitation services are limited or unavailable.[7] The World Health Organization Assembly has established a resolution to assist member states in developing policies on disability and rehabilitation, especially for those who are poor.[8] It has become a global priority to develop technologies for early disease detection, improved treatment, and high-quality assistive living.

While a limited standard of living in developing countries is a huge barrier to technological investment and development, making a wireless leap is easier in some ways. For example, most developing countries have few or no hard-wired telephone systems. Rather than investing in a network of landlines, these countries have gone straight to using mobile technology. As such, mobile service in some developing countries may actually be more available and reliable than in some developed countries. The challenge of making phones widely available to the general public should improve with time as the cost of this technology continues to decrease. The actual cost of the phone itself is small relative to building and maintaining communication networks. Nevertheless, it is likely that mobile, wireless phones, rather than personal computers, hold the promise of boosting entrepreneurship and economic development in poor countries.[9] Where economic development is increased, better public health will follow.

1.3 The healthcare model: Transforming with technology

There are many ways to review and discuss the interaction of technology with healthcare: by persons involved (patients, caregivers, clinicians), by the technologies and methods employed, by time of use (acute versus chronic care), and by locus of care. We have chosen to organize this review by locus of care for two reasons. First, presenting technologies by locus of care cuts across most domains, such as disease or condition requiring care or category of technology that might be used by a patient or clinician. Second, it allows us to describe major challenges and opportunities in a manner that reflects the sequence and location of care that an individual might experience when challenged with a health problem over time. Although our organizing principle here is locus of care, the reader should keep in mind that the ultimate goal of the ideal healthcare system is to ensure optimal functioning and quality of life of patients. Implicit in this assumption is that most individuals would choose to maintain their health in their own residences; if they require care outside of the home, most individuals would also choose to minimize their time in institutional facilities. Thus, the ultimate health system is "homeless" and able to provide care and ensure health regardless of location. This is the promise of pervasive healthcare.

1.3.1 Home- and community-based health

The vast majority (84 percent) of Americans over sixty-five live independently in single-family homes or apartments.[10] They arguably spend most
of their time in the home, including approximately 364 minutes (just over
six hours) spent sleeping or at least in bed at night.[11] The home environment
is usually where the realization that someone is sick first occurs, as well as
where most acute and chronic illnesses are managed. This means that among
the most important classes of technology that may be brought to bear in this
setting is in the realm of monitoring. Many homes may have a few simple
health monitoring technologies such as thermometers, scales, or blood pressure machines. Some health conditions or concerns lead to specialized home
devices or kits, such as home blood glucose monitors for diabetes, CPAP
machines for obstructive sleep apnea, and nebulizers for asthma treatment.
By and large the general penetration of technology into the home setting for
healthcare so far has been rather rudimentary, reactive, and underdeveloped,
particularly for devices aimed at self-diagnosis or decision making (as to
whether more specialized help is needed).

The initial point of care at home is most likely the telephone, as the
conduit for determining the need to leave the home for testing, consultation,
and treatment. It is suspected, although not proven, that most initial outside
medical advice received in the home is provided by informal contact with
relatives and friends through phone calls. Growth of the Internet and home
computer use has resulted in a greater opportunity for individuals to access
multiple sources of healthcare advice. As of September 2005, 30 percent of
adults over sixty-five reported having Internet access,[12] up from 15 percent
in 2000.[13] In recent surveys of Internet users, 80 percent state they have used
the Internet to obtain health or medical information, while 6 percent state
that they had done so "yesterday."[14] However, Internet access is less accessible for those with lower incomes. Only 54 percent of households with
annual incomes below $30,000 have access to the Internet, compared to 78
percent of those with annual incomes of $30,000–$49,999, and at least 87
percent of those with $50,000 or greater report being online.[12]

There is a huge variety of health-related sites available through Internet
or Web portals ranging from those that are disease specific to more general
or comprehensive medical sites. A major challenge for the consumer is to
determine the validity of the information provided. It has been pointed out
that the consumer's risk of accessing an inadequate site is a function of both
the amount of inadequate information on the Web and the ability of the
individual to determine the quality of the information. This has made it
difficult to objectively determine the impact of the over six thousand (and
growing) health-related Web sites or Web pages available.[15]

Many nonprofit disease-oriented lay organizations (e.g., National
Cancer Society, American Heart Association, the Alzheimer's Association,
etc.), as well as governmental agencies (e.g., NIH, FDA, VA) occupy this
Web space, and their Web sites are generally considered more reliable and

less biased. Yet the fact that these Web sites are nonprofit or governmentally sponsored does not guarantee that the information is unbiased or useful or that the user will interpret the data appropriately.

A growing number of informational sites are offered through the healthcare system that an individual belongs to and a few even offer the opportunity to e-mail a clinician with a question, such as the Kaiser health plans or the Department of Veterans Affairs (My Health eVet). This type of interactive capability appears, at least on the surface, to offer an efficient procedure with no "phone tag" and allows for fully documented and enhanced communication among stakeholders (as others involved in the individual's care could be automatically notified). How this online interactive capability will develop is uncertain. Major forces that will impede or enhance the development of these Internet systems include policy development regarding reimbursement for providing these services[16] as well as liability concerns around providing care advice without a direct examination. More discussion of disease management through electronic media is discussed in Section 1.3.3 on hospital-based care.

Determining patient needs in the face of symptoms is a reactive approach to healthcare that is embedded in current U.S. care models. In these models, reimbursements are highest for acute care medicine. Aside from the sometimes convoluted economics that drive our care models, to some degree the institutional or clinic-based care model also is a product of the need to bring the person to a place where the tools and technologies for diagnosis and treatment reside rather than to provide these capacities at home. This is a tremendous opportunity for technology innovation and creativity, both from the standpoint of bringing these tools to bear in the home, and perhaps more important, for providing systems of care that are proactive in detecting a problem before it leads to major disability.

Proactive healthcare through home monitoring is not new. Pervasive or minimally obtrusive systems for home monitoring have been described for over a decade[17] and are further discussed in Chapter 5. In the home healthcare arena there has been a natural convergence of technologies and systems for monitoring and assisting people with special needs (e.g., low vision, hearing loss, limited mobility). In general, these systems consist of a few common elements, such as control devices for the automated modulation of the environment, health monitoring devices for disease management, and motion sensors for assessing general activity. Unfortunately, these integrated systems have yet to "catch on" for a number of reasons including cost and lack of reimbursement, complexity, and scalability. Most important, the vast majority of these systems have not been tested in rigorous clinical trials, therefore there is a lack of evidence that they are effective (and will not cause harm) in a generalized setting.

As the clinical efficacy of home-based monitoring and intervention systems is tested and refined, there will be certain areas that may especially benefit from these technologies. One such area is medication taking. Medication errors in the home usually occur when patients take the wrong drug, take the incorrect dose, misunderstand instructions, or are not made aware of potential drug interactions.[18] Medication errors are a particular threat to safety to all but are a special challenge for older patients, as they take more medications and are at greater risk for memory problems that may lead to adherence problems.[19] Pervasive technologies can provide checks for correct drug administration or reminders to take medications. For example, smart medicine cabinets equipped with radio frequency identification (RFID) capabilities can help ensure the correct medication is chosen from the shelf, personal digital assistants can provide timely reminders when medications are due, and pill boxes that record dose administrations can provide a record of medication compliance.[20,21]

Despite the current challenges of proactive home monitoring, the principle has strong face validity that is grounded in the power of preventive medicine and early detection. There are obvious advantages to detecting illness at an early stage or identifying patients at risk rather than managing advanced illnesses or patients in crisis. Home-based systems are growing in use and offer promising potential for facilitating wide-scale proactive monitoring of diagnosed diseases. The use of various technologies for disease management with an identified medical problem is generally called telemedicine. Although the practice of telemedicine focuses on the home, it depends on an outside clinical monitoring support system. Because telemedicine applications depend on having received a diagnosis or recognized treatable medical condition, outpatients are prime candidates for telemedicine systems, as we discuss in the Section 1.3.2 on outpatient care.

The integration of these formal telemedicine systems with more individually driven, clinician-independent home health monitoring programs is an important area for future development. Such integration may more effectively address the knowledge gap between professionally driven programs and well meaning but could potentially be harmful with unsupervised, informal home care. The blurring of professional with personalized lines of healthcare may require a new social contract with regard to responsibility for health outcomes. This inevitably will be followed by new regulations to apportion diagnostic and treatment responsibilities. Chapter 9 looks at some ways to assess and regulate pervasive healthcare, such as telemedicine.

1.3.2   Outpatient care

With the maturing of home-based systems, as well as evidence that diagnosis and treatment can be safely and efficiently augmented or provided through home-based care, one can envision that traditional clinic- or office-based care will evolve accordingly. This will be true for both primary care as well as specialty care.

Once the decision is made that a person needs medical attention beyond self-care, the current model is to seek an appointment with a clinician for an assessment. The location of this assessment is by and large an office or a clinic; therefore these constitute the center of outpatient diagnosis and treatment initiation. One of the first major challenges for outpatient care is the transfer of the information needed to conduct the medical interview and provide appropriate treatments. As medical history is the keystone to a diagnosis, the front end of each outpatient clinical encounter is in essence a data-gathering operation. Unfortunately, in current office practice much of this information gathering is time-inefficient, with most of the time taken up by reviewing with the patient what they can best recall of their symptoms, medical history, and relevant medications. Because patients may not seek care except when they feel unwell, they are often not at their best in terms of being prepared for the visit or remembering important information. Efficiently obtaining accurate information about a sick patient at the point of care is at a minimum a potent time saver. It also has important implications for quickly reaching the correct diagnosis and initiating proper treatments, ranging from ensuring that a logical sequence of diagnostic tests is pursued to prescribing treatments that are the most successful and safe. The latter issue of safety is not only a matter of advising an appropriate treatment, but also about avoiding lost time from ineffective treatments and preventing frank medical errors.

These challenges should diminish as electronic health records (EHRs) and information transfer systems mature. However, unless they are part of an integrated health system (such as those used within the Veterans Administration or Kaiser health systems), EHR systems are still rather scattered and not able to seamlessly transfer information. Outside of an integrated health system, EHRs rarely extend beyond the confines of the medical office to include accurate pharmacy or laboratory data. Nevertheless, even fully integrated EHR systems will only improve the information flow between the traditional office or laboratory settings. A truly integrated and pervasive health information system would expand a patient's health records to include information acquired daily at home when the patient feels healthy and is going about his or her daily routines. One can clearly see the advantage of marrying the home-based monitoring information noted above with the office visit. A clinician would have a better idea of a patient's overall health by including information obtained from the patient when in their typical environment and could also more accurately monitor the progress of any therapy instituted after the clinical visit.

Remote care capabilities are continually increasing as new technologies and systems are implemented. The remote monitoring and management of health is currently accomplished through a number of channels (e.g., phone, Internet, interactive video). This form of health delivery, now usually subsumed under the category of telemedicine, has received much attention during the last decade. Unfortunately, despite many applications being proposed and reported on, there is currently little systematic evidence for the efficacy of these programs.[22,23] Among home-based applications, the most highly studied area has been monitoring of blood sugar in patients with diabetes mellitus. Other applications have been pursued for a number of chronic disease management areas ranging from blood pressure to mental health. By and large these studies have been too small and of limited duration and thus statistically underpowered to provide definitive conclusions as to how effective and generalizable they are compared to conventional procedures. Despite the lack of clinical trials for most of these applications, there are a growing number of companies that offer commercially available remote care products, complete with online monitoring of data of many types in the users' homes (e.g., weight, blood pressure, temperature, pulse and cardiac rhythm, oxygenation or pulse oximetry, blood glucose, lung capacity). Examples of products that are currently available include HomMed (Honeywell), ViTel Care, and MedStar (CybernetMedical). Key to such developments is not simply attempting to optimize remote data acquisition, but also supporting the effective integration of the data, the inference of the individual's health state, and the ability to differentiate between salient outcomes.

In addition to applying telemedicine to home health, there is also increasing use of telemedicine approaches in the interface between office- and hospital-based medical practice. In this arena, most studies have suggested the benefits of teleradiology or the transmission and interpretation of images. Other related areas have been applied to patient interviewing or examinations, such as teledermatology. Obviously, the prefix "tele" can and will be added to any specialty using this technology. Again, as with the home-based telemedicine applications, the efficacy of telemedicine methods for office- and hospital-based practices has yet to be proven. Compared to pure clinical efficacy, there is even less evidence that telemedicine is cost-effective.[24] In part, this lack of evidence is the result of too few systematic studies. However, this also raises the important need to establish common standards or benchmarks to measure efficacy and assess costs in this field.

1.3.3   Hospitals

Although many, if not most, illnesses can be managed at home or in the outpatient setting, there are clearly medical conditions that require more intensive care and treatment in a hospital. Generally, patients are either brought to an emergency or urgent care department for acute diagnosis and management or a nonurgent procedure is scheduled weeks or months in advance at the outpatient office. These latter procedures are not necessarily acutely time-sensitive but require the diagnostic and operating facilities and technology only afforded in the hospital or surgical care unit. Contemporary hospitals have an exceptional variety and density of technologies. These range from the information systems needed to operate the facility to the medical devices required to diagnose, monitor, and intervene in acute medical illnesses. It is notable that most of the technologies within this array of systems and devices are stand-alone and are not integrated with one another. In part, this is the result of the institutions themselves. Hospitals are built and continually renovated or expanded in a piece-meal fashion. Additionally, their organizational structure typically results in semiautonomous departments and services that are not coordinated in terms of long-range planning or purchasing.

On average, today's hospitals spend approximately 2.5 percent of their operating budgets on technology, while hospitals investing in more advanced clinical information systems spend 3 to 5 percent or more.[25] A comprehensive review of all the technologies in hospitals and related facilities is beyond the scope of this chapter, but selected examples of technologies and their functions are presented in Table 1.1. Within this wealth of technology, it is worth emphasizing the crucial and dramatic changes in the hospital setting resulting from the huge prevalence of information technology (IT).

IT pervades the entire health institutional ecosystem from the moment of entry (e.g., registration, health record creation, or update), to traditional testing (e.g., laboratory and diagnostic procedures), treatment (e.g., medical or surgical procedures), and discharge from the hospital. In the best of settings these activities are tightly integrated with work flow and allow clinicians to instantly tap into the chain of events to retrieve any information needed at the point of care. This information-rich chain creates the opportunity for expert or intelligent ("machine") agents to facilitate not only the process, but the actual care provided as well. This ranges from tasks such as the appropriate interpretation of diagnostic results to the proper institution of medications and therapies. However, this ideal has yet to be achieved, mostly because hospital systems tend to operate separately. One only needs to observe the work of a typical hospital-based nurse or attending physician for a few hours to conclude that the practice of medicine has absorbed pieces of technology, such as computerized order entry or reference to drugs on a PDA carried in a pocket, but rarely achieves the full power of information integration that more pervasive healthcare technologies could deliver.

Table 1.1 Examples of Technology Applications in the Institutional Healthcare Setting

Function	Examples
Core information technology services	Patient registration, accounting, facilities management
Clinician documentation	Electronic health record, computerized order entry, ASR transcription
Patient safety	Bar coding, "smart" beds, alarm and monitoring systems
Digital imaging and diagnostics	PACS (picture archiving and communication system), advanced imaging, personal medicine
Computer-aided medicine	Computer-aided diagnostics, computer-aided surgery, robotics
Health research	Evidence-based medicine, quality assurance, virtual or simulated surgery
Telemedicine	Remote monitoring, patient–health system networking, education

Improved healthcare through new technologies or better application of existing systems has been slow to develop. For example, the Institute of Medicine issued its landmark report, *To Err Is Human: Building a Safer Health System*, in 2000, yet it is sobering that since that time there has been disappointing progress in the area of patient safety.[26] For instance, medication errors can lead to adverse drug events, causing unnecessary hospital admissions and deaths. An estimated one out of 854 inpatient deaths can be attributed to medication errors.[27] While not all medication errors cause direct harm to patients, they cost healthcare consumers and provider systems millions every year. Standard healthcare delivery systems present many opportunities for errors to occur. The most common hospital-related medication errors are the prescriptions of an incorrect dose, prescriptions of a drug that will adversely interact with an existing prescription in the patient regimen, administrations of an incorrect dose, and administrations of an incorrect medication. EHR systems are capable of providing warning alerts that identify potential medication errors before they happen by requiring the verification of prescribed doses and calling attention to possible contraindications. Bar-coding strategies have been developed to ensure the correct medications are administered to the correct patients, where the bar code on hospital patients' admission wristbands must match the bar code on the prescription bottle.

The slow adoption of patient safety-related measures is certainly not to be blamed solely on technology. However, technologies have a great deal of unmet opportunities to improve healthcare in the hospital environment. Although computerized order entry systems may reduce errors, they nevertheless have not been proven to reduce preventable injuries.[28–30] This highlights the need to consider the application of technology interventions as any other treatment: unproven until rigorously tested and benchmarked by meaningful outcomes. Additionally, while the cost of implementing such systems may quickly be regained in prevented events and admissions, a lack of start-up resources can be a significant barrier.[31]

Despite these challenges, the healthcare delivery system recognizes the promise and power of more effective technology solutions such as wireless connectivity and intelligent agents. Through 2010, IT spending among U.S. healthcare providers is projected to experience a 7.4 percent five-year compound annual growth rate.[32] When applied to a critical issue like medication errors, effective pervasive computing technologies would be a tremendous force for data consolidation, clinician coordination, and error prevention, thus avoiding unnecessary morbidity, mortality, and healthcare costs. Current and future trends in pervasive technology within the hospital are presented in Chapter 3.

1.3.4   *Assisted-care communities and institutions*

Unless individuals return to their homes, patients leave the hospital or acute care venue to enter an alternate care setting. This transition to institutional care takes many forms, specific to each person's needs. In general, the choice of an alternative living situation is related to the level of assistance or dependent care needed. In the best of situations these transitions are temporary and after a period of rehabilitation, the person returns home. However, returning to prior levels of independent living may not always be possible. Thus an individual may come to reside permanently in assisted living, adult foster care, or a nursing home. According to the National Center for Health Statistics for 1999, 4.3 percent of adults over sixty-five live in nursing facilities, with such facilities operating at 87 percent capacity.[33] Increasingly, people approaching the certainties of disability with advancing age have chosen to live in continuing care retirement communities (CCRCs) that provide the full spectrum of care in an integrated living environment. The precise number of elders residing in CCRCs is not readily available, as residents living independently within the communities are reported as living in single-family homes.

From the perspective of pervasive healthcare technology, these living situations run the spectrum of applications and opportunities already outlined in Section 1.3.1 on home-based care and Section 1.3.3 on hospital care. CCRCs and assisted living facilities more closely mirror the needs of a person living freely in the community, while facilities with a greater emphasis on skilled nursing care or rehabilitation reflect the types of technology applied in the hospital setting. Because the focus of people in any of these residences is to optimize independence, this emphasis has appropriately transferred to a focus on the design and use of assistive technology. These assistive technologies range from mobility-assistive devices to speech synthesizers and automated cognitive remediation. Current and future trends in pervasive assistive technology are discussed in greater detail in Chapter 8.

As in both the home and hospital environments, there has yet to be a seamless integration of technology to its full potential in assisted living environments. This is likely to change as the care industry is increasingly implementing and relying on technological solutions for common care problems. For example, many nursing homes have commercially available ubiquitous alarm or warning systems to protect residents who are at risk of wandering and becoming lost. One notable assistive living residence, Oatfield Estates in Milwaukie, Oregon, has gone farther than most in integrating pervasive technology for monitoring and care of elderly residents. In this model, bed and chair sensors track residents' sleep habits, weight, and daily movement. RFID badges track patient location continuously, including night-time behavior. Residents who are prone to wandering are discouraged from leaving the grounds by a motion-activated in-ground sprinkler system, rather than by locks or alarms on entrance or exit doors. The system is used not only to assess the status of the resident, but also as a way to determine the effectiveness of the staff. Recently, a "family portal" was added to the system to enable family members to log on to a secure Web site and check on the status of their loved one.

1.4 Conclusions

This overview of healthcare, disease, and disability has outlined several dominant themes relevant to pervasive computing in healthcare. In the coming years, the burden of disease in developed as well as developing countries will be driven by the demographics of an aging population. There will be a continuous shift toward chronic conditions and age-related illness as the dominant forces driving both care needs and financial obligations. Although older populations may dominate healthcare from a population perspective, obviously there will continue to be younger individuals in need of care as well. Many of the technical principles developed for the application of care to seniors will be applicable to children as well.

Small, nonrandomized studies of isolated devices or methods of care using technology have been conducted. These suggest that the power of technology and pervasive computing to assist in facilitating diagnosis and treatment while easing the burden of care appears to be real, but in many areas has yet to be proven. In part, this is the result of the strong need for adequately powered, replicated clinical efficacy and outcomes studies. One barrier to conducting these needed research studies is cost. Large-scale studies of cutting-edge technologies being used in typical clinical settings come with large price tags. However, the issue is not simply cost. There is not yet a research infrastructure designed to carry out these kinds of ecologically valid clinical studies. As a result of issues such as these, current research does not usually progress past small feasibility studies, where a few subjects are observed for a brief period of time. Larger, more in-depth studies with multiple subjects in realistic settings must be conducted with new technologies to ensure that the technologies are reliable, safe, and useful.

In addition to more substantial testing, there is a critical need to create standards and metrics so that measured outcomes using new technologies are valid, reliable, and comparable across individuals and populations. Additionally, most pervasive computing technologies inherently can collect continuous, real-time data. This results in a different frame of reference for health assessments than the episodic, data-poor measures available in current practice. Standardized methods must be developed to incorporate and compare conventional measures to new ones. As recently stated in an Institute of Medicine report, ultimately there is a need for a comprehensive and universal system for measuring and reporting healthcare quality.[34] Pervasive computing and new healthcare technology benchmarks must be integrated into the fabric of these developing measurement and reporting systems.

These are some of the challenges that face a transition from today's care methods to more technologically rich and pervasive ones. People's well-being and their lives are at risk when healthcare is mismanaged. Therefore pervasive care solutions must be robust, reliable, and proven. As such, the development and implementation of these systems can be slow and the cost high. Nonetheless, pervasive technology offers a promising method of augmenting care across the healthcare spectrum.

References

1. Guyer, B., Freedman, M., Strobino, D., and Sondik, E., Annual summary of vital Statistics: Trends in the health of Americans during the 20th century, *American Journal of Pediatrics* 106 (6), 1307–1317, 2000.
2. Health, United States, 2005 with chartbook on trends in the health of Americans, National Center for Health Statistics, 2005.
3. Ezzati-Rice, T., Kashihara, D., and Machlin, S., Health care expenses in the United States, 2000, Agency for Healthcare Research and Quality, Report No. 04-0022, 2004.
4. Litan, R., Great expectations: Potential economic benefits to the nation from accelerated broadband deployment to older Americans and Americans with disabilities, New Millennium Research Council, 2005.
5. Philipson, T., Posner, R., and Page, M., The long-run growth in obesity as a function of technological change, *Perspectives in Biology and Medicine* 46 (3), S87–107, 2003.
6. Moriarty, D., Kobau, R., Zack, M., and Zahran, H., Tracking healthy days: a window on the health of older adults, *Preventing Chronic Disease* [serial online], 2005.
7. Disability, including prevention, management and rehabilitation: report by the Secretariat, World Health Organization, 2005.
8. Disability and rehabilitation: WHO action plan 2006–2011, World Health Organization, 2006.
9. Mobile phones and development: Calling an end to poverty, *The Economist*, 2005.
10. Bennefield, R. and Bonnette, R., Structural and occupancy characteristics of housing: 2000, United States Census Bureau, Report No. C2KBR-32, 2003.
11. Tractenberg, R., Singer, C., and Kaye, J., Symptoms of sleep disturbance in persons with Alzheimer's disease and normal elderly, *Journal of Sleep Research* 14, 177–185, 2005.
12. September 2005 tracking survey, Pew Internet & American Life Project, 2005.
13. Fox, S., Older Americans and the Internet, Pew Internet & American Life Project, 2004.
14. December 2002 tracking survey, Pew Internet & American Life Project, 2002.
15. Eyesenbach, G., Powell, J., Kuss, O., and Eun-Ryoung, S., Empirical studies assessing the quality of health information for consumers on the World Wide Web, *Journal of the American Medical Association* 287 (20), 2691–2700, 2002.
16. Miller, R. and Sim, I., Physicians' use of electronic medical records: Barriers and solutions, *Health Affairs* 23 (2), 116–126, 2004.
17. Stefanov, D., Bien, Z., and Bang, W., The smart house for older persons and persons with physical disabilities: Structure, technology arrangements, and perspectives, *IEEE Transactions on Neural Systems and Rehabilitation Engineering* 12 (2), 228–250, 2004.
18. Thompson, C., Medications commonly cause problems for community-dwelling elderly, *American Journal of Health-System Pharmacy* 60 (8), 741–742, 2003.
19. Sorensen, L., Stokes, J., Purdie, D., Woodward, M., and Roberts, M., Medication management at home: Medication-related risk factors associated with poor health outcomes, *Age and Ageing* 34 (6), 626–632, 2005.

20. Mann, W. and Helal, A., Promoting independence for older persons with disabilities, in *Assistive Technology Research Series*, IOS Press, Amsterdam, The Netherlands, 2006, p. 244.
21. Wan, D., Magic medicine cabinet: A situated portal for consumer healthcare, in *First International Symposium on Handheld and Ubiquitous Computing*, Karlsruhe, Germany, 1999, pp. 27–29.
22. Hersh, W., Helfand, M., Wallace, J., Kraemer, D., Patterson, P., Shapiro, S., and Greenlick, M., Clinical outcomes resulting from telemedicine interventions: a systematic overview, *BMC Medical Informatics and Decision Making* 1 (5), 2001.
23. Roine, R., Ohinmaa, A., and Hailey, D., Assessing telemedicine: A systematic review of the literature, *Canadian Medical Association Journal* 165 (6), 765–771, 2001.
24. Whitten, P., Mair, F., Haycox, A., May, C., Williams, T., and Hellmich, S., Systematic review of cost effectiveness of telemedicine interventions, *British Medical Journal* 324, 1434–1437, 2002.
25. Reactive to adaptive: Transforming hospitals with digital technology, PricewaterhouseCoopers, Global Technology Centre, Health Research Institute, Report No. TC-02-07, 2005.
26. Leape, L. and Berwick, D., Five years after *To err is human*: What have we learned? *Journal of the American Medical Association* 293, 2384–2390, 2005.
27. Kohn, L., Corrigan, J., and Donaldson, M., To err is human: Building a safer health system, in *Institute of Medicine, Quality of Health Care in America*, National Academies Press, Washington, D.C., 2000, p. 312.
28. Bates, D., Pruess, K., Souney, P., and Platt, R., Serious falls in hospitalized patients: Correlates and resource utilization, *The American Journal of Medicine* 99, 137–143, 1995.
29. Garg, A., Adhikari, N., McDonald, H., Rosas-Arellano, M., Devereaux, P., Beyene, J., Sam, J., and Haynes, R., Effects of computerized clinical decision support systems on practitioner performance and patient outcomes: A systematic review, *Journal of the American Medical Association* 293 (10), 1223–1238, 2005.
30. Koppel, R., Metlay, J., Cohen, A., Abaluck, B., Localito, A., Kimmel, S., and Strom, B., Role of computerized physician order entry systems in facilitating medical errors, *Journal of the American Medical Association* 293 (10), 1197–1203, 2005.
31. Poon, E., Blumenthal, D., Jaggi, T., Honour, M., Bates, D., and Kaushal, R., Overcoming barriers to adopting and implementing computerized physician order entry systems in U.S. hospitals, *Health Affairs* 23 (4), 184–190, 2004.
32. Technology opportunities in the North American healthcare market, Datamonitor, Report No. DMTC1161, 2005.
33. The national nursing home survey, National Center for Health Statistics, 1999.
34. Committee on Redesigning Health Insurance Performance Measures and Performance Improvement Programs, Institute of Medicine, *Performance Measurement: Accelerating Improvement*, National Academies Press, Washington D.C., 2006.

chapter two

Computer science tools and techniques

Martha E. Pollack
University of Michigan, Ann Arbor, Michigan
Bart Peintner
*Artificial Intelligence Center, SRI International,
Menlo Park, California*

Contents

2.1   Introduction 22
2.2   Pervasive computing 23
    2.2.1   Principles of pervasive computing 24
        2.2.1.1   Decentralization 24
        2.2.1.2   Diversification 24
        2.2.1.3   Connectivity 24
        2.2.1.4   Simplicity 24
    2.2.2   Elements of pervasive computing 25
        2.2.2.1   Devices 25
        2.2.2.2   Standards and protocols 26
        2.2.2.3   Application services 28
    2.2.3   Applications to healthcare 28
        2.2.3.1   Reduced cost of current tasks 28
        2.2.3.2   Increased quality of care 28
        2.2.3.3   Peace of mind for caregivers 28
        2.2.3.4   Assistive technology 29
        2.2.3.5   Reduced risk for common activities 29
        2.2.3.6   Unimagined applications 29
2.3   Intelligent applications 29
    2.3.1   Representing knowledge 29
    2.3.2   Probabilistic reasoning 30

        2.3.2.1 Bayesian networks...31

        2.3.2.2 Dynamic models...33

        2.3.2.3 Inference in dynamic models 34

    2.3.3 Machine learning..34

        2.3.3.1 Supervised learning 35

        2.3.3.2 Unsupervised learning 35

        2.3.3.3 Reinforcement learning 36

    2.3.4 Automated planning 37

2.4 Privacy and security..39

    2.4.1 Privacy 40

    2.4.2 Authentication 41

    2.4.3 Authorization...41

    2.4.4 Integrity 42

    2.4.5 Caveat user...42

2.5 Summary..42

2.6 To learn more 43

References 43

2.1 Introduction

This chapter provides an overview of *pervasive computing* techniques that are being used to develop the advanced healthcare systems described in the remainder of the book. Pervasive computing—sometimes also known as ubiquitous computing—can be seen as the third wave of computing paradigms. In the first wave, which started in the 1960s and held sway through the 1980s, computers were large mainframes shared by multiple users. The second wave centered on personal computers with individual users. The third wave of computing, still emerging now at the start of the twenty-first century, is one in which each user has continual access to a large number of networked computers. However, these are not the traditional desktop-style computers as we have come to understand them but instead may include devices such as mobile phones, personal digital assistants (PDAs), and various kinds of sensors. Users do not interact with these computers using a keyboard, mouse, and operating-systems commands; instead interaction is much simpler, and sometimes even automatic, with the pervasive computing system sensing what is needed and acting accordingly. Of course, traditional computers can also play a role in a pervasive computing environment, when they are connected with devices that allow them to support "everywhere, anytime" access to data.

Thus, in the vision of ubiquitous computing, computer-based systems anticipate and respond to the needs of their users to control everyday environments such as homes, workplaces, and automobiles. For example, in a home environment, a pervasive computing system could be used to support a variety of tasks, including turning lights on and off as people move from one room to another, automatically generating shopping lists as items are removed from the refrigerator, and learning a person's favorite television programs and recording them automatically. Within the area of healthcare, pervasive computing techniques could lead to systems in which devices like thermometers and sphygnometers immediately transmit information to a patient's electronic medical records. They could provide a physician in the hospital with instant access on a secure PDA to the up-to-date records of a current patient. Pervasive computing could make it possible for a cognitively impaired person to have his activities monitored and to receive a reminder if he forgot to take his medicine or needed directions to get back home should he get lost walking around his neighborhood. Pervasive computing techniques could also make it possible for a physically disabled person to readily control her environment, with doors opening automatically as her wheelchair approached and the oven responding to commands spoken into a small, wearable microphone.

This chapter will discuss the kinds of computational techniques that make systems like the ones described above possible. We have divided the chapter into three main sections:

- **Pervasive Computing:** We begin by describing the overall pervasive computing enterprise in more detail, discussing the kinds of devices and protocols used.
- **Intelligent Applications:** Next we describe techniques, primarily from the field of artificial intelligence (AI), that are used to make pervasive healthcare applications behave "intelligently."
- **Privacy and Security:** Because of the nature of the data they are using and the services they are providing, it is essential that pervasive computing systems ensure privacy and security. In this section, we describe how privacy and security are ensured.

It is impossible to provide a detailed tutorial on each of these topics. Instead, the goal of this chapter is to ensure that the remainder of this book is accessible to a wide range of readers, who may have different amounts of background in computer science. We highlight key computational challenges in the design of pervasive computing systems and we sketch the types of approaches used to meet them, providing references for readers who want to follow up on particular topics more extensively.

2.2 Pervasive computing

As suggested by the examples in the introduction, pervasive computing is a broad effort to make information-centric tasks simple, mobile, and secure. The effort aims both to uncover new applications and to improve the efficiency of current computing applications by making them available in more places, more often, and with more convenience for users.

One of the most well-known embodiments of pervasive computing is the mobile phone. The mobile phone in its simplest form allows two people to communicate at a distance, the same basic service that the standard phone has provided for decades. The mobile phone has made this service available in more places and—because we now rarely leave home without it—more often. The mobile phone is also an example of a new application made possible by pervasive computing. Current state-of-the-art mobile phones contain digital cameras, games, GPS services, and the ability to communicate in other ways, such as e-mail and text messaging. We do not think of our mobile phone as a computer, but rather just as a tool we use in our daily lives. In fact, this is one of the goals of pervasive computing: to seamlessly integrate information technology into our daily lives.

2.2.1   Principles of pervasive computing

The pervasive computing slogan "everywhere and anytime" summarizes its four basic principles: decentralization, diversification, connectivity, and simplicity.

2.2.1.1   Decentralization
In contrast to the earlier waves of computation where centralized computers performed a wide variety of tasks, the pervasive computing approach has many devices that perform more specialized tasks. Applications and services are provided by multiple networked devices that coordinate their resources.

2.2.1.2   Diversification
The pervasive computing paradigm employs a wide array of devices, all of which can interoperate and share information. Therefore, services are designed in such a way that current devices will be able to communicate with future devices. This focus aims to ensure that infrastructures created today do not limit future possibilities.

2.2.1.3   Connectivity
Interoperability requires communication protocols that must be agreed on by all parties that develop pervasive computing devices and services. For example, the communication protocols used by mobile phones are rigidly standardized to enable phone manufacturers to create phones that communicate and share services with the phones of other manufacturers. Computer languages and protocols that work on different types of hardware are needed to ensure that basic services do not have to be redeveloped for each new device.

2.2.1.4   Simplicity
Pervasive computing has a heavy focus on issues of human–computer interaction. The goal is to design systems for which there is almost no "learning

curve" (i.e., systems that are so intuitive that people can use them as easily as they use their telephones and toasters). In part, the motivation for simplicity is the fact that small, mobile devices cannot support the same kinds of interactions as do computers with large screens, keyboards, and mice; however, simplicity also recognizes that mobile devices and applications must be easy to learn and use if they are to be more useful than the current way of doing things. Being simple is not the same as being primitive—pervasive computing systems can and do have complex functionality. Rather, simplicity focuses on creating effective, easy, and enjoyable methods of interacting with computer-based systems.

2.2.2 Elements of pervasive computing

The tangible results of pervasive computing can be categorized in three main areas: the devices that run pervasive computing applications, the protocols that allow devices and applications to communicate, and the applications and services pervasive computing provides.

2.2.2.1 Devices

We commonly think of pervasive computing devices as small portable devices, such as PDAs, mobile phones, pagers, and mobile digital music players. However, other, more stationary devices also play a part in seamlessly integrating technology into our lives. For example, home entertainment systems and game consoles are now able to connect to the Internet, providing access to digital content and a mechanism for playing games against remote players. Other devices provide behind-the-scenes support for pervasive applications, including network routers and modems, mobile phone towers, and wireless access points.

Many advanced applications, including a number of healthcare applications, rely heavily on sensors that can monitor aspects of an environment and on actuators that can take actions to physically change an environment. For example, patient monitoring is currently an active research topic. The goal is to recognize how well patients have slept, whether they have taken their medicine, whether they are eating and drinking, and so on. Various devices are being explored for these purposes. One approach makes use of a wide variety of environmental sensors, such as motion detectors, which sense a person's movement through the home; contact switches, which can indicate whether a cabinet, refrigerator door, or closet door has been opened; flush sensors on toilets, flow sensors on faucets, load sensors on beds, and so on. Networks of these sensors can be installed in a person's home or in an assisted living facility and can wirelessly transmit information that is useful in recognizing a person's activities.[1,2] An alternative approach is to use radio frequency identification (RFID) technology: here one attaches tiny RF sensors to household objects, and the user wears an RF reader (e.g., installed in a glove or on a piece of jewelry). When the person nears one of the tagged objects, the RF reader is triggered and wirelessly transmits information about the object.[3] When activities can be identified, then assistance can be provided (e.g., in the form of reminders about activities that need to be performed).[4,5] In an outdoor environment, global positioning systems (GPS) can be used to sense information about a person's location and to provide him with directions should he become disoriented and lost.[6] In addition, pervasive healthcare systems may use biosensors, which can range from a stationary device, such as a sphygnometer in the home that transmits a blood pressure reading to a doctor's office, to devices that are worn by a user to measure her temperature, heart rate, and so on, once again transmitting that information over a wireless network.[7] The information sensed in a pervasive computing environment can also trigger actuators that change the environment, such as stoves that automatically turn off if left on for too long without a person nearby, faucets that turn off when the water level is too high, or alarms that are activated when a person forgets to perform an important activity.

2.2.2.2   *Standards and protocols*

Standards and protocols are essential in allowing diverse devices to share information, and are required on many levels. At the hardware level, they are needed to specify which devices can "talk" and also to ensure that arbitrary devices "understand" each other. One level higher are network protocols, which define exactly how to address messages, package information, and ensure an entire message arrives at its destination. At the application level, interfaces must be defined to allow diverse applications to access the same functionality. Standards also define how certain aspects of security are handled, such as how to ensure a user or a device is properly identified, how to ensure a user or a device has permission to access a resource, and how to prevent others from "listening" to private messages. Security is discussed in more detail in Section 2.4 of this chapter.

Important current hardware-level standards include UPnP™ (universal plug and play) and the Tivoli® device management system. Using these protocols, device manufacturers can describe the kinds of information that their hardware produces and accepts—essentially, their I/O (input/output) patterns. By clearly defining how their devices communicate externally, designers enable their devices to be readily integrated into heterogeneous networks. Most computer users are familiar with the convenience of installing plug-and-play peripheral devices: after a new piece of hardware, such as a memory stick or an iPod, is plugged into their computer's USB port, the computer automatically identifies the device and downloads and installs drivers (programs that manage communication between the devices) for it. With plug-and-play, all the user has to do is simply plug in the new device and it is ready to use. This feature is possible because both the peripheral device and the computer make use of the same device management protocols and hence the peripheral is able to "tell" the computer about what it is, as well as about the name and location of the drivers that it needs. This type of essentially automatic network configuration is at the heart of the pervasive computing paradigm, where the goal is to make it possible for arbitrary mobile devices to be *seamlessly integrated* into existing networks whenever and wherever they are needed.

Network protocols have existed for quite a while. Possibly the best-known one is the TCP/IP (Transmission Control Protocol/Internet Protocol), which was developed in the mid- to late 1970s—practically ancient by computer-science standards! TCP/IP is built in to many operating systems and is also the protocol used by the Internet. TCP/IP is one way of ensuring that messages are correctly sent from one location to another in a network. It provides a way of specifying information about the network address to which the message is going, about the size of the message, and so on. It also includes a set of standard techniques for ensuring that the entire message is received at its intended destination and for message retransmission in case of error. Other protocols have been built "on top of" TCP/IP. An important example is Hypertext Transfer Protocol (HTTP), a communication protocol for transferring files of data on the World Wide Web. HTTP is a server–client-based protocol where one system called the client (e.g., a Web browser) makes a request to another system called the server (e.g., a Web server) for particular files, expressing the request according to HTTP protocol. The server replies, also using HTTP protocol to ensure the client can understand the response. By using the same protocol, systems that have completely different applications are able to successfully communicate with each other because they have rules regarding how to structure and interpret communications.

A number of other protocols have been developed in recent years to allow communication among cellular and other wireless devices. These include protocols for cellular telephone communication (there are a large variety of these, such as GSM, TDMA, CDMA, GPRS, and SMS); Bluetooth, a standard that supports communication devices that use a particular kind of small, inexpensive radio chip; Wireless Application Protocol (WAP), which can be thought of as a kind of HTTP for cellular devices; and X10, a protocol that allows home automation devices to communicate using 110V wiring.

Above we mentioned the importance of standardization at the programming and applications level. One of the main motivations for the development of the Java programming language in the early 1990s was to enable programs that were written and compiled on one machine to be run, with identical behavior, on any other machine. To achieve this goal, Java programs are not compiled directly into the machine language for any specific computer, but are rather compiled into a machine language called "Java bytecode." Java bytecode can be run on any computer that has a Java bytecode interpreter, and it will produce the same results, regardless of the machine the code is run on. Thus, although a Java bytecode interpreter must be constructed for each type of computer, once a computer has such an interpreter it can run any Java program that was compiled into Java bytecode on any machine. In a similar fashion, Hypertext Markup Language (HTML) was designed as a standard language for specifying Web pages in a way that (in principle) allows them to look the same on different machines and in different browsers. eXtensible Markup Language (XML) is a standard that outlines how to describe the data that is included on Web pages, enabling different devices to interpret it.

2.2.2.3   Application services

The purpose of developing devices and standards is to provide the necessary infrastructure for application services. Many basic services fall into the general category of communication including e-mail, telephones, text messaging, and video conferencing. Other services deal with managing information, such as remote database access and file transfer. Many services make secure transactions, such as streaming music, trading stock, or placing merchandise orders. A fundamental belief in pervasive computing is that new, previously undreamed-of services and applications will arise when devices, standards, and current services mature. In Section 2.2.3 we briefly mention some of the types of applications that are emerging from the use of pervasive computing in healthcare, and, of course, the remaining chapters of this book provide many more examples of such applications.

2.2.3   Applications to healthcare

How do advances in pervasive computing affect healthcare? Such advances are predicted to lead to profound changes in healthcare, including the following significant effects.

2.2.3.1   Reduced cost of current tasks

Reducing the heavy load of paperwork carried by medical professionals would free up time for these highly trained individuals to concentrate on their primary, caregiving tasks. Mobile devices, if secure and easy to use, could be used as methods for entering much patient data only once, automatically propagating relevant data to the many forms that must be completed. In addition to saving clinicians' time, digital information could be automatically processed, significantly reducing labor costs.

2.2.3.2   Increased quality of care

Pervasive devices could give doctors real-time access to medical records and available research. This could help doctors make more informed decisions about care. Automated reasoning about medical histories and proposed treatments could possibly reduce clinician errors.

2.2.3.3   Peace of mind for caregivers

Sensors and other mobile devices in a home can provide information about elderly or disabled individuals to their caregivers—for example, reducing a caregiver's worries about whether her loved one has fallen or has failed to carry out important activities.

2.2.3.4 Assistive technology

The same sensors and devices used to give caregivers peace of mind can also be used to provide information to people recovering from an injury or living with a disability. For example, for someone with mild memory impairment, sensors could provide enough information to allow an application to issue reminders to users about important items, such as activities that they need to perform.

2.2.3.5 Reduced risk for common activities

Smart sensors and actuators can reduce the risk of injury for those in assisted living environments. Stoves or faucets that reason about when to shut off would reduce the risk of fire or other damage.

2.2.3.6 Unimagined applications

Of course, with new technologies come new ideas for applications. Just as we did not imagine all of the uses for mobile phones before they became part of our everyday lives, we cannot foresee all of the possibilities for healthcare applications and services.

2.3 Intelligent applications

The devices and protocols for pervasive computing make it possible to collect and transmit data to healthcare applications such as those outlined at the end of the previous section. The applications themselves incorporate a wide range of computational techniques, including many that come from research on artificial intelligence (AI). This section surveys those techniques, providing a "gentle" introduction to the interested reader. However, complete understanding of these concepts is not required for reading the chapters in the remainder of this book.

2.3.1 Representing knowledge

Pervasive computing applications tend to be knowledge intensive and make use of a great deal of information about the domains in which they operate. For example, a system that tracks the activities of a person would need to "know" a lot about those activities. It would have to know that making lunch typically occurs around noon, that it involves being in the kitchen and preparing certain kinds of foods, and that preparing food may involve things like opening cabinets and the refrigerator door. A system that helped a user get back home if he became lost would have to know about the normal travel habits of its user as well as understanding maps and bus routes. A system that automated much of the paperwork in a doctor's office would have to know about diagnosis codes and insurance-approval processes and billing procedures.

The field of AI has long been concerned with knowledge-intensive applications and has developed many techniques for representing and reasoning about knowledge. Researchers in the field speak about the design of *ontologies*. Ontologies are common vocabularies used for representing knowledge about some domain that also organizes the vocabularies in some specified way. For example, the early medical expert system MYCIN,[8] which could diagnose blood diseases, adopted a rule-based ontology where it modeled its knowledge in terms of rules that related concepts to other concepts, specifically symptoms to hypotheses, and hypotheses to other hypotheses or to diagnoses of disease. To reason with this type of ontology, MYCIN started with observed symptoms and applied "backward-chaining": first finding rules that matched the symptoms, applying those rules to derive intermediate hypotheses, then finding rules that matched the intermediate hypotheses and applying those, and so on. This style of backward-chaining is commonly used in many AI systems. An alternative style of ontology can be seen in INTERNIST,[9] an early medical AI system that performed diagnoses for problems that would typically be encountered in a general internal-medicine practice. INTERNIST used a frame-based ontology where each disease was associated with a "frame" that included all the information about prototypical presentations of that disease. The frame was structured into slots (e.g., "symptoms") and fillers (e.g., "rash"). Reasoning then occurred by having the system employ a pattern-matching algorithm to find the frame that most closely matched the observed symptoms.

Modern intelligent systems, including the kinds of pervasive healthcare systems that are the focus of this book, tend not to use either pure rule-based or frame-based ontologies. As we describe in more detail in the next section, this is because most of these systems require the ability to reason extensively about uncertain information, something that the rule-based and frame-based ontologies handled only in somewhat ad hoc ways. However, rich ontologies still play an important role in these modern systems and are integrated with probabilistic reasoning techniques. For example, the designers of the ILSA[1] system have developed an extensive ontology for use in systems that support the caregiving process (e.g., systems that provide reminders to cognitively impaired people or that issue alerts to the human caregivers of cognitively or physically impaired patients). This ontology consists of a hierarchically structured network of more than 1,000 terms that can describe many caregiving activities.

2.3.2   *Probabilistic reasoning*

Many interesting pervasive computing applications require a method to make decisions or recommendations based on incomplete or incorrect information. For example, a doctor may want to enter a list of symptoms into a portable device and have it search databases for rare conditions that match or partially match those symptoms. Or perhaps an application that monitors the location of a patient may need to locate her using sensor information that is insufficient or simply inaccurate, especially because most current sensor systems are "noisy." Applications like these require tools for reasoning about uncertainty—in other words, a way of converting incomplete or inaccurate information into "beliefs" about the situation being sensed or reasoned about. The field of probabilistic reasoning focuses on this problem and has provided tools for compactly representing and reasoning about uncertain knowledge and relationships.

At the core of probabilistic reasoning is the *probability distribution*. Given a set of possible situations (also called outcomes or states), a probability distribution defines the probability that each possible situation is the true situation. Consider a case in which a patient's symptoms indicate three possible conditions: a cold (denoted "C"), influenza ("I"), and food poisoning ("F"). One probability distribution could be {C = 0.1, I = 0.7, F = 0.2}. Given such a distribution, a doctor would strongly suspect that the patient has influenza, but further tests might be used to rule out food poisoning.

With only three outcomes, a probability distribution is easy to manage. However, distributions with millions of elements are very common and require a more structured representation. Imagine a distribution that defines a probability for every combination of symptoms and diseases. With only fifty conditions and twenty symptoms, the distribution would have more than fifty million elements.

2.3.2.1   Bayesian networks

The Bayesian network (BN) provides a compact, graphical way to represent large probability distributions. Its strength comes from its ability to represent *conditional independence*. For example, we might imagine that the probability that a person has a cough does not depend in any way on the probability that he has high blood pressure, and vice versa. If this is true, then the two conditions (cough and high blood pressure) are said to be conditionally independent of one another. The BN representation allows one to encode this conditional independence and thus reduce the number of elements that need to be included in the probability distribution: for example, we would not need to include a separate value for the probability of cough given high blood pressure, because that is the same as the probability of a cough without high blood pressure.

Figure 2.1(a) shows a BN that illustrates the problem mentioned above, in which there are fifty conditions and twenty systems. The BN factors the probability distribution over symptoms and conditions into twenty-one nodes: one node, labeled *Conditions*, represents all fifty conditions, while each of the other nodes each represent whether a particular symptom exists. In other words, the node labeled *Conditions* can be assigned any value that represent(s) the condition(s) a person might have. For example, it could be set to "cold" or "food poisoning" or even "cold and food poisoning." Each of the nodes labeled *Symptom* represents a particular symptom, so the node labeled *Symptom 2* might represent the fact that the person has a cough.

In this BN, the arrows represent cause and effect and thus point from condition to symptom, because the condition causes the symptoms. We can use the BN in Figure 2.1(a) to take a set of known symptoms in a given

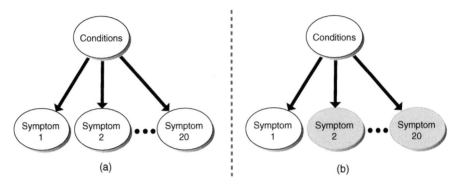

Figure 2.1 (a) A Bayesian network representing a probability distribution over conditions and symptoms. (b) A subset of the nodes set in evidence before inference.

patient and set them into evidence. Thus, suppose that a patient exhibits symptoms 2 and 20, as shown in Figure 2.1(b), where they are indicated by filled-in circles: if symptom 2 is "cough" and symptom 20 is "fever" then setting these in evidence represents the case where the patient has a cough and a fever and no other symptoms. Using the evidence that the patient has only symptoms 2 and 20, a probability distribution over all conditions can be calculated. It would tell us, in this example, that the probability that a person has a cold is relatively high, that the probability that he has a cold and influenza is relatively low, and that the probability that he has food poisoning is very low.

Calculating a probability distribution using evidence is known as *probabilistic inference*. Many different types of inference methods for Bayesian networks have been developed in recent decades. Some of these aim to calculate exact distributions, while others attempt to quickly calculate a distribution that approximates the true one.

While the example in Figure 2.1 illustrates the use of BNs for medical diagnosis problems, BNs are also frequently used for sensor interpretation. For example, consider the problem of activity recognition, which was mentioned in the previous section. A sensor network may detect that a person has entered her kitchen, opened a particular cabinet, used the electric can opener, turned on the faucet, and turned on the stove. What activity is she engaged in? If it is around noon, it may be fairly likely that she has opened a can of soup for lunch, but it is also possible that she is instead getting food and water for her cat. Just as medical conditions probabilistically give rise to symptoms, the performance of certain activities probabilistically gives rise to sensor firings. The connection is probabilistic, not only because people do not always do things the same way every time but also because the sensors themselves may be noisy, sometimes firing (or failing to fire) erroneously.

2.3.2.2 Dynamic models

BNs only represent static situations: they assume the relationships and situations they model do not change over time. However, often the real situations we want to model do change over time. Imagine we want to know which room or hallway an elderly resident of an assisted living facility is in by using hallway sensors that detect when the patient is nearby. The hallway sensors alone may be insufficient to tell us which room the resident is in, but the sequence of hallway sensor readings could give us a much better idea.

To model a changing world such as this one, the hidden Markov model (HMM) maintains a probability distribution over all possible defined variables (in the example just given, this would be the rooms, hallways, and sensor readings) *for each point in time*. The resulting distribution is referred to as the *state*. At each point in time, an HMM uses the current state to estimate the likelihood of an *observation* occurring, such as the occupant's presence in a particular room, causing the observation of a particular sensor firing. Figure 2.2 shows a graphical representation of an HMM. The figure shows a series of identical *time slices* with two nodes each. The top node represents the state at a specific point in time and the bottom node represents the corresponding observation. The arrows model causality in the same manner as they did in the BN represented in Figure 2.1. For our resident location example, each time slice would represent the point in time when a sensor fired. It is the resident's being in a certain room (state) that causes the sensors to fire (observation) with some specified probability.

HMMs can incorporate and exploit information about impossible circumstances. For example, the current location of a person restricts the possibilities for her next location, as a person cannot instantly jump from one part of the building to another. If a person is in a room with a single door, the next location must either be within the same room or in the hallway outside of that door. Therefore, HMMs enable the construction of a much more accurate model than could be achieved by a simple BN, which, in the case of the tracked resident, would only be able to describe a static relationship between hallway sensors and location.

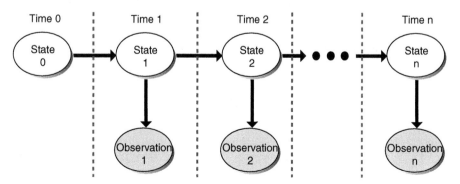

Figure 2.2 A generic hidden Markov model.

Often, the size of the state or the size of the observation can be quite large. For example, imagine that we now want to track not only the location of the resident, but also her heart rate, walking speed, and the activity she is performing. Imagine that we also had several additional sensors that give us extra information about these additional aspects of the patient's state. Just as adding more symptoms to our probability distribution in the BN example caused the size of the distribution to sky-rocket, so does the size of the state and observation nodes explode when we try to monitor these additional elements.

A dynamic Bayesian network (DBN) extends an HMM in exactly the same way that BNs extend a probability distribution. DBNs allow the state and observation nodes in an HMM time slice to be broken into smaller pieces, much like the BN breaks a probability distribution into smaller pieces. The result is a more compact representation that requires more complex algorithms for inference.

2.3.2.3 *Inference in dynamic models*

The goal of inference in dynamic models such as HMMs and DBNs is to use a sequence of observations to infer a sequence of states. In other words, the observation nodes in each time slice are used as evidence and the probability distributions for the state nodes in each time slice are the desired result.

Many types of inference methods exist for HMMs and DBNs, both exact and approximate, but the concepts presented in this book can be understood without knowledge of how they work. However, we will mention one approximate method that is both intuitively appealing and often used. This method is called *particle filtering* and it is used to estimate the probability distribution for the sequence of state nodes that best explain the observations by using a set of "guesses." Each "guess" (called a particle) is a sequence of states. In our resident-monitoring example, a particle would be a sequence of rooms and hallways (with no probabilities) that indicate the resident's path. Each particle is given a weight based on how well it matches the recorded observations. By taking a weighted average of these particles, a distribution over each state can be obtained. The complexity of the algorithm lies in how each guess is obtained and how the set of guesses is maintained.

Probabilistic reasoning methods have been used in a number of very successful applications, both inside and out of the healthcare arena. For instance, Bayesian inference techniques have been widely used in the design of automated medical diagnosis systems for more than twenty years (see references 9, 10, and 11), while much of the recent work on activity recognition for healthcare intervention and delivery makes use of HMMs and DBNs (see references 2, 3, and 6).

2.3.3 *Machine learning*

The discussion above describes some ways we can represent uncertain knowledge and gives hints about how to reason about such knowledge. The techniques above all require a large number of probabilities to relate the information they are reasoning about, such as conditions to symptoms or actions to sensor firings. But where do the probabilities come from?

Probabilities can be estimated when there are many examples of each situation (or outcome) that is being modeled. For example, if we record the condition of thousands of patients that enter a hospital, we might easily be able to define a probability distribution over the most common conditions. However, for relatively rare conditions, thousands of patients may not be enough to create an accurate distribution. In addition, for many applications it is infeasible to obtain enough relevant examples, even if they exist in principle. Although we will not delve into the details of machine-learning techniques, we will briefly discuss the broad classes of techniques developed by the field: supervised learning, unsupervised learning, and reinforcement learning.

2.3.3.1   Supervised learning

One goal of machine learning is *classification*; it aims to sort input objects into one of several categories. For example, it may be desirable for a computer to learn to classify different skin diseases by using photographs. Classification problems are often solved by *supervised learning* techniques, which learn how to classify inputs using a set of example inputs and classifications. In other words, the algorithm is given "the answers" for an ideally large set of examples. A supervised learning technique attempts to build a model that maps known input to the correct classifications. Once this model is created, it can be used to classify other inputs for which the answers are unknown.

Supervised learning has been very successful and has been applied to a very wide range of problems, both inside and outside healthcare. As just a few examples, supervised learning techniques have been used to recognize credit card fraud,[12] to recognize potentially problematic traffic intersections,[13] and to learn the preferences of computer-system users.[14] In the medical domain, these techniques have successfully classified pregnant women who are at high risk for C-sections,[12] predicted recurrence of prostate cancer,[15] and learned to recognize potentially dangerous conditions in patients being monitored in intensive care.[16]

2.3.3.2   Unsupervised learning

As humans, we often have difficulty making sense out of mountains of data. For example, imagine recording the hallway sensor data in our resident-monitoring example for several months. A person looking at the raw data would most likely be overwhelmed by the sheer volume of data and would not be able to extract any useful information.

The goal of unsupervised learning is for a computer-based system to autonomously discover patterns in data. Unlike the supervised case, in which the answers for a training set of data are given in advance, the answers and even the forms of the answers are left for the unsupervised learning techniques to discover. Months of raw data for our resident's movements may reveal that the resident typically spends most of her time in one wing of the assisted living facility or that she is frequently up and about late at night. Such patterns could potentially be used to help improve the resident's quality of life, such as encouraging her to participate in social activities that take place in different parts of the facility or by correcting her medicine to help her sleep better at night.

Both supervised and unsupervised learning can be used to learn the parameters of an HMM or DBN, such as one that describes a resident's movements. In the supervised case, examples of room sequences and associated observations would allow the model's probabilities to be estimated quickly. If such room sequences do not exist, unsupervised techniques have been developed that can estimate the probabilities using only sequences of observations. The basic idea underlying these techniques is to identify natural clusters of data that may correspond to significant features of whatever is being learned. It has been found that by iteratively forming clusters (i.e., "best guess" grouping of data) and then using these clusters to reanalyze the data, one can converge upon good models that turn out to correlate well with phenomena of interest.

2.3.3.3 *Reinforcement learning*

Reinforcement learning techniques are useful when a system must choose an action to perform in a given situation. The goal here is to choose the action that leads to the best possible outcome, or *reward*. Imagine a system designed to notify a nurse when specific indicators for a patient change. The action of notifying the nurse should only be taken when the change is significant enough that it outweighs any inconvenience or irritation a false alarm may cause the nurse. If notifications are too frequent, the nurse will eventually pay less heed to them. If they are too rare, significant changes may go unnoticed. Rather than encoding precisely what is significant and what is not, reinforcement learning can be used to incrementally improve its evaluation of significance, based on whether the nurse gives it positive or negative feedback for notifications it makes or fails to make.

The reinforcement learning framework involves an "agent" (in our example, the system that learns when to issue a notification to the nurse) situated in an environment from which it may sometimes receive feedback. In the simplest formulation, the agent–environment interaction is formulated as a *Markov decision process*. In a Markov decision process there are discrete steps, and at each step the agent first senses its environment, chooses an action to perform (e.g., issue a notification or do nothing), and then receives a payoff based on the action the agent took and what the outcome of that action was. Note that payoffs may be delayed or omitted, and thus the agent has at best incomplete knowledge about the consequences of its actions. The agent has to use its interaction with the environment to learn a near-optimal *policy*. A policy is the strategy for deciding what action to take in each situation. Reinforcement learning algorithms have to address the fundamental challenge of temporal credit assignment (i.e., determining which subset of a long and complex sequence of actions was responsible for the good or bad long-term performance of the agent). Deciding which action to take can become difficult, especially in more complex reinforcement learning problems where the agent's sensors do not give it complete state information because of problems with noise and there is a variable, undeterminable amount of time between actions.

Until recently, reinforcement learning had been successfully applied only to a relatively narrow range of problems. Perhaps most notably, reinforcement learning was used in the design of a very successful computer backgammon game.[17] However, as algorithms and computing power improve, reinforcement learning techniques have begun to be used much more widely, including in certain pervasive healthcare applications. An example of this is a recent study that explored the feasibility of using reinforcement learning to create policies for deciding when to issue reminders to cognitively impaired people.[18]

2.3.4   Automated planning

We just described reinforcement learning, a technique for inferring correct policies for action through a process of (informed) trial and error. However, sometimes we want our systems to directly decide what actions they should perform by reasoning about actions whose outcomes systems already know, rather than by trying out alternatives to see what happens. For instance, a decision support system in a hospital might have knowledge about the various procedures that have been recommended for a patient and might want to schedule those procedures based on current information about the availability of different equipment and personnel. Usually it is not possible to schedule the procedures in any arbitrary order and certain procedures may need to precede other ones. Moreover, there may be contingent dependencies as well, with the need to schedule a specific procedure only if some other procedure has a particular outcome.

Automated planning techniques enable a system to construct plans of actions that achieve specified goals. In these systems, the computer typically has knowledge of a basic set of actions that can be performed, where each action is specified in terms of its *preconditions* (i.e., the thing or things that must be true before an action is done) and its *effects* (i.e., the thing or things that will become true if an action is performed). Given a specification of the current situation and of the goals to be achieved, both of which are expressed in the same language as the action preconditions and effects, the planning system then constructs a plan that is guaranteed to achieve the goal when performed in the initial state. Although this is quite simple when there are only a small number of possible actions and a small number of possible goals, planning can quickly become very time-consuming when the number of actions and goals grows, as in real-world situations. Additionally, the technique described above assumes that the computer's environment is completely observable—that is, it knows with certainty what the state of the world is and what the probable results of its actions are. This is not the case in the real world where noise and other sources of interference can cause input to be unreliable. The problems encountered when applying planning techniques are exacerbated when the planning system also has to reason about the temporal duration of actions or constraints on the times at which the goals are achieved, about actions whose effects are conditional on the circumstances in which they are performed, or about actions whose effects are uncertain. Significant research has gone into developing efficient algorithms for all these cases.

Many of these planning algorithms are based on the idea of *searching*. In a *state-space search*, the planning system first considers the initial state (i.e., the state of the environment at the time the plan will begin to be executed). It then reasons about what actions could be performed in that state by identifying all the possible actions whose preconditions are true in that initial state. The planning system reasons about what the effect will be of applying each possible action, decides which effects look most promising (using what is called an *evaluation heuristic*), and then repeats the process, essentially stringing together sequential actions into a plan. At any point in time, it can temporarily or permanently abandon the sequence it is currently considering and turn to another partial sequence. In another variation, a state-space search proceeds backward by starting with the goal state, considering what actions could lead to that state, and then reasoning about what preconditions must have been true for each of those actions to have been performed. The planning system will repeat this until the initial state is reached, essentially constructing the sequence of actions that form the plan by starting with the last action in the sequence and adding successively earlier actions.

A commonly used alternative to a state-space search is a *plan-space search*. Here, the constructed plans are not complete sequences but rather are only partially ordered with respect to one another (i.e., the plan includes constraints that each specify when some action must precede another). Other approaches to planning include the use of a structure called a *plan graph*, which approximates the effects of sequences of actions to facilitate more rapid plan construction. This technique translates the problem into one of logical inference where the system generates a proof that a sequence of actions will achieve the goal and then uses the sequence that was constructed as part of the proof to decide which actions to take.

When it is important for planners to reason about the times at which actions are performed or goals achieved, *constraint-satisfaction processing* is also used. A constraint-satisfaction problem (CSP) consists of a set of variables (V) and a set of domains (D), one for each variable, and a set of constraints on the ways in which the values from the domains can be legally assigned to the variables. An extremely simple example is shown in Figure 2.3. Here, V has two variables. b represents the time breakfast is finished and m represents the time frame in which medicine must be taken.

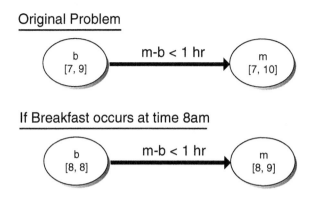

Figure 2.3 A very simple CSP and the result of propagating a constraint in the CSP.

The domain of b is [7, 9], representing the fact that breakfast can be eaten any time between 7 a.m. and 9 a.m. The only constraint is that medicine must be taken within one hour of eating, therefore $m - b < 1$. Consequently, the domain of m is [7, 10]. Should the system recognize that breakfast is finished at 8 a.m., then constraint propagation will cause the domain of m to shrink to the interval [8, 9] indicating that medicine must be taken by 9 a.m. In this case, we say that the legal solutions to the problem are those that assign a value between 8 and 9 to m. Again, this is a simple example and when there are many variables and many constraints, it is very difficult to find a legal solution to a CSP. Indeed, it can even be difficult to decide whether a valid solution exists. However, a great deal of research in computer science has produced algorithms that often perform well in practice on these types of problems.[19]

2.4 Privacy and security

The "everywhere, anytime" philosophy of pervasive computing is a double-edged sword: although it makes information available to users in more places and at more times, it also provides added opportunities for those who wish to steal or corrupt information. In fact, in many areas of information technology a trade-off exists between security and convenience. Typing in passwords or waiting for authorization to use resources is often an unwelcome hassle, but one we accept for a certain level of security. This trade-off is extremely important in pervasive healthcare applications. Busy healthcare professionals do not have time to constantly enter passwords or perform other types of validation when using technology. People who are ill or cognitively impaired may be unable to deal with onerous means of ensuring security. On the other hand, patients and their advocates will not accept the use of any information system that places their personal information at risk to those who could use the information against them.

Given the importance of security in healthcare applications, we cannot expect to simply "add security" once all the other problems are solved.

Security must be a principle concern designed into devices and services from the beginning. The computer industry has learned tough lessons about the dangers of viewing security as an afterthought.

2.4.1 Privacy

When most people think of security, they think of the need for privacy: an assurance that a piece of information is not read by entities other than those who are intended to have access to it. The privacy of information is most vulnerable when it is traveling from one place to another. Information in pervasive computing applications usually travels through media that are shared by others: through the air, through fiber optic cables, or through other devices that direct the information to its destination. In any of these media, packets of information can be read by anyone with knowledge of the information infrastructure and the standards to which they adhere. Thus, in another example of the trade-off between convenience and security, the same standards that allow diverse devices to communicate also allow third parties to eavesdrop on that communication.

The general approach to providing many aspects of security is called *cryptography*. Cryptography is the process of converting a message into a code before transmission (encryption) and converting the code back into the message at the destination (decryption). If two parties have a secret "key" for encrypting a message on one end and decrypting the message on the other end, then the message will not be interpretable by a third party if intercepted. Thus, much of the research on privacy revolves around developing complex security keys that are difficult to "break." Other research focuses on methods for distributing keys and for enabling two parties to decide which key to use. If a third party intercepts the message that contains the key, then they will know how to decode messages that were encrypted with that key. The key itself could be encrypted, but with what key?

There are two main types of key strategies used in cryptography: symmetric and asymmetric. Symmetric strategies require a single key for both the sender and the receiver devices. The sender encrypts the message using the key and an encryption algorithm and the receiver decrypts it using the same key and algorithm. These approaches are usually very fast and are good for sending large amounts of data. The drawback is that the key somehow must be determined and distributed securely.

Asymmetric strategies use two keys: one for encryption called the *public key* and the other for decryption called the *private key*. The problem of determining and distributing keys is avoided in this scheme by allowing each device to have its own public and private keys, which are mathematically related in such a way that a message encoded using the public key can only be decoded using the matching private key. The public key is distributed freely, while the private key is only known by its owner. Using this system, if device X wants to send a message to device Y, X encrypts the message using Y's public key. When the message reaches Y, Y uses its private key to decrypt it. Because the message can only be decoded using an undistributed private key, third parties can only understand intercepted messages if they guess the private key.

Guessing the key sounds impossible, but because computers excel at quickly trying different combinations, the key must have many digits to make trying all possible combinations next to impossible. The longer the key, the harder it is to guess. As computers become faster and better at cracking codes, the keys must grow in length. If longer is more secure, you may ask why the keys are not made as long as possible. Asymmetric strategies are much slower than their symmetric counterparts and their speed depends on the length of the key. Therefore, a trade-off between speed and security must be made.

2.4.2   Authentication

Authentication is the process of proving an identity. Secure systems require this process to prevent third parties from accessing resources by using another's identity. Authentication is most commonly achieved through passwords. Systems assume that if you know your username and password, you are who you say you are. However, passwords, like keys, can be guessed, either by trying all possible passwords or by using knowledge about the person who chose the passwords. Passwords are often required to be at least some minimum length (e.g., eight characters) to make the first method more difficult. Nonalphabetic characters, like punctuation marks or numbers, are often required to prevent people from simply using their dog's name or other easily guessable passwords.

A password system can be strengthened by asking additional personal questions. However, this adds inconvenience. If the protected resource is very important, such as a bank account, then the inconvenience may be tolerated. For often-used resources like an e-mail account, the extra security may not be worth the trouble. Cryptography is (and should be) used in the authentication process as well. If passwords are not encrypted, then they are vulnerable to theft. Encryption also prevents those who administrate computing systems from viewing stored passwords. Biometric data provides an alternative to passwords. Physical characteristics of a person, including their fingerprints, voices, and eye patterns, cannot be guessed or easily faked.

2.4.3   Authorization

While privacy deals mostly with ensuring that communication is secure (i.e., that unknown people are not able to access information), authorization aims to guarantee that resources are only used by those authorized to use them. Authorization occurs after authentication: once a system determines who the user is, it can determine which resources the user can access.

When a system has many users and many resources, it becomes unwieldy to define which users have access to which resources. Therefore, authorization research focuses on allowing flexible specification of authorization policies. Under these policies, users and resources are categorized into groups, allowing an administrator to allow or deny entire user groups access to groups of resources.

2.4.4   Integrity

When one party sends a message to another, it is often important to verify that the message sent is the same one that is received. Even though a third party may not be able to understand a message (thanks to encryption), the third party may still be able to corrupt the message. Even without a malicious third party, messages may get lost or corrupted by the network infrastructure or by other programs on the devices simultaneously sending and receiving. Therefore, a secure communication channel must have built-in ways to ensure the integrity of a message.

Integrity, like privacy, may be ensured using a "keylike" system. Before a message is sent, the message is run through a *hash function*, which computes a single number based on the entire message. For example, a simple hash function might count the number of vowels in the message and multiply that by the number of time the letter "t" occurs. This number is tacked on to the end of the message before transmission. When the message is received, it is run through the same hash function. If the number calculated by the hash function matches the number at the end of the message, then the message is believed to be intact. Of course, some hash functions are better than others. The function described above is probably more likely to be a better identifier for a message than one that simply counts the number of characters before the first nonalphabetic character.

If a third party knows the hash function, it could modify the message and modify the hash number as well. Therefore, the hash function itself must be distributed securely before messages are transmitted.

2.4.5   Caveat user

A computer system is never completely secure. It is always possible for a third party to guess a key, a hash value, or a password. However, we can control the likelihood of security being broken by choosing a longer key, choosing a longer password, or adding more hurdles before access is granted. In general there is a trade-off between security and convenience. Where a system needs to be in this trade-off depends on many factors, including the cost of a security break and the cost of the added inconvenience.

2.5   Summary

The field of computation is changing rapidly. Since they were first invented, computers have inevitably become smaller, faster, and cheaper. However, we have now moved into an era in which they have also become pervasive. They are no longer devices that sit in machine rooms or even on desktops but are instead embedded throughout our environments, where they are connected to one another either through wires or, more frequently, wirelessly. When combined with the power of advanced algorithms for automated reasoning, planning, and learning, as well as techniques for ensuring privacy and security, these pervasive devices will lead to intelligent applications that will radically change the way healthcare is delivered.

2.6 To learn more

There is a large amount of technical literature available on all the topics described in this chapter. Several books provide good overviews on pervasive computing,[20] wireless sensor networks,[21] and network protocols.[22] There are also numerous books available with details about each of the different protocols—for example, on TCP/IP,[23] HTTP,[24] and Bluetooth.[25] A widely used textbook provides a comprehensive survey of AI techniques.[26] There are also a number of good texts on probabilistic reasoning,[27,28] machine learning,[29,30] automated planning,[31] and constraint satisfaction processing.[19] Discussion of computer security techniques can be found in several sources.[32,33]

References

1. Karen Zita Haigh, Liana M. Kiff, Janet Myers, Valerie Guralnik, Christopher W. Geib, John Phelps, and Tom Wagner, "The Independent LifeStyle Assistant™ (I.L.S.A.): AI Lessons Learned." In *The Sixteenth Innovative Applications of Artificial Intelligence Conference*, pp. 852–857, 2004.
2. Emmanuel M. Tapia, Stephen S. Intille and Ken Larson, "Activity Recognition in the Home Setting Using Simple and Ubiquitous Sensors," *Proceedings of PERVASIVE*, pp. 158–175, 2004.
3. Matthai Philipose, Kenneth P. Fishkin, Michael Perkowitz, Donald J. Patterson, Dirk Hahnel, Dieter Fox, and Henry Kautz, "Inferring ADLs from Interactions with Objects," *IEEE Pervasive Computing* 3:50–56, 2004.
4. Martha E. Pollack, Laura Brown, Dirk Colbry, Colleen E. McCarthy, Cheryl Orosz, Bart Peintner, Sailesh Ramakrishnan, and Iaonnis Tsamardinos, "Autominder: An Intelligent Cognitive Orthotic System for People with Memory Impairment," *Robotics and Autonomous Systems*, 44:273–282, 2003.
5. Jennifer Borger, Pascal Poupart, Jesse Hoey, Craig Boutilier, Geoff Fernie, and Alex Mihailidis, "Decision-Theoretic Approach to Task Assistance for Persons with Dementia," *Proceedings of the 9th International Joint Conference on Artificial Intelligence*, pp. 1293–1299, 2005.
6. Donald J. Patterson, Lin Liao, Krzysztof Gajos, Michael Collier, Nik Livic, Katherine Olson, Shiaokai Wang, Dieter Fox, and Henry Kautz, "Opportunity Knocks: A System to Provide Cognitive Assistance with Transportation Services," *Sixth International Conference on Ubiquitous Computing*, Nottingham, England, 2004.

7. Bodymedia, www.bodymedia.com.
8. Bruce G. Buchanan and Edward H. Shortliffe, editors, *Rule-Based Expert Systems: The MYCIN Experiments of the Stanford Heuristic Programming Project*, Addison-Wesley, 1984.
9. Randolph A. Miller, Harry E. Pople, Jr., and Jack D. Myers, "Internist-1: An Experimental Computer-Based Diagnostic Consultant for General Internal Medicine," *New England Journal of Medicine*, 307(8):468–476, 1982.
10. Michael Schwe, B. Middleton, David E. Heckerman, Max Henrion, Eric J. Horvitz, and Gregory F. Cooper, "Probabilistic Diagnosis Using a Reformulation of the INTERNIST-1/QMR Knowledge Base I: Probabilistic Model and Inference Algorithms," *Methods of Information in Medicine*, 30:241–255, 1991.
11. David E. Heckerman, and B. N. Nathwani, "An Evaluation of the Diagnostic Accuracy of Pathfinder," *Computer and Biomedical Research*, 25:56–74, 1992.
12. Tom M. Mitchell, "Does Machine Learning Really Work?" *AI Magazine*, 18(3):11–20, 1997.
13. Saso Dzeroski, Nico Jacobs, M. Molina, and C. Moure, "ILP Experiments in Detecting Traffic Problems," *Proceedings of the 10th European Conference on Machine Learning*, pp. 61–66, 1998.
14. Geoffrey I. Webb, Michael J. Pazzani, and Daniel Billisus, "Machine Learning for User Modeling," *User Modeling and User-Adapted Interaction*, 11:19–22, 2001.
15. Bla Zupan, Janez Demar, Michael W. Kattan, J. Robert Beck, and Ivan Bratko, "Machine Learning for Survival Analysis: A Case Study on Recurrence of Prostate Cancer," *Artificial Intelligence in Medicine*, 20(1):59–75, 2000
16. Katharina Morik, Michael Imboff, Peter Brockhausen, Thorsten Joachims, and Ursula Gather, "Knowledge Discovery and Knowledge Validation in Intensive Care," *Artificial Intelligence in Medicine*, 19(3):225–249, 2000.
17. Gerald Tesauro, "Temporal Difference Learning and TD-Gammon," *Communications of the ACM*, 38(3):58–68, 1995.
18. Matthew Rudary, Satinder Singh, and Martha E. Pollack, "Adaptive Cognitive Orthotics: Combining Reinforcement Learning and Constraint-Based Temporal Reasoning," *21st International Conference on Machine Learning*, July 2004.
19. Rina Dechter, *Constraint Processing*, Morgan Kaufmann, 2003.
20. Uwe Hansmann, Lothar Merk, Martin S. Nicklous, and Thomas Stober, *Pervasive Computing*, 2nd ed., Springer-Verlag, 2003.
21. Feng Zhao and Leonidis Guibas, *Wireless Sensor Networks*, Morgan Kaufmann, 2004.
22. James F. Kurose and Keith W. Ross, *Computer Networking: A Top-Down Approach Featuring the Internet*, 3rd ed., Addison-Wesley, 2004.
23. Behrouz A. Forouzan, *TCP/IP Protocol Suite*, 2nd ed., McGraw-Hill, 2003.
24. Brian Totty and David Gourley, *HTTP: The Definitive Guide*, O'Reilly, 2002.
25. Robert Morrow, *Bluetooth Operation and Use*, McGraw-Hill, 2002.
26. Stuart Russell and Peter Norvig, *Artificial Intelligence: A Modern Approach*, 2nd ed., Prentice-Hall, 2002.
27. Judea Pearl, *Probabilistic Reasoning in Intelligent Systems*, Morgan Kaufmann, 1988.
28. Arnaud Doucet and Nando de Freitas, editors, *Sequential Monte Carlo Methods in Practice*, Springer-Verlag, 2001.
29. Tom M. Mitchell, *Machine Learning*, McGraw-Hill, 1997.

30. Richard S. Sutton and Andrew G. Barto, *Reinforcement Learning: An Introduction*, MIT Press, 1998.
31. Paolo Traverso, Malik Gallab, and Dana Nau, *Automated Planning: Theory and Practice*, Morgan Kaufmann, 2004.
32. Abraham Silberschatz and Peter Baer Glavin, *Operating Systems Concepts*, 5th ed., John Wiley and Sons, 1999.
33. Charles P. Pfleeger and Shari Lawrence Pfleeger, *Security in Computing*, 3rd ed., Prentice-Hall, 2002.

section two

Architectures, systems, and technologies for pervasive healthcare

chapter three

Pervasive computing in hospitals

Jakob E. Bardram
University of Aarhus, Aarhus, Denmark
Heribert Baldus
Philips Research Laboratories, Aachen, Germany
Jesus Favela
Centro de Investigación Científica y de Educación Superior de Ensenada, Ensenada, Mexico

Contents

3.1 Introduction 50
3.2 Contemporary computer technology in hospitals 52
    3.2.1 Patient monitoring 52
    3.2.2 Hospital information systems and
            electronic patient records.. 53
    3.2.3 Picture archiving and communication systems (PACS)........... 54
3.3 Challenges for computer technology in hospitals.............................. 55
    3.3.1 Nomadic work... 55
    3.3.2 Collaboration and coordination.. 56
    3.3.3 Mobility among heterogeneous devices..................................... 57
    3.3.4 Rapid context switching 58
    3.3.5 Integration of the digital and the physical world 59
3.4 Current trends in pervasive computing research
    for use in hospitals 59
    3.4.1 Mobile and pervasive computing 60
    3.4.2 Wireless sensor networks for patient monitoring.................... 64
    3.4.3 New human–computer interaction technology......................... 65

    3.4.4   Hospital groupware and collaboration support.......................66
    3.4.5   User authentication and security 69
3.5   Discussions and conclusions..69
References 75

3.1   Introduction

Pervasive computing offers a compelling vision of unobtrusive, proactive aids to our daily life that vary with and are appropriate to our location and activities. In many respects, pervasive computing aims at moving beyond the desktop computer to weave itself into the very fabric of our surroundings, being available for use when needed. Taking the working conditions for clinicians in a hospital into account, this vision is particularly appealing. Hospital clinicians work in a setting that is fundamentally different from the office—they are extremely mobile and often do not even own desks or personal computers; they work collaboratively and seldom on personal tasks on personal computers; their jobs are intrinsically tied to the physical domain of the patient and not to digital material in computer systems; and they constantly alternate between different working contexts both in terms of physical location and the task at hand. Mobility, collaboration, interruptions, ad hoc problem solving, and physical work are fundamental aspects that characterize the work of nurses, physicians, surgeons, radiologists, and so on.

In such settings, pervasive computing concepts and technology seem to offer attractive solutions. One can easily extrapolate Weiser's original ubiquitous computing vision into a hospital setting where pads (equivalent to PDAs), tabs (equivalent to tablet PCs), and liveboards (equivalent to interactive Smart Boards) would be available in large numbers everywhere.[29] These devices would all be seamlessly connected in a wired and wireless network to provide location and context awareness, available for all to use anywhere, anytime. For example, a nurse would pick up a pad in a patient's room and use it for small, quick jobs like documenting medicine handout; a physician would pick up a tab when arriving at the ward, be automatically identified, and then use this pad during the ward round; a radiologist would be able to access and present radiology images in medical-grade quality on an arbitrary liveboard in any of the conference and meeting rooms in the hospital; collaborative software systems for colocated and distributed cooperation regarding patients and their treatments would be available on all three devices. Location and context awareness would ensure that the proper information would be available for easy access in all places and would reduce the chances of relevant medical information being overlooked.

However, for historical reasons, pervasive computing research has focused almost exclusively on the well-known office and home environments. Taking pervasive computing into the hospital could substantially improve the understanding of requirements for nonoffice environments and would benefit hospital procedures and thus patients and clinicians. Hence, by moving pervasive computing into the hospital, we would improve the working conditions of clinicians and thereby help them do what they are good at—treating and caring patients.

Moving pervasive computing into hospitals, however, puts forth a range of challenges. The idea of publicly available devices originally pursued in the visions of ubiquitous computing would need to be supplemented with proper security and privacy-enabling technologies. A hospital is a rugged environment that has little resemblance to office and meeting settings normally studied in pervasive computing research. A hospital environment would put hardware and software under an extraordinary degree of wear, especially physically but also in usability—there is little room for unresponsive systems or complicated user interfaces in life-critical situations.

This chapter introduces some of the main computer-based systems used in hospitals today, discusses some of the key challenges for the use of computing and networking technology in today's hospitals, and examines how current software and hardware technology is evolving to meet some of these challenges.

Section 3.2 describes current computer technology used in hospitals, including hospital information systems (HIS), picture archiving and communication systems (PACS), and intensive care monitoring systems. These three types of systems work to a different extent as the computational backbone in all hospitals and, hence, these systems are candidates for change.

Section 3.3 discusses the core challenges to computer technology as deployed in a hospital setting. The central argument is that contemporary computer technology has evolved during the last forty years to primarily support personal office work at a desk. Because clinical work in a hospital is so fundamentally different from office work, many of the modern software and hardware components fit poorly into a hospital environment. The discussion of these fundamental challenges is based on ethnographic research on the issues faced by the hospital users of modern computers.

Pointing into the future, Section 3.4 introduces examples of current research aimed at applying pervasive computing technology in hospitals. There are numerous efforts that attempt to address the challenges. These efforts include research on mobile computing in hospitals, location- and context-aware computing, wireless communication, software infrastructures, support for cooperation and social awareness, and multimodal interaction with computers (e.g., during a surgical operation).

Based on these research examples, we discuss some key considerations to take into account when designing, developing, and deploying pervasive computer technology in hospitals. These considerations range from security issues, such as user authentication, to fundamental software infrastructure and operating systems concerns, to the integration of digital systems as representations of the physical world, a vital consideration for the success of digital-based pervasive technology in physical-based clinical work.

The chapter concludes with a summary of the key points that were presented and contemplates the future direction of pervasive computing in hospitals.

3.2 Contemporary computer technology in hospitals

Computer technology is continuously evolving in hospitals. Incorporating stand-alone solutions for monitoring and diagnosis, such as ECG monitors and x-rays, communication technology allows distributed hospital information systems (HIS); connected picture archiving and communication systems (PACS); electronic patient records (EPR); and networked intensive care monitoring systems. Such types of systems are very common in all modern hospitals today.

Hospital communication is moving toward wireless solutions, which are increasingly important for both administrative and clinical applications. Wireless and portable technologies help increase efficiency by allowing the equipment to be brought to the patients and not vice versa. Wirelessness enables the complete mobility of devices and persons, a new degree of flexibility of caregiving, and seamless access to all different types of patient data wherever and whenever needed, resulting in an increased quality of care, operational efficiency, and convenience. Wireless patient monitoring, for example, allows recovering patients to be mobile instead of being confined to a hospital bed close to a stationary monitor. Wireless access to patient data and real-time vital signs allows a clinician to quickly and efficiently assess the patient situation, without the need for immediate physical presence in noncritical situations.

3.2.1 Patient monitoring

Patient monitoring systems can provide real-time, continuous, or intermittent assessments of critical physiological parameters. This availability of a wide range of vital-sign parameters plays an important role in the clinical benefits that patient monitoring can provide. An unfortunate side effect is the required cables, which tether patients to stationary monitors, compromise patients' movements, and hinder access by the caregiving staff.

Traditional monitoring systems track different types of biological measurements (like ECG, arterial oxygen saturation (SpO_2), and blood pressure) using dedicated sensors that are cabled to a measurement server (like a bedside monitor); via backbone connectivity these sensors can be linked to central surveillance systems. Building on this technology, today's telemetry solutions (i.e., devices used to measure and transmit data) provide a certain degree of patient mobility by connecting cabled sensors to a body-worn telemetry-relay system that forwards the data to bedside monitors and surveillance systems. Enabled by a hospitalwide wireless infrastructure, the coverage of telemetry systems is being extended from isolated, departmental solutions to hospital-wide systems. Usually, the computational infrastructure for the monitoring and communication of clinical data is separated from that used for hospital information and administration to ensure reliability and security.

Along with wireless monitoring, location-based solutions are receiving increased attention. Patient-centric monitoring can be combined with a location estimation service, which can reliably calculate the room where a user or device is located. This could be achieved by automatically deploying wireless communication signal information from an existing telemetry infrastructure. This would make it possible to locate any patient using telemetry immediately in the event of an alarm. By using the same technology for caregivers, the system could also help to locate and alert the closest emergency responder to the patient.

3.2.2 Hospital information systems and electronic patient records

Hospital information systems encompass all the software and applications used in the administrative and business settings, as well as clinical environments in integrated delivery networks, acute hospitals, secondary hospitals, primary care centers, free-standing medical centers, diagnostic centers, and group practices. Today, approximately 40 percent of HIS data is administrative data, while clinical data have a share of approximately 60 percent.[15]

Examples of administrative functions include:

- Materials and supply chain management
- Assets management
- Equipment maintenance
- Financial management
- Human resource management

Clinical systems cover:

- EPRs and patient clinical management systems
- Medicine management systems
- Referral and booking systems
- PACS information management systems
- Laboratory and pharmacy information systems
- Patient management, such as order tracking, reporting, physician access, and charting
- Care management, such as care protocols and care schemes

Historically, different types of systems have been represented by separate computer systems and this is still the general picture. Therefore, considerable effort is spent trying to create an integration standard that would enable different types of systems from different vendors to integrate and cooperate. However, different standards exist and so far there is no generally accepted standard. HL7 is the U.S.-based industrial standard that is used worldwide. However, there is a European standard (HISA), and the ISO 18308 is a standard set for electronic health record (EHR) reference architectures. Hence, system interfaces to modalities and legacy systems (e.g., diagnostic devices or monitoring devices, either for networked access or for complete information system integration) are an ongoing challenge.

Nonetheless, wired and wireless networks are increasingly allowing seamless access to individual systems, providing point-of-care information access, and order and clinical data entry. System access can be performed remotely from fixed terminals or mobile devices, such as tablet PCs or PDAs. Increasingly, access to specific information systems is also being integrated into other hospital functions, such as into monitoring systems. In the future, we expect to see an increase in the integration of differing data sets from various sources, particularly in clinical decision support systems. Mobile access to these services will allow efficient, high-quality caregiving.

3.2.3   Picture archiving and communication systems (PACS)

PACS started as the computerized replacement of conventional radiological films, opening the path toward a filmless clinical environment, where hospital users can acquire, store, transmit, and display images digitally. Today's installations are typically small and link the intensive care unit with the radiology department and a few workstations.

However, advances in medical digital imaging technology are making x-rays, computer tomography (CT) scans, ultrasound (US) scans, magnetic resonance imaging (MRI) scans, and other radiological data increasingly effective. Also, technology for PACS has significantly evolved over the last fifteen years. PACS were mainly developed to archive, manage, and save diagnostic images obtained with different modalities. Evolving from stand-alone radiology systems, PACS are being integrated into the health-care information system environment and cover the following areas:

- Image acquisition—interfacing with digital acquisition devices (CT, MRI, US, etc.)
- Image processing—either at the acquisition phase or post acquisition (e.g., to enhance image quality or analyze images for diagnostics)
- Image viewing—at diagnostic, reporting, consulting, and remote workstations
- Archiving—on short- or long-term storage devices
- Communication—primarily via hospital local area networks but also via wide area networks or public communication services
- Integration—offering one integrated system to the user that includes modality interfaces and gateways to healthcare facilities and departmental information systems

Driven by technology advances and medical needs requiring high-resolution digital images for diagnoses, increasing image resolution and quality result in an extremely high amount of data to be processed, transported, and archived. To fulfill these requirements, there is an increasing trend to merge proprietary medical systems with industry-standard data storage and network equipment. New storage solutions currently allow hospitals to manage more than a terabyte of digital images. High-speed networks, such as transmission via a fiber channel, are used to transport data.

New caregiving procedures increasingly require that all relevant medical data be available at the point of care, namely by moving devices and data to the patients rather than by moving patients to different sets of stationary equipment. This implies a broad coverage of a reliable, fixed PACS infrastructure throughout the hospital. In addition, mobile and handheld devices are being used more frequently by clinicians, imposing new challenges to efficient data management, such as the ability to display images on small screens and the need for reliable, secure, and efficient wireless communication.

3.3 Challenges for computer technology in hospitals

Computers first appeared in hospital environments more than forty years ago, mostly in support of administrative and accounting tasks. These information systems still constitute the computational backbone of many hospitals. However, with the adoption of EPRs, clinical decision support systems (CDSS), and computerized physician order entry (CPOE) systems, the focus of computers in hospitals is increasingly moving away from assisting in administrative tasks to supporting health workers in patient care, where computers are expected to reduce medical errors, improve quality, and reduce costs.

Work in a hospital, however, is very different from work in an office, for which traditional desktop computers were developed. Hospital work demands close coordination and collaboration among specialists distributed in space (i.e., different locations within the hospital) or time (i.e., working different shifts). Hospital workers are constantly moving to locate colleagues, evaluate and care for patients, access information, and obtain other resources. Thus, mobility is a characteristic element of hospital work. In addition, physicians and nurses experience frequent interruptions and often need to change the context of their work based on their locations, the patients they are currently attending, the notification of new lab results, or sudden changes in the states of patients. These working conditions call for a new computing paradigm that is designed for hospital work: one that supports collaboration and coordination, mobility, seamless interaction with heterogeneous devices, and frequent task switching.

In the remainder of this section, we describe results from field studies that characterize hospital work and discuss how current computer technology fails to support it, thus introducing the need for nontraditional forms of computer support.

3.3.1 Nomadic work

Most hospital workers need to move continuously to perform their daily work to access people, knowledge, and resources.[7] In addition to moving between patients' beds, physicians and nurses need to change locations to find artifacts (e.g., patient records, x-ray images, and medications) or locate specialists. Indeed, most hospital workers do not even have desks.

A recent study, conducted in the internal medicine area of a midsize teaching hospital, estimated that hospital staff walk more than one kilometer during each work shift.[23] Physicians in particular spend about 50 percent of their time in "base locations," 35 percent of their time in bed wards, and the rest of their time in other areas such as hallways. Physicians move constantly from one area to another spending, on average, fewer than ten minutes in a given place before changing location.

In contrast with other working environments, information in hospitals is generally not concentrated in a single place but is distributed among a collection of artifacts in different locations. Patients' records are maintained and used in coordination with data on whiteboards, computers, or binders that are located in rooms, labs, common areas, and offices. The entire hospital can be seen as an information space that is "navigated" by hospital staff to get the information required to perform their work effectively.[11]

Physicians and nurses frequently need to locate colleagues. A physician might require the opinion of a specialist to confirm a diagnosis; a nurse might need to contact the doctor in charge of a patient who is showing discomfort or pain; a resident physician might need a couple of free hands to help with an intervention. This does not necessarily require people to move within the hospital, but it very often does.

Hospital workers make decisions and act in ways that are highly influenced by their locations, the locations of others with whom they collaborate, and the locations of relevant artifacts, such as patient records or specialized equipment, required to perform their daily work.[24] All of these locations are essential in determining the type of information staff might require.

In almost all situations, instant information access is required. The adoption of desktop-based computer technology under these circumstances introduces important challenges. Currently, the most common solution is placing PCs throughout the hospital. However, the process of moving to the nearest computer, logging into the system, locating relevant information, and logging out is a rather cumbersome process in many instances if only one record needs to be consulted or one item entered into the system. A more recent trend is the widespread adoption of handheld computers, which increasingly provide access to hospital information such as EPRs through wireless networks. It has been estimated that 40 percent of practicing physicians in the United States used PDAs in 2004; this is more than four times the overall consumer adoption rate.[12]

3.3.2   Collaboration and coordination

Work in a hospital setting is characterized by a high degree of collaboration among specialists, such as physicians, nurses, and pharmacists.[26] A study conducted in a public hospital reported that 70 percent of the medical staff consulted clinical records with other doctors and nurses at least once a day, and all of those who participated in the study reported sharing this information at least once a week.[14] It would be difficult to conceive of a hospital where physicians and nurses do not communicate and coordinate their actions. Nurses need doctors' medical notes to plan and conduct their activities, while doctors' diagnoses and clinical decisions are based on the analysis of patient information obtained and captured by nursing staff. Exchanges of information are intense, occurring face to face, over telephones, or asynchronously using different kinds of messaging systems. The patient is the center of work activity in the hospital and his or her care is the responsibility of several specialists throughout the day.

Care cannot be interrupted, nor should it be affected, by shift changes. Hospital workers make use of shared information artifacts that are flexible enough to convey information across group boundaries and to coordinate work.[26] These artifacts serve as containers of relevant patient data, as well as a channel of communication with other individuals. The most notable example of this is the patient record, which is used by physicians, nurses, and other hospital staff to record the work done on a patient, to extract information relevant to clinical work, and to support clinical decisions. Artifacts used in a hospital to convey information should support diverse needs while still conveying information common to all users.

The coordination of work in a hospital ward utilizes a wide range of different artifacts (whiteboards, notes, patient records, Post-its, etc.), which are highly interdependent and supplement each other in providing an appropriate view on the current status of work.[8] Artifacts such as whiteboards hung on walls communicate information regarding patients' conditions and locations.[28] Mobile artifacts, such as clipboards with an individual patient's records, provided greater detailed patient records including information on medication administered to him or her.

The effectiveness of information artifacts depends on their location but also on their ability to provide adequate information to the user. Due to their different professional backgrounds, hospital personnel are likely to experience problems when defining and agreeing on the most useful way to represent information.[26] Therefore, in order to be effective, an information artifact has to be versatile and able to present large amounts of information in a way that is meaningful to the diverse needs of all the artifact's users.

3.3.3   Mobility among heterogeneous devices

Clinicians are highly mobile and use many different computers and devices as part of their daily work. Often it is difficult to keep track of where a user was in an interrupted task or to transfer a user's session between different computers. As a result, the computational support for clinical tasks must be manually reestablished continuously during a working day. For instance, a nurse might access a desktop computer to record the medications she has given to a patient. She gets interrupted to attend an urgent call and when she returns to her previous task, she might need to log in on a different computer to resume her interrupted task. The nurse might have to retrace several steps to get to the point she was at before the interruption took place.

The problem with current desktop computer technology is that applications run in isolation on homogeneous devices. Computer interaction is designed with the assumption that there is only one user for each computer, which is clearly not the case in hospitals. It is difficult to move a set of applications or services from one computer to another, and it is even more difficult to move them between different kinds of devices, such as from a desktop computer to a PDA. Even though portable devices like PDAs and tablet PCs are increasingly used in hospitals, there is still a need to use different computers during a working day. The use of thin client technology is attempting to address this challenge; however, current solutions still suffer from some of the same limitations. Most clinical applications do not run as Web-based solutions accessible from browsers, because they often require the responsiveness that only a native platform application can currently provide. The use of technology such as Citrix, which provides remote access to desktop computers, does not help to resolve the limitations found in desktop-based computing because although Citrix executes remotely, it still provides the user with a standard desktop PC functionality.

3.3.4   Rapid context switching

Hospital workers are involved in many concurrent activities and they constantly alternate between these activities. Interruptions are a frequent part of hospital work as physicians and nurses need to take care of several patients and attend to different chores.

Constant interruptions and task switching is not exclusive of hospital work. Almost all workers interleave their attention across different areas of concern.[17] As people switch from one task to another they need to integrate resources associated to the new task. A daily clinical round, for instance, involves the evaluation of several patients. To do this, the physician in charge and a group of residents move from bed to bed, gathering information from medical records and direct patient examinations to discuss and decide on the plan of care for the day. The context and resources associated to each case are specific to each patient and often patients are treated more or less in parallel.

Modern computer operating systems are application, rather than task, oriented. They do not integrate well the different information resources that are associated to the care of a patient; even if one is able to display several documents on the screen, switching to a new patient involves manually closing each file that is currently displayed and then opening the files related to the new case in each application of interest. This additional work can be quite time-consuming, creates the potential for errors (e.g., if information from two different patients is accidentally open in different applications), and breaks up the seamless transition desired for task switching.

3.3.5 Integration of the digital and the physical world

Clinical work is inherently tied to the physical world. In contrast to office-style work, which uses text documents, technical drawings, spreadsheets, and bank accounts, the main object of clinical work—the patient—is hard to represent digitally and even harder to treat and care for by using computers. Even though many aspects of clinical work can be represented digitally, such as medical records, radiology images, and basic patient information, manipulating these digital documents does not cure the patient. As many doctors say, "No patient has ever been cured by writing in the record." Core clinical work inevitably lies outside of the computer and in the realm of the physical world. In this case, the physical world includes the patient and all his or her physical parts, as well as items such as medications, test samples (e.g., blood and urine samples), and surgical instruments.

Most contemporary computer systems and clinical computer applications ignore this physical dimension of clinical work and are used merely for retrospective record keeping. In a clinical setting, computer systems may provide a greater benefit to the healthcare profession if they are more closely tied to the physical world by being able to sense it, affect it, and augment it. Technologies for sensing and adapting to the physical world can be used to create a link between a (physical) patient and his or her medical record or between (physical) medication and relevant online information. Technologies for affecting the physical world could include actuators for moving samples around and for adjusting surgical tables and lighting in an operating room. Technologies could, for example, project digitally analyzed three-dimensional images onto the body of the patient before and during an operation to guide the surgeon.

3.4 Current trends in pervasive computing research for use in hospitals

So far, we have read how a hospital is saturated with computing technology yet there are fundamental challenges for contemporary technology in supporting the nomadic, collaborative, interrupted work of clinicians. The basic ideas, concepts, and principles of pervasive and ubiquitous computing offer a new kind of technology that moves beyond the limitations of the personal, desktop computer and into the users' surroundings. Going back to the original work on ubiquitous computing at Xerox PARC, support for mobile and collaborative computing was at the core of the hardware and software developed there.[29] The mobile Tab and Pad and the LiveBoard worked seamlessly together as a common infrastructure. Location- and context-aware systems helped users to locate each other and provided relevant information tailored to the users' work contexts. Different applications were able to support remote and colocated collaborations. These original ideas seem to be very relevant for a hospital environment and thus several academic and industrial research groups have worked to design reliable and effective pervasive computing for a hospital or clinical environment. In this section we give some examples from this research to provide an overview of this effort and to illustrate the core challenges that this research tries to address.

3.4.1 Mobile and pervasive computing

As discussed in Section 3.3, one of the core challenges to hospital work is the mobility of clinicians, patients, and clinical devices. This need for mobility must be supported by computer technology if it is to be used in clinical work. The current strategy for supporting mobile work in a hospital is to use mobile computer technology (PDAs, tablet PCs, and laptops) with a wireless local area network. Many hospitals now have WiFi (IEEE compatibility standards for wireless local area networks) and many clinicians, especially physicians, use PDAs as an information repository that provides access to handbooks on topics such as medicine, departmental procedures, clinical guidelines, and referral forms. In many hospitals laptops are used during ward rounds to access medical information in the EPRs.

However, there are still many challenges in the current use of mobile computers in hospitals. These challenges relate to the following aspects:

Hardware—Mobile hardware designed for office use, such as laptops, tablet PCs, and PDAs, perform poorly in a rugged environment like a hospital. These devices break if you drop them on the floor, are difficult to sterilize or keep sterile, cannot be cleaned with alcohol and are fragile regarding liquids in general, and are still rather heavy to carry around. As a consequence, it is not uncommon to see laptops mounted on wheeled tables around the hospital.

Information overload—The amount of information that clinicians handle is enormous. They treat hundreds of patients and each patient can have voluminous medical records. Hence, navigating this information on the small screen of a mobile device while working in a high-paced environment presents a core challenge to the use of mobile equipment in hospitals.

Heterogeneous devices—As mentioned in previous sections, clinicians do not use a single device. They constantly switch to different computers, depending on the purpose of their current tasks and locations. There is a need to move users between devices. With current mobile technology, the user must reestablish their user session each time they move to another device.

Isolated devices—Clinicians often meet and engage in ad hoc, colocated cooperation while discussing information, such as when a nurse and a physician discuss a patient's medical treatment. Because current mobile devices are inherently isolated, there is no technological support for a colocated exchange of medical information, such as using two tablet PCs together to share views and work together.

Wireless communication—The application domain demands medical-grade quality of communication. However, today's systems still struggle with problems like interference between wireless technologies, limited bandwidth capabilities, and no network roaming within or between wireless technologies.

Several research initiatives address these challenges. The use of context-aware technology is a way to mitigate information overload. In general, medical context-awareness systems characterize situations by identifying "who, where, and what." Context-aware systems will help locate and present relevant information to users by taking into account contextual information, such as a user's identity, role, location, device used, time, and status of an information artifact (e.g., the availability of lab results). Based on the combination of this information, applications could adapt to different situations and behave optimally according to the specific situation, thus improving caregiving procedures. For example, when a physician carrying a PDA linked to a context-aware system is near one of his patients, the system would be able to automatically display a clinical record for that specific patient.[4,24] Or—as illustrated in Figure 3.1—medical images that are relevant to an ongoing operation may be shown on large displays in the operating room. By making a context-aware PACS client for the operating room, the system can more or less automatically provide access to relevant images based on the knowledge of the patient, the type of operation, the surgeon, and the progression of the operation.

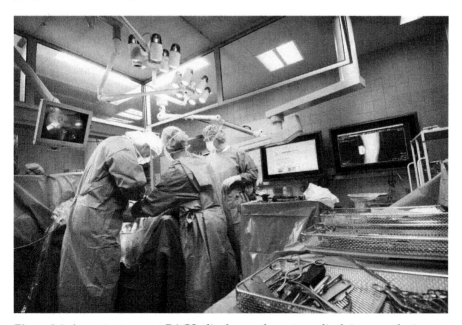

Figure 3.1 A context-aware PACS displays relevant medical images during an operation.

Context awareness can also be used for communication, enabling hospital staff to send different types of messages depending on environmental conditions. For example, a physician could send a message to the doctor responsible for a patient in the next shift when he evaluates the patient and the laboratory results are ready. In this case, the time of day (next shift), the location of the physician (near the patient), and the availability of the results are the conditions that trigger the delivery of the message. Context awareness is also essential for wireless patient monitoring. One application of this is an emergency alarm that is transmitted to the central information center if an at-risk patient is ambulating alone. In the case of patient transport, the alarm would first be sent to the attending clinician. Realizing such an application requires not only information on patient mobility, but also on the current situation in which the patient is mobile.

Application roaming[10] addresses the challenge of having multiple heterogeneous devices in a hospital environment, some of which are mobile. Application roaming supports the transfer of a user session from one device to another, such as from a desktop PC to a PDA. Application roaming requires that the session be adapted to the devices during the transfer (e.g., the smaller screen on a PDA and its limited network access is taken into consideration). It is important to realize that mobile devices in a hospital are not isolated but need to blend into the existing technological infrastructure beyond mere network access. Mobile devices need to work seamlessly with EPRs by not only providing access to clinical data, but also by cooperating with any applications running on different computers in the hospital.

Device composition tries to address the challenge of isolated mobile devices. A composite device is simply a logical device that is made ad hoc from several distinct physical devices.[25] For example, if the nurse and physician, each carrying a tablet PC, met in the ward, they can put these two tablet PCs next to each other and the PCs could merge as one tablet PC with an extended display surface. The nurse and physician can now drag and drop files, objects, and applications between the two displays and interact with them directly. A related technique is to support information transfer between heterogeneous devices, such as the transfer of medical information from a computer with multiple users to a personal PDA.[14] Or a PDA could be used as a remote access point to a display, like an interactive wall-size display, as illustrated in Figure 3.2. Here two physicians collaborate by using one device each. More PDAs can be connected to the large display to support a medical conference.

Looking further ahead, a parallel strategy for supporting mobile computing in hospitals is to embed computers everywhere in the hospital, thereby enabling clinicians to gain access to whatever information they need, regardless of location. This strategy seeks to move away from personal computers to publicly available computers by allowing all clinical personnel to go to an arbitrary computer and start using it as if it were a personal computer. Examples of such computers in the hospital of the future would include large

Figure 3.2 Device composition between a PDA and a large interactive display that is supporting a medical conference.

wall-based displays for conferences, displays build into the tables and walls all over the hospital (especially in bed wards), displays built into patient beds, and mobile displays the size of today's PDAs and tablet PCs. The mobile displays would still be publicly available; such devices would not be personal but could be handed over from one user to another.

Because all contemporary operating systems—namely Windows, Linux, Mac OS—are designed for personal computers, the ability to have seamless context-aware systems requires a new infrastructure layer or some fundamental new operating systems technologies. There are needs for smooth user identification and authentication, for migrating the user's sessions between devices, for adapting to the heterogeneous devices used, and for supporting collaboration while moving around. Several research projects work with these challenges, which are not isolated to a hospital environment. Some projects look specifically at operating systems support for smart spaces, like the Interactive Room project at Stanford[21] and the Gaia project at the University of Illinois at Urbana-Champaign.[27] Projects like Aura at Carnegie Mellon[16] and Activity-Based Computing (ABC) at the Centre for Pervasive Healthcare in Denmark[6,13] research how computers may support human activities and tasks on a higher conceptual level than on the level of files and applications, which is the case in present operating systems. The ABC project specifically focuses on the challenges of the hospital environment.

3.4.2   *Wireless sensor networks for patient monitoring*

Cables connecting a patient to related equipment can be bulky, intrusive, inconvenient, and can even hamper access to the patient by clinicians. To get rid of the various cables at the patient body, the next phase goes beyond telemetry through the introduction of wireless sensors and wireless sensor networks, to provide nonobtrusive monitoring and unlimited mobile monitoring. Medical body area sensor networks consist of intelligent, wireless sensors; transducers; and devices that communicate with one another or within the immediate vicinity of the patient's body. A simple example would be an ID tag, three ECG electrodes, and a bedside monitor with a measurement module to receive the sensor signals. The electrodes would transmit their data at regular intervals, along with an identifying code received from the ID tag, by sending low-power radio signals to the monitor for display. If it became necessary to monitor the patient's blood pressure, for example, the nurse could just add a self-contained blood plethysmometer that combined its data with ECG data to calculate and transmit the blood pressure to the corresponding module.

As this example illustrates, rather than delivering specific, single wireless solutions for different, individual measurements, the core concept of medical body sensor networking is a peer-to-peer, self-configuring platform. Peer-to-peer operation makes the network failure tolerant, as it is not dependent on any particular component. Self-configuration makes it possible to integrate components into the network simply by placing them on or near the patient. Apart from sensors, this type of network could include data recorders or illness-specific processing components. For example, a diabetic processor could combine data from glucose and ECG sensors to provide a better and more immediate clinical picture.

In practice, the medical sensor network will mean more than just getting rid of the cables. The ability of the sensors to communicate with nearby devices means that if the patient needs to be moved, the caregiver could just clip a transport monitor to the head of the bed and wheel the patient off, without having to disconnect, reconnect, or reconfigure cables or sensors. As a patient's condition improves, a caregiver would only need a single telemetry unit to cover the whole range of parameters, as even telemetry would become cableless, with a seamless transition between systems as the patient physically leaves and enters the range of their bedside monitor.

There is, however, a new and interesting challenge associated with wireless monitoring systems. If there are no physical wires connecting the patient to external devices, then how do you know which device is associated to a certain person and which signal comes from which patient?[1] This problem calls for secure and robust automatic identification solutions that operate anytime and anywhere, regardless of the devices or sensors that are attached to patients. This capability is of vital necessity, otherwise the patients' data could get mixed up, causing treatment problems and, quite possibly, life-threatening situations.

Finally, medical sensor networks not only enable flexible hospital monitoring, but also enable the seamless transition toward "care everywhere." Relaying the sensor network to a mobile phone can transform the phone into a new personal healthcare device that not only provides local display and storage, but also can automatically forward alarm messages in critical conditions to emergency centers.

Along with the advances of research in smart sensor technology, external sensors (e.g., located in places such as the patient's bed) will be increasingly available and can be flexibly integrated to a patient sensor network, thus allowing completely unobtrusive monitoring. The absence of wires in a sensory system means the systems could also be built into items of clothing, while advanced low-power wireless technology and intelligent power management systems would ensure that they operated for months or years on small batteries. As a result, patients will be able to go freely about their normal daily lives while being well looked after.

3.4.3   New human–computer interaction technology

Looking at computer use in hospitals, some fundamental human–computer interaction challenges become evident. The use of a keyboard and mouse requires a desk, and even using a laptop requires the user to place the laptop on a horizontal surface. In hospitals, it is not uncommon to see a laptop being placed on the patient's bed as the clinician has no other flat surface to use. Moreover, in clinical situations there is little time and room for tedious clicking and typing. Moving to more acute and intense clinical situations, like those found in operating rooms, it is difficult, if not impossible, for the surgeon to use a keyboard and a mouse. In many ways, new human–computer interfaces are needed in hospitals.

One promising technology for hands-free interaction is speech recognition, which is already in use in several hospital settings. Speech recognition is often used for transcribing continuous speech and is most notably used for dictating to the medical record. The use of a specialized language in the different medical specialties helps to train a speech recognition engine, which can achieve very accurate recognition rates. Speech recognition is also used for executing commands. For example, in the Stryker Endosuite system (www.stryker.com) for endoscopic surgery most of the instruments and tools can be controlled by voice commands from the surgeon. However, voice commands are used less for clinical applications such as EPRs.

A promising technology being explored for data entry in clinical settings is the use of a digital pen that records the user writings on special paper forms. The information can then be transmitted to a computer wirelessly by using the pen's Bluetooth transmitter or by placing the device in its cradle. The special paper allows the pen to register the location where the user writes, and through predesigned paper widgets it can be determined whether, for instance, a box was checked, indicating that a medicine was administered to the patient. The technology, developed by Anoto (www.anoto.com), allows for a seamless transition between the physical and digital domains. This dual representation facilitates data capture while allowing the information to be shared and processed.

An interesting research topic within the field of pervasive computing support for hospitals is the development of multimodal interaction technologies for medical applications. A multimodal interaction is the idea of using more than one way of interacting with a system. For example, allowing a user to point to an object and say "copy this." Multimodal interaction techniques are being researched in an operating room context.[18,19] Today it is very difficult for a surgeon to access medical information while operating on a patient. If she needs to take a closer look at an x-ray, she needs to suspend the operation and walk to a PC, which is typically located in the corner of the operating room. Because the surgeon cannot touch the computer for fear of contamination, she must ask the nonsterile operation nurse to find and display the image. If context awareness was combined with multimodal interaction techniques, the surgeon could access patient data, such as x-rays, directly perhaps by using hand movements and voice commands to manipulate the data. This research is still at a very preliminary and challenging stage. There is, however, no doubt that helping clinicians to access clinical data while their hands are busy is of central importance in hospitals.

3.4.4   Hospital groupware and collaboration support

As discussed in previous sections, collaboration is a crucial part of patient care. However, little computer support for intra-hospital collaboration exists today, as most forms of support are designed to assist coordination. For example, most HIS incorporate basic support for interdepartmental referral and requisitions, like requesting a blood test for a patient. Similarly, workflow types of systems, like computer support for clinical guidelines, have been developed and implemented in hospitals, supporting the distribution of work tasks among a set of cooperating nurses and doctors.

Sharing information is central to cooperation and collaboration. This sharing of information can occur silently without explicit communication, such as when different care professionals examine a patient's medical records. Because the patient is usually the focus of collaborative care efforts, the patient record is a powerful and central tool for implicit communication and coordination among care professionals. Research into cooperation via common information spaces reveals that additional computer support should be embedded into electronic medical records. For example, it should be possible for a clinician to subscribe to events and be notified when they occur (e.g., available lab results). It should also be possible to add meta-comments to the record for each user and to reflect an individual's needs for seeing data in specific ways, while not causing the data to be inaccessible to others.

Furthermore, due to the many medical conferences taking place in hospitals, groupware technology that enables online, real-time conferencing, including video and shared access to medical records, could become quite beneficial. For example, a physician and a radiologist may want to discuss an x-ray image using a conferencing system. Currently, collaborative group-ware support exists outside the EPR and is not integrated with medical applications. As a consequence, clinicians need to go to special teleconference rooms and launch special teleconferencing applications. This means that a collaborative session must somehow be arranged beforehand so that the collaborators can reserve and physically go to a specific conference location. Furthermore, in current groupware technologies you start a session between specific computers, not users. Hence, a potential collaborator must know which computer his colleague is in front of so that he can contact that computer and start a collaborative session.

Some research projects have looked at these challenges. In the Intelligent Hospital project,[22] multimedia conferences were established between users, not machines. The infrastructure routed the conference session between the computers used by collaborators or the ones closest to them. If the user was moving while engaged in a teleconference, the session would roam to follow him or her around. In the ABC project,[6] support for collaboration was designed to be inherent to the runtime environment—basically all sessions can be collaborative and can have a range of participants. In an EPR implemented on top of the ABC environment, activities regarding the treatment of a patient can be shared among a set of participants. If, for example, a physician is engaged in prescribing medicine for a patient and a nurse completed this activity as well, the physician and nurse would be given the option to choose to communicate and cooperate directly or to continue working individually. In the latter case, however, ABC allows them to maintain a peripheral awareness of the actions of one another.

Supporting social awareness among collaborating colleagues in a hospital is another central research area within pervasive healthcare. Field studies have shown that clinicians maintain a peripheral and social awareness of one another in order to keep up to date with the flow of work and to align their own actions to those of others.[3,7,9] For example, people involved in an operation monitor the status and progress of the operation in order to be ready when needed.[28] The AWARE architecture[9] is designed to help develop clinical applications that support these kinds of social awareness. Building on top of a context-aware infrastructure that collects information about clinicians' locations, statuses, and current activities, the AWARE architecture supports the development of different client applications that can help clinicians maintain a mutual social awareness. For example, the Aware-Phone has been built on top of the architecture and enables clinicians to use their mobile phones to look up the statuses, calendar information, and locations of colleagues before calling them. By gaining access to this basic information about a colleague's current working context, the clinician is able to reduce the number of interruptive calls. Hence, she can decide whether it would be more appropriate to call another, less busy, colleague or to postpone the call to a more suitable time. Similarly, the AwareMedia application supports social awareness within an operating ward. AwareMedia runs on large interactive displays and shows the location, status, and current activity of all surgical personnel; the operating schedule for each operating room; the people located inside the operating room, including the patient; and a video feed from inside the room. Figure 3.3 shows the use of the AwarePhone and AwareMedia in an operating ward.

It has been estimated that hospital workers spend up to 5 percent of their work shift time tracking the locations of colleagues and hospital assets.[23] This has motivated the development of location estimation solutions for hospitals. Versus Technology offers hospital tracking systems that use active RFID in badges worn by clinicians and tags attached to artifacts. An alternative solution is the one commercialized by Ekahau, which estimates location based on the intensity of the RF signals emitted by WiFi access points and that are registered in laptops or PDAs. Although less accurate, this technology has the advantage of not requiring additional infrastructure if the hospital already has a wireless network and hospital staff carry mobile computing devices. When integrated with context-aware mapping applications, location estimation can help hospital workers find nearby colleagues to consult a clinical decision or assist in a procedure.[24] Vocera Communications offers WiFi-based badges that estimate location and support two-way voice communication between users. These solutions foster collaboration in highly mobile working environments and are increasingly being adopted in hospitals.

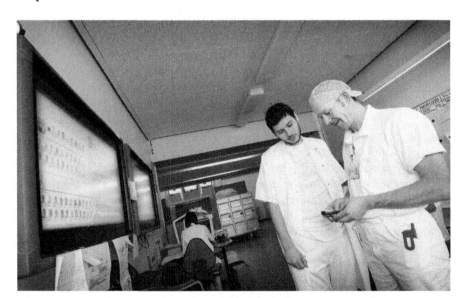

Figure 3.3 The AwarePhone and AwareMedia in use at a surgical ward.

3.4.5   User authentication and security

Studies of hospital work have revealed that frequent user log in and log out is the source of many usability problems.[5] First, users log in often; sometimes a nurse will log in thirty to fifty times during a day shift. This is caused by the combination of the nomadic nature of hospital work and the use of desktop PCs that force the nurse to log in every time she returns from a task. Second, conventional user authentication mechanisms do not support the cooperative nature of medical work. Often clinicians will sit together and work on the same patient. However, because there is no such thing as a "shared log in" only one of them can be logged into the medical record at a time. Hence, they often need to take turns, logging on and off one at a time. This is often time-consuming and annoying, because they are collaborating about the same patient. Finally, combining many log-in and log-out events with the typical user authentication mechanisms of typing usernames and passwords makes logging in in a hospital setting really frustrating. It is not surprising that studies have shown that user authentication is circumvented to a large degree at many hospitals.

Because computer security and access control to personal medical data is of the highest importance in medical computer applications, studies regarding the trouble of logging in should be taken very seriously in the design of clinical applications. It is important to create secure user authentication mechanisms that support frequent log in and log out at a fast pace with minimal manual and cognitive overhead. Hence, technologies like smartcards[2] and biometric systems that use methods such as fingerprint recognition or iris scanning[20] seem like appealing alternatives for this setting, and indeed, they have been deployed in several hospitals around the world. These technologies, however, also have their limitations. For security reasons, a user on a smartcard system still has to type in a PIN code, making this technology only marginally better than the username and password mechanism. Fingerprint recognition is ill suited in a hospital environment where the staff often use latex gloves, and iris scanning is still a rather invasive and expensive technology. Furthermore, these user authentication mechanisms merely address the identification and verification of a single user. Log-in and log-out mechanisms that acknowledge the collaborative and mobile work environment of a hospital ward are still needed.

3.5   Discussions and conclusions

In this chapter we discussed some of the more fundamental challenges with respect to hospital work and we presented current trends in research into pervasive computing technology that address these challenges. Most of the topics covered in this chapter are summarized in Table 3.1.

Table 3.1 A Summary of Challenges and Technologies

Challenge	Description	Technology Support	Issues	Section
Mobile and nomadic work	Hospital work is highly mobile and most users do not have desks to place computers on.	Mobile devices	Portable and small Robust hardware Wireless communication Small displays Limited resources Isolated devices Requires easy access	3.3.1, 3.4.1
		Shared embedded devices	Stationary and large Powerful devices Large displays Networked Requires easy access Requires application roaming	
Collaboration and coordination	Most work in hospitals is highly collaborative and requires much coordination across time, space, and organization.	Coordination, booking, scheduling, workflow	Explicit coordination Work scheduling and booking (e.g., operation plans) Support for workflow (e.g., treatment of certain diseases)	3.3.2, 3.4.4
		Teleconferencing	Real-time videoconferencing across distance Often requires special setup and equipment	
		Shared medical records	Using data entry for coordination Subscribing for change events Leaving traces	

	Social awareness	Mutual awareness of the status of common work (e.g., an operating schedule)	3.4.4
		Rendering visible issues in work that may be relevant to others	
		Monitoring the surroundings for cues on status	
	Device composition	Merging separate devices for shared use in an ad hoc manner	3.4.1
		Easy sharing of resources, like screen, memory, files, etc.	
Heterogeneous device roaming	Web-based applications	Easy deployment	3.3.3
No device fits all tasks in a hospital and therefore users need to move between heterogeneous devices often.		Standard technology	
		Limited functionality and responsivity	
	Application roaming	Adaptation to heterogeneous devices	3.4.1
		Requires elaborate middleware	
		Exploits local resources	
Patient monitoring	Wireless monitors	Mobile patients	3.2.1, 3.4.2
Monitoring of vital body signs are increasingly done wirelessly while being integrated with other systems.	Ad hoc medical sensor networks	Patient identification	
		Wireless communication	
		Sensor fusion	
		Patient location	
		Wireless device association	
		Context-sensitive monitoring	
		Resource-efficient security for wireless sensor networks	

(Continued)

Table 3.1 (*Continued*)

Challenge	Description	Technology Support	Issues	Section
Rapid context switching		Context-aware systems	Sensors to sense the context (e.g., location) Device adaptation based on context Information retrieval based on context Reminders based on context Tagging data entry with context information	3.3.4, 3.4
Integration of the digital and the physical		Physical linking and bookmarking	Making links between digital material and the physical world Tagging using barcodes or RFID	3.3.5
		Augmented reality	Overlaying or displaying digital material on the physical world Merging the digital and the physical in one view	
User authentication		Smartcards	Easy access Requires manual interaction, often PIN	3.4.5
		Biometrics	Easy access Mixed usage in a hospital environment (e.g., fingerprint reading is hard when wearing gloves)	3.4.5

The recurrent theme in this chapter has been the inadequacy of using computer technology designed and developed for office use in a nonoffice setting like a hospital. Two of the most obvious differences are the highly mobile—or nomadic—nature of medical work in a hospital and the degree of intense cooperation taking place. Therefore, computer support for hospital work must consider the fact that most hospital workers are highly mobile and require information throughout the hospital premises. Appropriate technological solutions demand mobile computing and wireless networking solutions, through which physicians and nurses can remotely access and update clinical information. Coverage by a wireless network should enable access from multiple and diverse locations where information is required (e.g., bedside, ward, meeting rooms, etc.) while guaranteeing security.

Although it is the current state-of-the-art solution, consulting information on mobile devices can be unfeasible at times for several reasons: the user might not have two free hands to hold the device and type or scribble on it; there might not be an appropriate surface on which to place the device; the screen size might not be suitable for consulting images or lab results; or it might just not seem socially acceptable to consult a handheld device in front of a patient. These and related issues call for augmentation with other forms of computer devices within the hospital, including traditional desktop computers as well as large displays. The locations of these devices should be such that they are easily accessible to potential users while offering guarantees for privacy. More generally, most computing technology in hospitals should be designed to be publicly available with easy access for everybody while providing the proper mechanisms for maintaining privacy, access control, and security.

No one single device will address all requirements for information access. Indeed, there is a strong demand for the seamless transfer of data and applications from one device to another. With the implementation of data transfer over heterogeneous devices, a physician might receive a notification indicating the availability of a patient's x-ray images on his or her PDA or SmartPhone, move to a large display, and easily access the images to visualize, analyze, and possibly discuss them with a colleague. Future infrastructures should allow the user to log in only once and then transfer information from one device to another quickly and effortlessly. This calls for environments that have built-in support for application roaming and device composition, as discussed in Section 3.3.2.

Collaboration is absolutely vital in getting work done in a hospital and, while there is still room for improvement, functional teleconferencing technologies to support collaboration are already deployed and used in many hospitals. Workflow systems for coordinating work according to clinical guidelines are also being implemented. While these applications are useful, in this chapter we wanted to also draw attention to other kinds of support for clinical cooperation, namely support for cooperation within clinical applications and for social awareness and implicit coordination in hospitals. When designing computer applications for clinical use, it is important to consider how these applications could support these more indirect types of cooperation.

Much coordination in hospitals is done ad hoc and implicitly. For example, even though an overall plan for the flow of operations during a day exists, this plan is constantly adapted and adjusted to the changing circumstances in workflow.[3] Hence, methods for ad hoc adjustment must be available and mechanisms for implicitly conveying the updated plan to all personnel involved must be devised. Large whiteboards hanging on walls in all hospitals are extensively used for ad hoc rescheduling, for displaying the updated schedule to all involved, and for maintaining status information in a prominent way that allows everyone to be aware of the current status of the plan with a glance. This is an example of implicit coordination that does not require explicit communication regarding the adjusted plan and its status. Because of the large number and diverse backgrounds of people on a medical team, implicit mechanisms for coordination and cooperation that represent data of interest to many people are evident in many aspects of hospital work life. Computer systems designed to support hospital work need to take the necessity for diverse collaboration into account. Computational support for implicit coordination could also help distribute and update plans more effectively than whiteboards, making plans visible to interested users anytime and anyplace.

Turning to the physical context, we find that clinicians engage in many parallel work tasks while moving physically in the hospital and using a number of different devices. Technologies for hospital work need to take into account that a user's location and task, and thus computational needs, are constantly changing. Building context-aware technologies into clinical applications can provide a considerable benefit by helping the users locate each other and patients, by retrieving medical data that is relevant to the current work situation, by adapting a device to the situation, and by tagging medical entries with contextual information.

Although the discussion in this chapter mostly focused on software, there are also substantial challenges regarding the hardware to be used in hospitals. The hardware in hospitals needs to be resistant to shocks and liquids, must have limited ventilation in sterile areas, and must preferably be embedded in cabinets or housing in the wall or in separate rooms, thereby containing them as much as possible and making the visible parts easy to clean. Large displays, which are often needed for overviews and for looking at large amounts of medical data, should be developed so that they can be installed discreetly all over the hospital. The unique demands of the hospital environment is driving the development of more appropriate technologies than the standard desktop and laptop PCs.

In this chapter we discussed challenges and research involved with new technology for use in hospitals. Although many technologies are being developed to support work in a hospital environment, they would be beneficial in other settings. It is not difficult to imagine how successful technologies that support the presentation of context-aware information, move user sessions across heterogeneous devices, and provide device composition would have huge ramifications in both general businesses and the home. Hence, we argue that the challenging environment presented by a hospital results in technological solutions that will possess many properties that are relevant in other settings as well.

References

1. Baldus, H., Klabunde, K., and Muesch, G. Reliable Set-Up of Medical Body-Sensor Networks. In *Lecture Notes in Computer Science,* vol. 2920, Springer-Verlag, 2004.
2. Baentsch, M., Buhler, P., Eirich, T., Höring, F., and Oestreicher, M. Javacard—From Hype to Reality. *IEEE Concurrency,* Oct.–Dec. 1999: 36–43.
3. Bardram, J.E. Temporal Coordination—On Time and Coordination of Collaborative Activities at a Surgical Department. *Computer Supported Cooperative Work. An International Journal,* 9 (2) 2000: 157–187.
4. Bardram, J.E. Applications of Context-Aware Computing in Hospital Work—Examples and Design Principles. In *Proceedings of the 2004 ACM Symposium on Applied Computing,* ACM Press, 2004, pp. 1574–1579.
5. Bardram, J.E. The Trouble with Login—On Usability and Computer Security in Ubiquitous Computing. *Personal and Ubiquitous Computing,* 9 (6) 2005: 357–367.
6. Bardram, J.E. Activity-Based Computing: Support for Mobility and Collaboration in Ubiquitous Computing. *Personal and Ubiquitous Computing,* 9 (5) 2005: 312–322.
7. Bardram, J.E. and Bossen, C. Moving to Get Ahead: Local Mobility and Collaborative Work. In *Proceedings of the Eighth European Conference on Computer Supported Cooperative Work (ECSCW 2003),* Netherlands, Klüwer Academic, 2003, pp. 355–474.
8. Bardram, J.E. and Bossen, C.A. Web of Coordinative Artifacts: Collaborative Work at a Hospital Ward. In *Proceedings of the International ACM SIGGROUP Conference on Supporting Group Work (GROUP 2005),* ACM Press, 2005, pp. 168–176.
9. Bardram, J.E. and Hansen, T.R. The AWARE Architecture: Supporting Context-Mediated Social Awareness in Mobile Cooperation. In *Proceedings of the 2004 ACM Conference on Computer Supported Cooperative Work,* ACM Press, 2004, pp. 192–201.
10. Bardram, J.E., Kjær, T.K., and Nielsen, C. Supporting Local Mobility in Healthcare by Application Roaming among Heterogeneous Devices. In *Proceedings of the Fifth International Conference on Human Computer Interaction with Mobile Devices and Services,* vol. 2795 of *Lecture Notes in Computer Science,* Udine, Italy, Springer-Verlag, September 2003, pp. 161–176.
11. Bossen, C. The Parameters of Common Information Spaces: The Heterogeneity of Cooperative Work at a Hospital Ward. In *Proceedings of the 2002 ACM Conference on Computer Supported Cooperative Work (CSCW 2002),* ACM Press, 2002, pp. 76–185.
12. Chin, T. Untapped Power: A Physician's Handheld. AMNews. January 17, 2005. Available at http://www.ama-assn.org/amednews/site/free/bisa0117.htm. Accessed on April 7, 2006.

13. Christensen, H.B. and Bardram, J.E. Supporting Human Activities—Exploring Activity-Centered Computing. In *Proceedings of Ubicomp 2002: Ubiquitous Computing*, vol. 2498 of *Lecture Notes in Computer Science*, Göteborg, Sweden, Springer-Verlag, September 2002, pp. 107–116.

14. Favela, J., Rodríguez, M., Preciado, A., and Gonzalez, V. Integrating Context-Aware Public Displays into a Mobile Hospital Information System. *IEEE Transactions on Information Technology in BioMedicine*, 8 (3) 2004: 279–286.

15. The European Hospital Information Systems Markets. *Frost and Sullivan*, 2004.

16. Garlan, D., Siewiorek, D.P., Smailagic, A., and Steenkiste, P. Project Aura: Toward Distraction-Free Pervasive Computing. *IEEE Pervasive Computing*, 1 (2) 2002: 22–31.

17. Gonzalez, V.M. and Mark, G. Constant, Constant, Multitasking Craziness: Managing Multiple Working Spheres. In *Proceedings of ACM CHI 2004*, ACM Press, 2004: 113–120.

18. Hansen, T.R. and Bardram, J.E. ActiveTheatre—A Collaborative, Event-Based Capture and Access System for the Operating Theatre. In *Proceedings of UbiComp 2005, Lecture Notes in Computer Science*, Toyko, Japan, Springer-Verlag, October 2005.

19. Hansen, T.R., Bardram, J.E., and Soegaard, M. Pervasive Interaction—Using Movement and Speech to Interact with Computers. In *Proceedings of the CHI 2006 Workshop: What Is the Next Generation of Human–Computer Interaction?* Montreal, Canada, 2006. Available at http://www.eecs.tufts.edu/~jacob/workshop/.

20. Jain, A., Hong, L., and Pankanti, S. Biometric Identification. *Communications of the ACM*, 43 (2), 2000, 90–98.

21. Johanson, B., Fox, A., and Winograd, T. The Interactive Workspaces Project: Experiences with Ubiquitous Computing Rooms. *IEEE Pervasive Computing* 1 (2) 2002: 67–74.

22. Mitchell, S., Spiteri, M.D., Bates, J., and Coulouris, G. Context-Aware Multimedia Computing in the Intelligent Hospital. In *Proceedings of the 9th ACM SIGOPS European Workshop*, ACM Press, 2000, pp. 13–18.

23. Moran, E., Tentori, M., Gonzalez, V.M., Favela, J., and Martinez-Garcia, A.I. Mobility in Hospital Work: Towards a Pervasive Computing Hospital Environment. *International Journal of Electronic Healthcare*, March 2006.

24. Muñoz, M., Rodriguez, M., Favela, J., Gonzalez, V.M., and Martinez-Garcia, A.I. Context-Aware Mobile Communication in Hospitals. *IEEE Computer*, 36 (8) 2003: 60–67.

25. Pham, T.L., Schneider, G., Goose, S., and Pizano, A. Composite Device Computing Environment: A Framework for Situated Interaction Using Small Screen Devices. *Personal Ubiquitous Computing*, 5 (1) 2001: 25–28.

26. Reddy, M.C., Dourish, P., and Pratt, W. Coordinating Heterogeneous Work: Information and Representation in Medical Care. In *Proceedings of the Seventh European Conference on Computer-Supported Cooperative Work (ECSCW2001)*. Netherlands, Klüwer Academic, 2001, pp. 239–258.

27. Romn, M., Hess, C., Cerqueira, R., Ranganathan, A., Campbell, R.H., and Nahrstedt, K. A Middleware Infrastructure for Active Spaces. *IEEE Pervasive Computing*, 1 (4) 2002: 74–83.

28. Xiao, Y., Lasome, C., Moss, J., Mackenzie, C., and Faraj, S. Cognitive Properties of a Whiteboard: A Case Study in a Trauma Centre. In *Proceedings of ECSCW 2001*, pp. 258–278.
29. Weiser, M., The Computer for the Twenty-First Century. *Scientific American*, September 1991: 94–104.

chapter four

Pervasive computing in the home and community

Donald J. Patterson
University of California, Irvine, California
Henry A. Kautz
University of Washington, Seattle, Washington
Dieter Fox
University of Washington, Seattle, Washington
Lin Liao
Google Inc., Kirkland, Washington

Contents

4.1 Introduction 80
4.2 Overview of research on assisted cognition 81
    4.2.1 Location, navigation, and wayfinding 81
        4.2.1.1 Nursebot...81
        4.2.1.2 IMP...82
    4.2.2 Wandering alert systems 82
    4.2.3 Home ADL tracking and support..82
        4.2.3.1 Smart homes...82
        4.2.3.2 COACH 83
        4.2.3.3 Autominder 83
        4.2.3.4 PDA-based reminding systems......................................84
        4.2.3.5 Wearable activity recognition systems 84
4.3 Opportunity Knocks: Assisting outdoor navigation 84
    4.3.1 A usage scenario 85
    4.3.2 System architecture...85
    4.3.3 The inference engine..88
    4.3.4 Status 91

4.4   Understanding home activities..91
      4.4.1   Sensing using RFID 92
      4.4.2   Modeling activities...92
      4.4.3   Coarse-grained ADL recognition 94
      4.4.4   Barista: Fine-grained ADL recognition.....................................94
      4.4.5   Modeling choices...95
      4.4.6   Accuracy experiments 97
      4.4.7   Improving robustness...97
4.5   Summary 100
Acknowledgments..101
References 101

4.1   Introduction

One of the major risks to independence for the elderly is the decline in the accomplishment of typical activities of daily living (ADLs) due to mild cognitive impairment (MCI), a precursor to Alzheimer's disease. (In this chapter, we use "ADL" to refer to basic ADLs [such as eating or dressing] as well as to instrumental ADLs [such as cooking or housework].) Lawton[1] measured the impact of impairment on instrumental ADLs and showed that the time spent on activities such as housework, shopping, and recreation declined 27 percent to 44 percent for impaired individuals. Furthermore, regardless of the cause of cognitive disabilities, research suggests that one of the best ways to prolong independence is to encourage the successful completion of ADLs.[2] A side benefit of such an improvement is an increase in the quality of life of family caregivers.

The term assisted cognition (AC)[3] has been coined to describe systems that use sensor data to determine the activities that a person is trying to perform and optionally provide prompts, warnings, or other kinds of interventions to help the person perform the activities safely and independently. Research in AC combines ideas from sensor networks and ubiquitous computing, artificial intelligence (AI), and human–computer interaction (HCI).

The number of sensors that can be used to track human activity, both in the marketplace and deployed in the environment, is quickly growing. They include familiar consumer-grade technologies like global positioning systems (GPS), Wi-Fi, and radio frequency identification (RFID) tags, as well as more specialized technologies such as MICA motes (small low-power CPUs optimized for sensing). Research in AI, some of which is described in this chapter, has developed methods for interpreting noisy sensor data in terms of hierarchical models of subjects' physical actions, activities, goals, and cognitive states. Work on HCI has shown that people with cognitive disabilities can effectively use interfaces that present simple choices and instructions and that employ simple text and meaningful photographic images.[4,5,6]

AC has the potential to be useful for many types of assistance:

- Logging: Logs of ADL performance provide valuable data for health-care professionals. Creating logs automatically would be more comprehensive and less expensive than relying on manual observations.
- Rating and trending: A system that provides direct assistance to a subject must be able to rate how well an activity is performed. Evaluating how performance changes over time, even in subtle ways, provides information about cognitive decline.[7]
- Guidance: Providing immediate feedback about the results of evaluation could help an individual successfully complete difficult ADLs. Guidance strategies must be carefully designed to avoid increasing the subject's cognitive load and should take into account the probability and associated costs of providing incorrect guidance.
- Actuation: In some situations, an AC system should simply perform an activity for the user, in order to improve immediate safety and security; for example, it might lock the doors at night after the user has gone to sleep.

This chapter will review academic and commercial research on assisted cognition systems and then consider two projects from the University of Washington in detail: an outdoor navigation system called Opportunity Knocks and an indoor ADL tracking system called Barista.

4.2   Overview of research on assisted cognition

There are a number of research groups that have progressed in developing assistive technology with an element of cognitive reasoning.

4.2.1   Location, navigation, and wayfinding

Location-sensing technologies, which are surveyed and evaluated in reference 8, are the foundation for wayfinding systems. More recently, the Place Lab initiative[9] is a project designed to make outdoor Wi-Fi localization ubiquitous through mass collaboration. A single sensor can be augmented with a user model, possibly learned, to improve its accuracy.[10,11,12,13,14,15] Complementary to this work, Liao et al.[16] presented a discriminative model to automatically classify significant places and activities based on the framework of relational probabilistic models.[17]

A popular class of location-aware applications is the tour guide,[18] such as Campus Aware[19] and GUIDE.[20] Such tour guides are precursors to the kind of wayfinding system with which we are concerned.

4.2.1.1   Nursebot

Nursebot is a project at Carnegie Mellon University that provides a robotic platform for delivering navigation assistance to the elderly. This robot is envisioned to operate under the auspices of a community living home and helps users make it to appointments on time and provides directions to get there as well.

Nursebot requires caregiver support to update its understanding of the world: for example, the remote location of people and the state of their schedules. Because it is a robotic platform it only makes use of sensors embedded on the robot, although in principle extending the robot's knowledge to a network of distributed sensors would be possible.

Nursebot's navigation is based on robotic mapping technology and laser-range finders. Potential destinations are identified on a known map. When an elderly individual indicates a potential destination, Nursebot plans a route to the location and executes the plan. The plan is updated according to real-time laser-range finder inputs, which help to regularly update the position of the robot and the robot's knowledge of the people in the environment.

4.2.1.2   IMP

Closely related to Nursebot is IMP,[21] a walker that is augmented with a laser-range finder and navigational reasoning. When a user wants to go somewhere in the mapped facility, he or she can indicate the destination on an attached computer and a path-planning algorithm will guide the user to the destination using an arrow. Onboard sensors monitor progress and assist the user in getting to the destination.

One of the design decisions that this system made was to navigate a person directly through the use of a displayed arrow. This puts a high burden on the navigational system and sensor suite to avoid leading users into dangerous environments that cannot be sensed by the walker. This is probably not a large concern for controlled environments such as nursing homes but would be a problem for outdoor transportation assistance.

4.2.2   Wandering alert systems

A number of companies have attempted, with varying success, to create wandering alert systems. Some have been stand-alone systems, such as Digital Angel, and others have been integrated into smart assisted living environments, such as Elite Care assisted living homes. The company Independent Living has an installable system that promises to monitor both ADLs and wandering and alert a caregiver when programmed parameters are exceeded.

4.2.3   Home ADL tracking and support

There are many research groups that apply ubiquitous and wearable computing to the goal of aging in place.

4.2.3.1   Smart homes

One of the first and best-known smart home projects is Georgia Tech's Aware Home.[22] This project experimented with many technologies for capturing low-level sensor data about human activities but did not focus on automated activity recognition.

The MIT House_n project developed an instrumented condominium for studying activity recognition in a naturalistic environment.[23] This space is designed as a living laboratory from which experiments in activity recognition can be conducted in a naturalistic manner. From this line of research a variety of activity recognition research has been published that explores how simple sensors can recognize activities as they are being performed.[24,25]

The University of Florida's Mobile and Pervasive Computing Laboratory has several projects that are directed at cognitive assistance for the elderly, including meal preparation assistance and preliminary research into using cell phones for cognitive assistance.[26]

The Bath Institute for Medical Engineering developed a number of prototype commercial applications of cognitive devices and collaborates with Dementia Voice, a dementia services center for the southwest of England, and Housing 21, a U.K. housing association. Some of their projects include a cooker monitor (an instrumented stove that monitors for dangerous situations such as gas leaks, smoke, or burning pans); a misplaced-object finder; and a tap monitor (an instrumented faucet that controls temperature and prevents flooding).

4.2.3.2 COACH

The COACH (Cognitive Orthosis for Assisting aCtivities in the Home) system aims to address all aspects of ADL performance, from recognition to guidance, for the specific tasks of bathroom activities.[27] The current system has been tested for assisting people with advanced dementia with handwashing.

COACH is an adaptive device that learns by using Markov decision processes for how best to guide a user through the process of washing hands. The sensor input for this task is an overhead camera located over the sink, which is processed primarily for the location of relevant objects and then becomes the source of information for the decision process. This work is noteworthy for its use of verbal prompts that increase in specificity as the user becomes less and less likely to achieve the goal of handwashing. Of all the systems mentioned here, it targets individuals with the most severe forms of dementia.

4.2.3.3 Autominder

Autominder[28] is a planning assistance system that is designed to help a user meet scheduling goals for day-to-day activities. Its main reasoning component is a temporal constraint satisfaction engine that can determine when activities should be performed in order to avoid conflicts with other activities. Autominder's prompting module tries to minimize the number of times it interrupts the user. For example, if the user is scheduled to perform two activities at around the same time (for example, taking meds and brushing teeth), Autominder will combine the prompts.

4.2.3.4   PDA-based reminding systems

PEAT[29] is a commercial product that has many of the same goals as Auto-minder. It is built on a PDA platform and its goal is to help individuals who experience cognitive difficulty when formulating and following plans. It allows a caretaker to enter information about the time and steps of each task to be performed and then uses sounds and graphical alerts to guide the user through the tasks. The user manually clicks the PDA after each step; no automated activity recognition is attempted.

The AbleLink company does research and production on practical computer-based systems for supporting people with cognitive disabilities. Like PEAT, AbleLink has developed and tested PDA-based task-prompting systems, mainly for users with mental retardation.[5]

4.2.3.5   Wearable activity recognition systems

There has been much recent work in using wearable sound, video, and acceleration sensors and computers for activity recognition, both individually and combined. Later we describe work on activity recognition using RFID-tagged household objects and a wearable tag reader that led to the Barista system.[30] A theme of much of this recent work is that "heavyweight" sensors such as machine vision can be replaced by large numbers of tiny, robust, easily worn sensors.[31]

4.3   Opportunity Knocks: Assisting outdoor navigation

We now turn to detailed case studies of two prototype assisted cognition systems, beginning with the navigation system Opportunity Knocks (here-after OK). Many ADLs, such as working, shopping, going to a doctor's office, or attending social events, require a person to move throughout his or her community. Individuals with cognitive disabilities must generally rely on rides from caregivers, use point-to-point taxi or shuttle services, use public transportation, or restrict their movements to places reachable by foot. The first two are often unavailable or affordable, and few cities are so compact that they can be easily traversed by foot, so we will concentrate on public transportation.

Public transportation provides a variety of cognitive challenges, such as remembering transit schedules; getting on the correct bus or train; determining when to get off; making changes between vehicles; and recovering from errors. These challenges are so great that many cognitively disabled individuals become housebound. However, if impaired individuals had effective compensatory cognitive aids to help them use public transportation, their independence and safety would improve, they would have new opportunities for socialization and employment, and stress on their families and caregivers would be reduced.

This idea of a personal navigation aid is substantially different from current commercial GPS navigation systems. Today's personal GPS devices are typically optimized for a particular mode of transportation, such as driving in a car or hiking cross-country. The personal navigation aid we propose would help a user perform complex transportation plans that involve moving between modes of transportation—for example, walking to a bus stop, riding on a bus, and then walking again. The system not only tracks the person's location, but also the mode of transportation and the status of the transportation plan.

Additionally, because our target audience is cognitively disabled, this solution should not require a user to explicitly program a device or to always take the initiative in using it. We will describe a system that learns a user's pattern of public transportation use, predicts the user's current transportation goals, infers user errors, and provides proactive assistance.

4.3.1 A usage scenario

In order to ground our system, we present a running example that will help illustrate the most important features of OK. The steps of the scenario are illustrated in Figure 4.1. John works at a cafeteria at the university. One day he leaves work to go home and is momentarily confused about where to go. He consults his OK system, which is running on his GPS-enabled cell phone. OK offers images of four destinations that he typically travels to after work: home, his doctor's office, and the homes of two of his friends (A). OK has learned two different plans for John to get home from work: he can walk to a bus stop and catch a bus home, or he can walk to the parking lot where he gets a ride with a work colleague. Because OK is uncertain about which plan is correct for the day, it asks John to choose between the two (B). He selects the bus icon, and OK provides walking directions to the bus stop (C) and instructs him to get on the #17 bus. Unfortunately, John erroneously gets on the #19 bus, which initially travels along the same route as the #17, but which ultimately goes to John's friend's house, rather than to his own home. However, once the #19 departs from the common portion of the route, OK recognizes the error (D). The system alerts John and immediately constructs a repair plan to get him home, using its general knowledge of the transit system (E). OK guides him to exit the bus, walk to a nearby stop where he can board the #17, and resume his journey home (F).

4.3.2 System architecture

Figure 4.2 diagrams OK's overall system architecture. The data flow of the system starts at a sensor beacon that is carried by a user. The sensor samples the environmental context of the user and forwards this information over a secure Bluetooth connection to the cell phone. The cell phone initially acts as a network access point and again forwards the context information to a remote server over the high-speed General Packet Radio System (GPRS) data network. The remote server, which is running the OK software, uses the sensor information in conjunction with Geographic Information Systems'

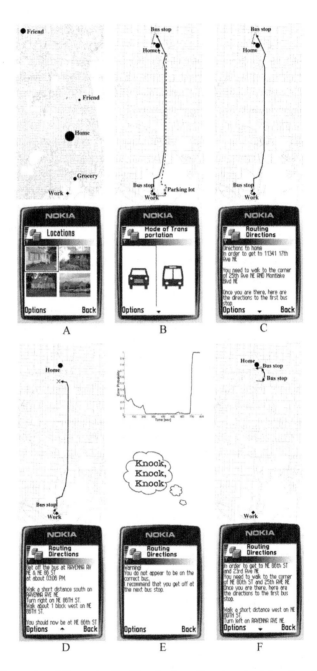

Figure 4.1 Scenario of Opportunity Knocks in action (see text for description).

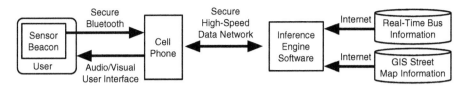

Figure 4.2 Architectural diagram of Opportunity Knocks.

(GIS) databases to localize the user. When the software has sufficient confidence in the position of the user, it is then able to suggest opportunities about which the user may want to know. These opportunities are sent back to the cell phone for display through the user interface. If an urgent opportunity, such as a plan for recovering from boarding the wrong bus, is recognized, the phone proactively alerts by making a door-knock sound; otherwise the phone remains passive with information available for reference by the user. If the user selects an opportunity, such as a route to a frequent destination, the cell phone requests supporting information from the server, which may require referencing real-time information about bus schedules.

We chose a cell phone as the client hardware because of its role as a de facto standard for a portable computing device. It has inherent value that is related to its primary function as a phone and for many people it is as common to carry as a wallet or a purse. As a result, it is likely to be a familiar, nonstigmatizing method of delivering assistive services. In the cell phone market products span from a traditional phone to a personal digital assistant (PDA). We opted for devices that were more like traditional phones rather than smartphones because of their ubiquity, simple interface, and limited maintenance requirements.

The system currently uses a Nokia 6600 cell phone. The Nokia 6600 phone is a GSM phone that has a wide range of features required by OK. First it supports the J2ME mobile information device profile (MIDP) 2.0 that provides support for secure networking, serial port connection support, and the application management system—a push registry that enables authorized applications to be launched remotely. Some model-specific features of the phone that we utilize include a high-resolution (176×208 pixels), high-color (16-bit) screen, a digital camera, Bluetooth support, and high-speed data network capabilities (GPRS).

When the user desires transportation assistance, he refers to the phone and observes up to four images of predicted destinations (later we describe how this selection is made). If he would like to go to one, he selects it. If the system has observed the user going to this destination in different ways (for example, by foot or by bus) it will prompt him for the method he would like to take. The previously observed route is then provided in text form. The system will not present destinations to which the user hasn't previously traveled, but it will allow the user to select a familiar destination even if it has never observed the user getting there from the current location. In this case OK presents a route that is based on a real-time bus route–planning service provided by the local transit authority. In the course of this interaction the user did not have to provide any information about where he was, and only a very small amount of information about where he wanted to go, yet the system was still able to route him effectively.

There are two occasions when the phone might become proactive and make a knocking sound. The first is when the system has high confidence that a novel or erroneous event has occurred. The second is when the system identifies that the user is at a new significant location and may wish to photograph it using the phone's camera. In the future, whenever the system wants to refer to that location, it simply uses the photo to identify the spot. Then the user does not need to input or recognize GPS coordinates or street addresses. A further advantage of this approach is that the user can decide what is visually meaningful about the location.

The current prototype system does not make use of voice prompts. Using voice is an obvious extension and is being incorporated in the next version of the system.

4.3.3 The inference engine

The inference engine that drives OK must learn and reason about the user's movements. As outlined earlier, the system must learn about its user's transportation routines in an unsupervised and unobtrusive manner, be able to predict likely destinations the user may want to go to at any given moment in time, and be able to recognize anomalous behavior. Because of the inherent uncertainties about human behavior as well as the possible errors from maps and GPS measurements, OK must reason probabilistically.

Ashbrook and Starner[11] have proposed using a second-order Markov model as a predictive tool for reasoning about likely destinations toward which a user may next travel. In contrast to our desired behavior, this model is not able to refine estimates of the current goal using GPS information observed when moving from one significant location to another. Because significant locations might be long distances away, this causes an unacceptable lag in noticing unusual behavior and significant amounts of GPS information are disregarded.

Patterson et al.[13] proposed using a dynamic Bayesian network (DBN) for inferring a user's transportation mode and location on a street grid from GPS data. Although this approach provided accurate tracking, it had poor predictive power, because the model had no representation of a user's destinations or transportation plans; information about the user's patterns of movement was only recorded as the probability of the user turning in a particular direction at each street corner.

In order to overcome these limitations, OK employs the new hierarchical DBN model representing transportation routines introduced by Liao et al.[32] The new model subsumes the capabilities of the previous models and bridges the gap between the raw sensor measurements and the abstract goal intentions of a user. A brief discussion of this model follows; refer to reference 32 for full technical details of the model structure, inference, and training.

Figure 4.3 shows the graphical structure of the new model. At the very highest level of this model, goals g_k (subscript k indicates the discrete time step) are explicitly represented as significant locations. Transitions between goals have specific probability distributions independent of the routes by which they are reached. Each goal destination influences the choice of which trip segment the user takes. Trip segments are sequences of motion in which the transportation mode is constant. Each trip segment t includes its start location t_{sk}, end location t_{ek}, and the mode of transportation t_{mk} the person uses during the segment. Each trip segment biases the expectation over the mode of transportation and the changes in location. The mode of transportation m, in turn, determines the location and velocity distribution of the user. At the bottom level, we denote by $x_k = <l_k,v_k>$ the location and motion velocity of the person. Edge transition τ_k indicates the next street when passing an intersection and data association θ_k "snaps" a GPS measurement onto some streets around it. The switching nodes $f_{kg}, f_{kt},$ and f_{km} indicate when changes in a variable's value can happen.

An efficient algorithm based on Rao-Blackwellised particle filters[33,34] has been developed to perform online inference for this model. At the lowest level, location tracking on the street map is done using graph-based Kalman filtering. At the highest level, the joint distribution of goals and trip segments is updated analytically using exact inference techniques. As a result, this

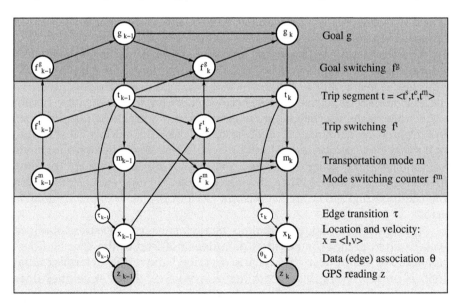

Figure 4.3 Hierarchical activity model representing a person's outdoor activities. The top level estimates the current goal, the middle layer represents segments of a trip and mode of transportations, and the lowest layer estimates the person's location on the street map. Figure reproduced from reference 32.

model makes it possible to reason about high-level goals (or significant locations) explicitly. The contribution of this model is that it considers not only previous significant locations visited but also the current location and the path taken so far to reason about likely destinations.

The parameters in the model are estimated in an unsupervised manner. This is a three-step process. In a first pass through the data, the possible goals for a user are discovered by observing when the user stays at a location for a long time. Then in a second pass, the usual parking spots and bus stops are inferred using an expectation-maximization algorithm.[35] Finally, the transition matrices at all levels are reestimated simultaneously using a second expectation-maximization procedure with the full model. The learning process does not require any labeled data and therefore requires no intervention from the user.

To detect abnormal events, the approach uses two models with different transition parameters. The first tracker assumes the user is behaving according to his personal historical trends and uses the learned hierarchical model for tracking. The second tracker assumes a background model of activities and uses an untrained prior model that accounts for general physical constraints but is not adjusted to the user's past routines. The trackers are run in parallel, and the probability of each model given the observations is calculated. When the user follows his ordinary routine the learned hierarchical model should have a higher probability, but when the user does something unexpected the second model should become more likely. To compute the probability of each model, we use the concept of Bayes factors, which are standard tools for comparing the quality of dynamic models based on measurements.

The above approach can detect unexpected events but cannot distinguish errors from deliberate novel behavior. An important contribution of OK, however, is the ability to differentiate these cases using knowledge of the user's destination. This is possible because there are times when the system knows where the user is going: for example, if the user asks for directions to a destination, if a caregiver or job coach indicates the "correct" destination, or if the system has access to a location-enabled datebook. In those situations we can clamp the value of the goal node in our model and reinterpret the low-level observations. When the observations diverge significantly from the clamped high-level predictions, the system is able to signal a possible error.

This model can spot anomalous behavior even if the user follows a well-trodden path, provided that path does not lead to the specified destination. For example, in the scenario described above, OK determines that John is on the wrong bus to get home, even though he sometimes does take that bus to go to his friend's house. A small graph illustrating this example of error detection when there is a clamped goal appears in step E of Figure 4.1.

4.3.4   Status

OK was built as a proof-of-concept prototype of an assisted cognition navigation system. It was successfully tested on real data gathered by students enacting various scenarios using models learned by using the system for approximately two hours a day for three weeks. However, the prototype was not robust enough for clinical trials. A particular problem of the architecture is that it required a continuous (wireless) Internet connection to a central computer server where inference was performed.

OK is currently being reengineered to be more robust and to perform inference on a Windows CE PDA carried by the user. Development leading to clinical trials for subjects with navigation difficulties due to brain injuries is being supported by the National Institute for Disability and Rehabilitation Research.

4.4   Understanding home activities

We now turn to our second case study with the assisted cognition system Barista. Indoor activity recognition has many potential benefits. Tracking the performance of ADLs is the first step in creating systems for ADL prompting and guidance. Furthermore, ADL monitoring is an ongoing, important activity in healthcare. For example, in the United States, any nursing home that receives Medicare funds has to record and report ADLs. Trained caregivers spend a great deal of time measuring and tracking ADL accomplishment for persons under their care. However, manual monitoring is time-consuming, error prone, and invasive. Automated aids that can address these issues and reduce the record-keeping burden on caregivers are of great interest.

Most systems that have been built to recognize home activities have been limited in the variety of activities they recognize, their robustness to noise, and their ease of use. In particular, most previous work on activity recognition has used sensors that provide only a very coarse idea of what is going on—for example, by detecting movement in a room, one might infer that an activity associated with that room is happening.[36] Also previous work required deployment of an extensive custom-sensing apparatus to monitor each task[27,37,38,39,40] or relied on solutions to deep technical problems such as machine vision.[41,42]

In this section we describe an approach to activity recognition that addresses these problems by combining the use of wearable RFID tag sensors to determine when a user is manipulating physical objects, with a simple and flexible probabilistic framework for modeling activities in terms of object touches.

4.4.1   Sensing using RFID

RFID tags are the size of postage stamps (including adhesive backing), have
no batteries, and can withstand day-to-day use for years. A tag reader sends
a radio frequency pulse to the tag, which responds with a unique identifier.
Depending on the power of the reader, a tag can be sensed from a few inches
to several yards away.

RFID deployment involves tagging tens to hundreds of objects in the
environment, then entering each tag identifier into a database. This can be
done incrementally; the more tags there are, the broader and deeper
the potential coverage of ADLs. Furthermore, market forces are pushing
toward the near-universal use of RFID tags on essentially all products. Such
preexisting tags could then be used for applications such as ours by using
a database to map tag IDs to types of objects.

Recently Intel Research and other companies have begun to develop
small wearable tag readers. Such readers can determine when a wearer
touches a tagged object. The work described in this chapter used a
short-range RFID reader built into the palm of a glove, with a Crossbow
Mica Mote radio, a USB-based power supply, and a rechargeable battery.
Intel Research has more recently developed a reader in the form of a small
bracelet.

ADLs that would be difficult or impossible to detect using either coarse
location sensors or state-of-the-art machine vision can often be recognized
on the basis of contact with a tagged object. For instance, consider trying
to determine if a person is reading. Location alone is clearly inadequate,
while reliably recognizing the act of "reading" from a video stream under
a wide range of orientations, positions, and lighting conditions is far beyond
the capabilities of machine vision for the foreseeable future. On the other
hand, if all the books, magazines, and newspapers in the home were tagged,
determining when a person was reading could be done quite reliably.

Although data from RFID tags are less noisy than many other kinds of
sensor data, any real-world data streams still contain extraneous readings
(for example, when the user's hand happens to brush by a tagged but unused
object) and missing readings. Therefore we propose to interpret the data
using probabilistic models that are robust in the face of noisy data.

4.4.2   Modeling activities

Representing ADLs in terms of the gross manipulation of physical objects
requires us to face the problem of developing a formal model that satisfies
a number of constraints: first, the model should easily express significant
properties of and distinctions between activities, while remaining robust to
unimportant variations in activity performance; second, the parameters
of the model should be easily estimated; and third, the model should be
implementable in a manner that supports efficient and scalable inference.

Descriptions of activities from a wide variety of sources, including healthcare literature, instructional manuals, and recipes typically break an activity down in a set of steps, where each step involves manipulating one or more objects over some period of time. Although textual descriptions usually present the steps in a total order, the underlying logical dependencies between steps often form only a partial order and include alternative and optional steps. The kinds of objects used in a step are usually flexible, and it is not difficult to form a coarse estimate on the probability of object use on the basis of the description. For example, while making a cup of tea (used as a running example in this section), we might estimate that the probability of using a spoon to stir the tea is 75 percent, allowing for cases where one uses a different utensil or none at all.

Figure 4.4 gives an example of modeling the activity "making tea" in four stages: getting out the supplies, heating the water, steeping the tea, and flavoring (i.e., adding sugar or lemon to the tea). The first stage consists of two steps that must both occur but in any order: this is indicated by a conjunctive arc across the first pair of outgoing arrows and the following pair of incoming arrows. This is an example of a partial ordering constraint. The disjunctive choice of which of two ways to heat the water (using the microwave or using the stove) is indicated by a set of plain arcs. The fact that the flavoring step is optional is represented by a disjunctive arc that bypasses the step.

Each step also has a Gaussian duration. Duration information can provide important constraints for distinguishing activities that use similar objects. For example, washing your hands at the kitchen sink takes about a minute, while washing dishes at the kitchen sink takes about ten minutes. Finally, each step includes a set of objects that is expected to be used. (In Figure 4.4 the duration and object information are only shown for the "boil water" step.) The value associated with each object is termed an object use probability and is the estimate of the probability that the object is manipulated at least once before the step completes. Also included in the model but not shown in the illustration are prior probabilities on each activity as a whole and on choice transitions within a model (such as the probability of including the optional "flavor tea" step); by default these are uniform across choices.

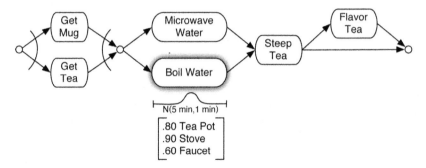

Figure 4.4 Making tea represented as an activity graph.

4.4.3   Coarse-grained ADL recognition

Matthai et al.[30,43] demonstrated that RFID sensors and probabilistic models of the form described earlier could reliably distinguish the performance of a large number of common ADLs. They chose fourteen ADLs to monitor, such as using the toilet, completing housework, using an appliance, taking medication, etc. (It is notable that the fourteen ADLs tested were eleven more than any other system has attempted.) One of the experimenter's homes was instrumented with 108 tags. Next, an activity model was created by hand to relate ADLs to tagged objects, using rough prior estimates of the timing and probability of object use, and a uniform prior estimate over all transitions from the end of one activity to the start of another.

Fourteen test subjects were each asked to perform the activities in any manner they liked and in any order, while wearing an RFID tag-reader glove. Activities took from twenty to sixty minutes, depending on the subject. A particle-filter inference engine computed the most likely sequence of activities performed by each subject, which was compared to a manual log of the actual sequence. The system did well on average, with a precision of 88 percent (correct labeling of recognized activities) and recall of 73 percent (percent of actual activities correctly recognized). To place these numbers in further context, we note that this was the first time any system had been created that could handle nine of the fourteen ADLs.

4.4.4   Barista: Fine-grained ADL recognition

Although some of the ADLs in the work described above used some objects in common, most could be uniquely distinguished by the manipulation of particular objects. The purpose of the Barista system was to determine if the same general approach could be used for fine-grained tracking of interleaved activities that shared many common objects.[44] In particular, we wished to find the simplest and most robust modeling methodology.

Barista focused on a morning routine in a small home. The following eleven activities were considered:

1. Using the bathroom
2. Making oatmeal
3. Making soft-boiled eggs
4. Preparing orange juice
5. Making coffee
6. Making tea
7. Making or answering a phone call
8. Taking out the trash
9. Setting the table
10. Eating breakfast
11. Clearing the table

To create the data set, one of the authors performed each activity twelve times in two contexts: Each activity was performed by itself twice, and then on ten mornings all of the activities were performed together in a variety of patterns. In order to capture the identities of the objects being manipulated, the kitchen was outfitted with sixty RFID tags placed on every object touched by the user during a practice trial. For example, in the bathroom the door knob, toilet handle, and faucet handle were tagged. In the kitchen, tags were placed on appliances, cookware, dishes, and food packages. Deploying the tags required less than two hours.

In this experiment the user simultaneously wore two RFID gloves (unlike in the first experiment in which one glove was used). The time and ID of every object touched was sent wirelessly by the glove to a database for analysis. The mean length of the ten interleaved runs was 27.1 minutes ($\sigma = 1.7$) and object touches could be captured at approximately ten per second. The mean length of each uninterrupted portion of the interleaved tasks was seventy-four seconds. Most tasks were interleaved with or interrupted by others during the ten full data-collection sessions.

The activities were not performed sequentially or in isolation from each other. Whenever there was a pause in an activity, progress was attempted in other activities (such as when waiting for water to boil) and some activities interrupted others at uncontrolled times (such as answering the phone).

4.4.5 Modeling choices

In order to justify the inference model that we ultimately developed we proceeded systematically by first focusing on accuracy and then on robustness. We developed the simplest possible probabilistic model, evaluated its performance, and then augmented it with features that were sufficient to disambiguate errors. In this section we present the techniques we used to improve accuracy by describing the two baseline models and another model that incorporated reasoning with aggregate features. The models increase in complexity by adding representational power. In subsequent sections we present abstraction techniques that we used to improve robustness.

The first baseline model consists of independent, single-state hidden Markov models (HMMs) for each activity. Used in a generative context, each state emits an object-X-touched event or a no-object-touched event at each tick of the clock. Each state's emission probability was trained on the twelve examples of a user performing the corresponding activity. After training, the probability of emitting a no-object-touched event was equalized across all HMMs so that the timing characteristic of the model was completely captured by the self-transition probability. To infer the activity being performed at each second, each HMM was presented with a seventy-four-second window of data (the average activity duration) ending at the query second. This produced a log-likelihood value for each model at each tick of the clock. The activity model with the highest log likelihood was used as the system's estimate of the current activity. This model was trained and tested on data in which object types were equalized so that there was no distinction made between spoon #1 and spoon #2, for example, but both appeared identically as a "spoon."

The second baseline model connected the states from the eleven independent HMMs of baseline A in order to be able to learn about and subsequently smooth the transitions between activities. We retrained this HMM using the ten examples of the user performing the eleven interleaved activities. The no-object-touched emission probability was again equalized across all states. This HMM was evaluated over the entire data window, and the Viterbi algorithm[35] was used to recover the activity at every time point given by the maximum likelihood path through the state space. Again, this model was trained and tested on data in which object types were equalized to eliminate distinctions between instantiations of objects.

For our third model, we chose to examine the effect of reasoning about aggregate information. The specific feature that we wanted to model was how many objects of a given type were touched during the course of the current activity. This aggregate can only be computed if globally unique object instances can be identified. This choice was motivated by the desire to differentiate activities that use the same object repeatedly from those that use many different objects of the same type. For example, consider the activities of setting the table and eating breakfast. The first involves single touches of several different plates, spoons, and cups, while the latter involves touching the same plate, spoon, and cup repeatedly.

Aggregate features can be handled in a DBN model by introducing variables that keep track of how many different instances of each object type are touched during the performance of an activity, in addition to a global variable whose value is the current activity. At the end of an activity (i.e., the value of the global variable changes), the final counts are treated as pseudo-observation. See Patterson et al.[44] for details.

The various features of these models are summarized in the following table:

	Exponential Timing Distribution	Interactivity Transitions	Aggregate Information
Independent HMMs			
Connected HMMs	✓	✓	
Aggregate DBN	✓	✓	✓

In this table, exponential timing distributions refer to the fact that the model expects the length of an uninterrupted portion of an activity to occur with a duration that is distributed according to an exponential distribution. The parameters of the distribution are learned from the data. This timing distribution is a result of the structure of the HMMs and DBNs used. Interactivity transitions refers to the ability of the model to represent the tendency of certain activities to follow or interrupt other activities more or less often.

Finally, aggregate information refers to the ability of the model to represent aggregations over individual objects.

4.4.6   *Accuracy experiments*

Our accuracy experiments were conducted with leave-one-out cross-validation across the ten interleaved runs. We calculated two accuracy metrics. The first was what percentage of the time the model correctly inferred the true activity. This metric is biased against slight inaccuracies in the start and end times and will vary based on the time granularity with which the experiments were conducted. We also evaluated our models using a string edit distance measure. In this case we treated the output of the inference as a string over an eleven-character alphabet, one character per activity, with all repeating characters merged. We calculated the minimum string edit distance between the inference and the ground truth. A string edit distance of one means that the inferred activity sequence either added a segment of an activity that didn't occur (insertion), it missed a segment that did occur (deletion), or it inserted an activity that didn't occur into the middle of an activity (reverse splicing). A perfect inference will have a string edit distance of zero. The string edit distance is biased against rapid changes in the activity estimate and is tolerant of inaccuracies in the start and end times of activities. The following table summarizes the results of the experiments:

	Time-Slice Accuracy (σ)	Edit Distance (σ)
Independent HMMs	68% (5.9)	12 (2.9)
Connected HMMs	88% (4.2)	9 (6.2)
Aggregate DBN	87% (3.1)	7 (2.2)

The independent HMM model performed badly because it rapidly and inaccurately switched between activities. The smoothing provided by the connected HMM model gave much better accuracy. However, Figure 4.5 shows that the connected HMM model confused the activity "eat breakfast" with the activity "clear the table." The aggregate DBN distinguished these properly and overall had a slightly improved accuracy and much better edit distance error measure.

4.4.7   *Improving robustness*

One of the concerns with the previous models is how well they will respond if someone used an object of a type that did not appear in the training data but was functionally similar to objects that did appear. For example, in our model we cooked oatmeal using a cooking spoon. Our inference should not fail if the user performed the same task using a tablespoon. Likewise, if the user makes tea in a cup rather than a mug, that should be a less likely but still plausible alternative. To solve this problem we introduce the concept of abstraction smoothing.

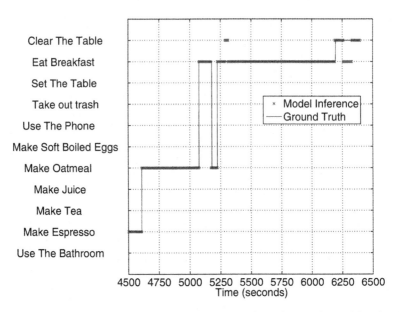

Figure 4.5 Sample connected HMM results. Ground truth is indicated by the thin line. Inference is indicated by the dots.

In order to perform smoothing over objects we created a relational model inspired by Anderson and Domingos.[45] Unlike the full power of that work, however, we used a single hierarchical object relation rather than a lattice. The hierarchy that we used was mined with supervision from the Internet shopping site Froogle (see Figure 4.6). The name of each object was entered into the shopping search engine and the hierarchy that was returned for that object was inserted into the global object tree. In the case of objects with multiple hierarchies, one was manually selected.

The semantics that we applied to the resulting tree were that objects that were close to each other in the graph were functionally similar. To specify the notion of "close," we weighted all edges on the graph equally and created an all-pairs functional equivalence metric according to the following formula,

$$P(O_i \Rightarrow O_j) = \frac{\exp\left(-\dfrac{Dist(O_i, O_j)}{2}\right)}{\displaystyle\sum_j \exp\left(-\dfrac{Dist(O_i, O_j)}{2}\right)}$$

where $Dist(O_i, O_j)$ is the shortest-path distance between O_i and O_j on the graph. This says that when object O_i is expected in the model, it will be substituted

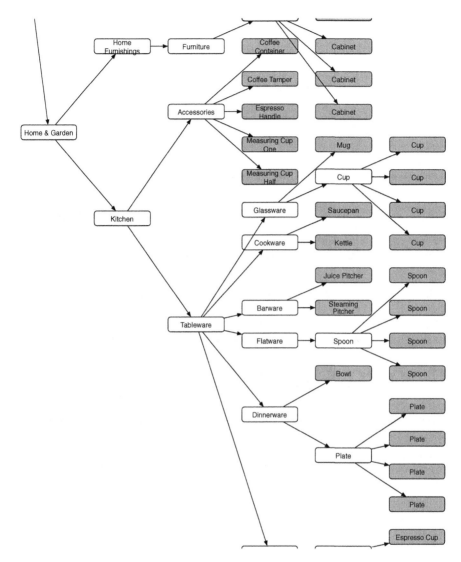

Figure 4.6 A portion of the object abstraction hierarchy mined from an Internet shopping site. Objects in our training data are shaded. Abstractions are not shaded.

by object O_j with probability $P(O_i \Rightarrow O_j)$. The likelihood of substituting one object for another falls off exponentially with distance in the hierarchy.

To validate how well this technique worked when objects were substituted, we reran our experiments with abstraction smoothing added to the aggregate DBN model. This resulted in an insignificant decrease in accuracy of 0.1 percent and −0.1 in edit distance.

Next, we reran our experiments with the same data streams, but with all instances of a particular object replaced by other instances of functionally similar objects of a distinct type. The following table shows our results for one scenario:

	Individual HMMs	Single HMM	Aggregates with Abstraction
Mean Accuracy	52.5%	77.4%	81.2%
Net Change	−15.1%	−10.9%	−6.4%
Mean Edit Dist.	24.7	35.6	8.8
Net Change	+12.7	+26.6	+1.1

Whereas abstraction smoothing doesn't greatly harm normal activity recognition, it greatly increases robustness to object substitution. The metrics from this table were generated by replacing a mug in all of the testing sequences with a cup. In two of the baseline models accuracy is dramatically lowered, but the abstraction model suffers a relatively modest decrease in accuracy, especially according to the string edit metric.

4.5 Summary

In this chapter we explored research on assistive technology for cognitive disabilities that combines advances in sensors and artificial intelligence to promote independence in the face of cognitive decline. By developing information systems that augment cognition in the same way that physical devices compensate for physical disabilities, we may be able to maintain higher quality of life for patients suffering from cognitive decline due to aging, trauma, or disease. At the same time, we can hope to reduce family caregivers' emotional and financial burden.

We looked at two possible avenues for assistance. The first is using an outdoor activity recognition system based on GPS to help people who make occasional cognitive errors recover safely. The second is an indoor activity recognition system based on a wearable computing platform and RFID tags that is designed to monitor which activities occur in a home.

In the outdoor case, we demonstrated that such a system could successfully be built now, that the reasoning that such a system can perform is both accurate and valuable, and that user–interface innovations make such a system usable without extensive user programming. In the indoor case, we saw that a single technology can subsume many previous activity recognition techniques in a way that is robust, easily deployable, and accurate at a fine level of detail.

An important challenge for work on assisted cognition systems is to develop and test effective user interfaces that decrease, rather than increase, a user's cognitive load. Opportunity Knocks began to explore the space of user interfaces, but the reliance on textual prompts would clearly be inappropriate for many users. The activity recognition system we described could be used as one part of a larger system that not only tracked activities but also provided prompts when necessary to help a user complete activities.

A significant complication in building interfaces for assisted cognition systems is that there is always the possibility that the system might misinterpret the data, so that what appears to be a user error might actually be a system (modeling) error. The problem of weighing the probabilities of system versus user errors, and the costs to the user of providing bad advice versus not providing good advice, is naturally formulated as one of decision making under uncertainty. Therefore we expect methods for solving such decision-theoretic problems to play important roles in future research in this area.

Acknowledgments

Preparation of this chapter was supported by grants from the National Institute on Disability and Rehabilitation Research (NIDRR), National Science Foundation (NSF), Defense Advanced Research Projects Agency (DARPA), and Intel. We thank Matthai Philipose for inspiration and technical leadership.

References

1. Lawton, M.P. Aging and Performance of Home Tasks. *Human Factors*, 32(5) 1990: 527.
2. Banaszak-Holl, J. et al. Predicting Nursing Home Admission: Estimates from a 7-Year Follow-Up of a Nationally Representative Sample of Older Americans. *Alzheimer Disease Association Disorders*, (2) 2004: 83.
3. Kautz, H. et al. An Overview of the Assisted Cognition Project, 2002. In *AAAI-2002 Workshop on Automation as Caregiver: The Role of Intelligent Technology in Elder Care*, 2002, 65.
4. Hopper, T., Bayles, K., and Kim, E. Retained Neuropsychological Abilities of Individuals with Alzheimer's Disease. *Seminars in Speech and Language*, 22(4) 2001: 261.
5. Davies, D.K., Stock, S.E., and Wehmeyer, M.L. Enhancing Independent Task Performance for Individuals with Mental Retardation through Use of a Handheld Self-Directed Visual and Audio Prompting System. *Education and Training in Mental Retardation and Developmental Disabilities*, 37, 2002: 209.
6. Inglis, E. et al. Issues Surrounding the User-Centred Development of a New Interactive Memory Aid. *Universal Access in the Information Society*, 2(3) 2003: 226.
7. Katz, S. Assessing Self-Maintenance: Activities of Daily Living, Mobility and Instrumental Activities of Daily Living. *Journal of the American Geriatrics Society*, 31(12) 1983: 721.
8. Hightower, J. and Borriello, G. Location Systems for Ubiquitous Computing. *Computer*, 34(8), 2001: 57.
9. Schilit, B., et al. Challenge: Ubiquitous Location-Aware Computing and the Place Lab Initiative. In *Proc. of WMASH 2003: The First ACM Intl. Workshop on Wireless Mobile Applications and Services on WLAN*, 2003, 29.
10. Krumm, J., Williams, L., and Smith, G. SmartMoveX on a Graph—An Inexpensive Active Badge Tracker. In *Proc. of the 4th International Conference on Ubiquitous Computing*, Springer-Verlag, 2002, 229.

11. Ashbrook, D. and Starner, T. Learning Significant Locations and Predicting User Movement with GPS. In *Proc. of the Sixth IEEE International Symposium on Wearable Computers*, Seattle, WA, October 2002, 101.

12. Cielniak, M., Bennewitz, G., and Burgard, W. Where is . . . ? Learning and Utilizing Motion Patterns of Persons with Mobile Robots. In *Proc. of the Seventeenth Intl. Joint Conf. on A.I.*, 2003.

13. Patterson, D.J. et al. Inferring High-Level Behavior from Low-Level Sensors. In *Proc. of UBICOMP 2003: The Fifth International Conference on Ubiquitous Computing*, A. Dey, A. Schmidt, and J.F. McCarthy, eds., Springer-Verlag, 2003, 73.

14. Schulz, D., Fox, D., and Hightower, J. People Tracking with Anonymous and ID-Sensors Using Rao-Blackwellised Particle Filters. In *Proc. of the Eighteenth Intl. Joint Conference on Artificial Intelligence (IJCAI)*, 2003, 921.

15. Hariharan, R. and Toyma, K. Project Lachesis: Parsing and Modeling Location Histories. In *Third International Conference on GIScience*, Aldephia, MD, 2004.

16. Liao, L., Fox, D., and Kautz, H. Location-Based Activity Recognition Using Relational Markov Networks. In *Proc. of the Nineteenth Intl. Joint Conf. on A.I.*, 2005, 773.

17. Getoor, L. et al. Learning Probabilistic Relational Models. In *Relational Data Mining*, S. Dzeroski and N. Lavrac, eds., Springer-Verlag, 2001, 307.

18. Abowd, G.B. et al. Cyberguide: A Mobile Context-Aware Tour Guide. *Wireless Networking*, 3(5) 1997: 421.

19. Burrell, J. et al. Context-Aware Computing: A Test Case. In *Proc. of the 4th International Conference on Ubiquitous Computing*, Springer-Verlag, 2002, 1.

20. Cheverst, K. et al. Developing a Context-Aware Electronic Tourist Guide: Some Issues and Experiences. In *Proc. of CHI 2000*, 2000, 17.

21. Morris, A.C. et al. A Robotic Walker That Provides Guidance. In *Proc. of the 2003 IEEE Conference on Robotics and Automation (ICRA '03)*, 2003.

22. Kidd, C.D. et al. The Aware Home: A Living Laboratory for Ubiquitous Computing Research. *Cooperative Buildings*, 1999, 191.

23. Intille, S.S. et al. The PlaceLab: A Live-In Laboratory for Pervasive Computing Research. In *Pervasive Computing, Third International Conference, PERVASIVE 2005, Video Program, Lecture Notes in Computer Science*, Springer-Verlag, 2005, 183.

24. Bao, L. and Intille, S.S. Activity Recognition from User-Annotated Acceleration Data. In *Proc. of Pervasive Computing, Second International Conference, PERVASIVE 2004*, A. Ferscha and F. Mattern, eds., Vienna, Springer-Verlag, 2004, 1.

25. Tapia, E.M., Intille, S.S., and Larson, K. Activity Recognition in the Home Using Simple and Ubiquitous Sensors. In *Proc. of Pervasive Computing, Second International Conference, PERVASIVE 2004*, A. Ferscha and F. Mattern, eds., Vienna, Springer-Verlag, 2004, 158.

26. Mann, W., Giraldo, C., and Helal, S. mPCA—A Mobile Patient Care-Giving Assistant for Alzheimer Patients. In *Proc. of UbiCog '02: First International Workshop on Ubiquitous Computing for Cognitive Aids*, Gothenberg, Sweden, 2002.

27. Mihailidis, A., Barbenel, J.C., and Fernie, G. The Efficacy of an Intelligent Cognitive Orthosis to Facilitate Handwashing by Persons with Moderate-to-Severe Dementia. *Neuropsychological Rehabilitation*, 14, 2004: 135.

28. McCarthy, C.E. and Pollack, M.E. A Plan-Based Personalized Cognitive Orthotic. In *Proc. of the 6th International Conference on AI Planning and Scheduling*, 2002.

29. Levinson, R. The Planning and Execution Assistant and Trainer (PEAT). *The Journal of Head Trauma Rehabilitation*, 1997.
30. Philipose, M. et al. Inferring Activities from Interactions with Objects. *IEEE Pervasive Computing: Mobile and Ubiquitous Systems*, 3(4) 2004: 50.
31. Hill, J.L. and Culler, D.E. Mica: A Wireless Platform for Deeply Embedded Networks. *IEEE Micro*, 22(6) 2002: 12.
32. Liao, L., Fox, D., and Kautz, H. Learning and Inferring Transportation Routines. In *Proc. of the 19th National Conf. on Artificial Intelligence (AAAI)*, 2004.
33. Bui, H. A General Model for Online Probabilistic Plan Recognition. In *Proc. of the Intl. Joint Conf. on Artificial Intelligence*, 2003, 136.
34. Doucet, A. et al. Rao-Blackwellised Particle Filtering for Dynamic Bayesian Networks. In *Proc. of the Conf. on Uncertainty in Artificial Intelligence*, 2000, 176.
35. Rabiner, L.R. A Tutorial on Hidden Markov Models and Selected Applications in Speech Recognition. In *Readings in Speech Recognition*, A. Waibel and K.-F. Lee, eds., Morgan Kaufmann, 1990, 267.
36. Glascock, A. and Kutzik, D. Behavioral Telemedicine: A New Approach to the Continuous Nonintrusive Monitoring of Activities of Daily Living, *Telemedicine Journal*, 6(1) 2000: 33–44.
37. Barger, T. et al. Objective Remote Assessment of Activities of Daily Living: Analysis of Meal Preparation Patterns. Technical Report, Medical Automation Research Center, University of Virginia Health System, 2002.
38. Mihailidis, A., Carmichael, B., and Boger, J. The Use of Computer Vision in an Intelligent Environment to Support Aging-in-Place, Safety, and Independence in the Home. *IEEE Transactions on Information Technology in BioMedicine*, 8(3) 2004: 238.
39. Shi, Y., et al. Propagation Networks for Recognition of Partially Ordered Sequential Action. In *Proc. of the 2004 IEEE Computer Society Conference on Computer Vision and Pattern Recognition, CVPR 2004*, 2004, 862.
40. Tran, Q. and Mynatt, E. What Was I Cooking? Towards Deja Vu Displays of Everyday Memory. Technical Report GIT-GVU-TR-03-33, Georgia Institute of Technology, 2003.
41. Jebara, T. and Pentland, A. Action Reaction Learning: Automatic Visual Analysis and Synthesis of Interactive Behaviour. In *ICVS '99: Proc. of the First Intl. Conf. on Computer Vision Systems*, Springer-Verlag, 1999, 273.
42. Moore, D., Essa, I., and Hayes, M. Exploiting Human Actions and Object Context for Recognition Tasks. In *Proc. of IEEE International Conference on Computer Vision 1999 (ICCV.99)*, 1999, 80.
43. Philipose, M. et al. Guide: Towards Understanding Daily Life via Auto-Identification and Statistical Analysis. In *Proc. of UbiHealth 2003: The 2nd International Workshop on Ubiquitous Computing for Pervasive Healthcare Applications*, Seattle, WA, 2003.
44. Patterson, D. et al. Fine-Grained Activity Recognition by Aggregating Abstract Object Usage. In *Proc. of the IEEE International Symposium on Wearable Computers*, 2005, 18.
45. Anderson, C., Domingos, P., and Weld, D. Relational Markov Models and Their Application to Adaptive Web Navigation. In *Proc. of the Eighth Intl. Conf. on Knowledge Discovery and Data Mining*, ACM Press, 2002, 143.

chapter five

Mobile and personal health and wellness management systems

Elina Mattila, Ilkka Korhonen, and Niilo Saranummi
Technical Research Center of Finland, Tampere, Finland

Contents

5.1 Introduction 106
5.2 Key building blocks.. 107
5.3 Stakeholders in health technology 108
5.4 Usage models 111
    5.4.1 Fitness and wellness 113
        5.4.1.1 Usage model of fitness and
            wellness technologies.. 113
        5.4.1.2 Fitness and wellness applications............................. 113
    5.4.2 Risk management and prevention 115
        5.4.2.1 Usage model of risk management and prevention
            technologies 115
        5.4.2.2 Risk management and prevention technologies....... 117
    5.4.3 Chronic disease management 119
        5.4.3.1 Usage model of chronic disease management
            technologies 119
        5.4.3.2 Chronic disease management applications................ 121
    5.4.4 Acute disease management and early discharge 124
        5.4.4.1 Usage model of acute disease management
            and early discharge technologies................................ 124
        5.4.4.2 Applications of acute disease management
            and early discharge technologies................................ 125

    5.4.5   Independent living...127
            5.4.5.1   Usage model of independent living technologies 127
            5.4.5.2   Applications of independent living 127
5.5 Discussion and conclusions 130
References 131

5.1 Introduction

The recent advances in information and communication technologies (ICT) enable technically the continual monitoring of health-related parameters with wireless sensors, wherever the user happens to be. Small, low-power sensors can, in principle, be embedded in almost anything in our surroundings: furniture, vehicles, wearable devices, and even clothes. Mobile phones or personal digital assistants (PDAs) with wireless networking capabilities may serve as gateways that process, store, and transfer measured parameters to clinicians for further analysis or diagnosis. This technology trend, also called mHealth (mobile health), is already visible in the market.

In parallel with the technological advances, there exists a need to reduce the costs of healthcare. The traditional care provisioning process, where the patient is a passive object cared for by healthcare professionals, has been challenged. In response, a new, proactive care paradigm is being introduced in health management, including management of risk factors such as being overweight, prevention of diseases and poor health, care for chronic diseases, and support for independent living.[1] In this model, the individual takes a more active role in the decision-making and management of his or her own health, while the health professionals step back into coachlike roles. Having individuals take more responsibility for the management of their health or illnesses allows a reduction in the workload of health professionals and provides an efficient opportunity for the prevention and management of diseases. This approach is especially suited for support by mHealth technologies.

The primary challenge for pervasive health technologies and mHealth technologies is to develop end-to-end systems that monitor relevant parameters in a way that is acceptable and usable (i.e., noninvasive, nonobtrusive, and easy to use), fits into the user's lifestyle, and produces valuable information. However, despite technological advances, the introduction of successful mHealth concepts into the market and into widespread use has proven complicated; therefore a full exploitation of these technologies is still to come. Many of the barriers of pervasive health technologies are nontechnical in nature. These barriers include the immaturity of the market for these products and the complexity of the stakeholder concept. The combination of the decision maker, payer, user, and beneficiary of a certain product or service may vary substantially from one region to another and also from one case to another.[2] It is often difficult to satisfy all the stakeholders' needs. Technological barriers include the lack of de facto standards (e.g., in communication technologies), the large number of incompatible parallel application platforms for mobile devices, and the lack of interoperability. Also, incorporating a long-lasting battery into a wearable device and even clothing, while still maintaining an agreeable form factor, is challenging. A number of experimental studies into applications of pervasive health technologies have been performed to try to remedy this problem, but a solution has not been reached yet. The future challenge is to bring together different healthcare professionals—namely technology providers, those in the clinical community, and those on the regulatory boards—to create a standardized network on which new applications can be launched.

In this chapter, we provide an analysis of mobile and personal health and wellness management systems from different perspectives. First, we list the general building blocks of typical mHealth systems. Then, we describe the stakeholders of health technologies. We present the main usage models of the technologies and the key user requirements for different application domains, from fitness to independent living. Then we introduce some examples of existing, mostly commercial, pervasive healthcare systems. Finally, future perspectives and the challenges of pervasive health technologies are discussed.

5.2 Key building blocks

Personal and mobile health and wellness management systems typically include several different pervasive computing technologies, the scale and variety of which depend on the application and its usage model. Upon looking at any personal mHealth system that is capable of monitoring or measuring some health or wellness-related parameter over the long term in a mobile context, it typically integrates the following key functions:[3]

1. A *sensor* to detect biomedical signals or parameters from the user or his or her environment. Such sensors can be noninvasive (e.g., intelligent biomedical clothing [IBC]), minimally invasive (biochips, lab on the chip, electronic nose), or embedded into a wearable or on-body device, such as a heart rate monitor with a belt-type electrode.

2. A *short-distance communication* method to route signals or data from the sensors for processing, storage, and user access. This transmission should ideally be wireless, such as the inductive link from a heart rate transmitter belt to a wrist unit in Polar Electro heart rate monitors.[4]

3. *Methods or algorithms to process the acquired data* into more usable information (e.g., the transformation of an electrocardiogram into a heart rate signal) and to reduce noise and other artifacts.

4. *Classification and decision-making methods* or algorithms to interpret parameters, diagnose a condition, and determine an adequate response (e.g., sending an alarm to a user who is going into anaphylactic shock

and recommending the proper corrective action, namely a shot of epinephrine).

5. A method of *long-distance communication* to transmit data to a third party in a remote monitoring site. This includes the transfer of data, parameters, alarm signals, onboard diagnosis, and recommended responses or interventions to offsite experts for validation and observation.

6. A *user interface* (UI) to interact both with the patient or user and the remote clinician/caregiver in a friendly, useful, and efficient manner.

7. A *back-end system* to enable data storage and access for data consumers (e.g., user, clinician). An example of a back-end system is the electronic patient record (EPR).

8. A *service* to provide the user with feedback, advice, and medical assistance based on the data.

In a real-world setting, the components that are employed are heavily dependent on the application itself. A successful application integrates all the required components, and no more, into a functional system in a cost-efficient and user-friendly manner. Currently, the major technical issues or constraints in designing a useful mHealth system are power consumption, true unobtrusiveness and acceptability, cost, long-term reliability and durability, and maintenance and calibration requirements. In most mobile and wearable solutions, powering all the components of the devices—sensors, actuators, communication, and UI—is often the bottleneck that hinders the shift of proof-of-concept prototypes of pervasive healthcare scenarios to real, practical systems.[5] Furthermore, in large-scale use, requirements such as automatic personalization of the system (including alarms, etc.), ease of installation in different platforms and environments, and adaptation to differing user needs have to be addressed. Thus there is an emerging need for approaches for the mass customization of applications and their intelligence and algorithms. These factors together make the transition from proof-of-concept prototype to a validated and commercially successful pervasive health technology solution rather challenging.

5.3   Stakeholders in health technology

Any successful technology must be efficient, effective, and satisfactory for the users' needs.[6] This implies that the technology must provide a high degree of usability, fit its purpose, and have some real benefits for the user (improved health or wellness, or interesting or entertaining information). These basic principles of user-centered design are of the utmost importance when designing mobile and personal health and wellness management systems. However, in the domain of health and wellness, technology will involve users and stakeholders who have different roles and needs (e.g., patients versus caregivers) and whose needs must be taken into account. We now briefly look at the different stakeholders and their roles in healthcare.

In consumer products the user concept is generally rather simple: the technology provider provides a technology product (or service) to the user, who makes the purchasing decision, pays for the technology, and uses it for his or her own benefit. In the healthcare domain the situation is usually more complex. This is illustrated in Figure 5.1, which represents the different stakeholders of the healthcare domain.

In the traditional health provisioning concept, a decision to purchase a given technology is usually made by medical or clinical professionals. These professionals may need to get approval for the purchase from the payer, which may be a public authority or private institution (e.g., an insurance company). The decisions may also be regulated by policymakers (e.g., through laws and regulations regarding compensation mechanisms for technology and care costs, privacy issues, etc.). The policymakers, in turn, represent citizens. It is the citizens who eventually pay for the service through taxes, insurance fees, or direct healthcare fees. The users of the technology include both the health professionals (doctors, nurses, and other care personnel) and the patients, although the latter typically have fairly passive roles. The beneficiaries of the purchase and use of the technology may vary: they may be patients (through improved care), health professionals (through more efficient or effective care or resource usage), or payers (through improved cost-efficiency). The assessment of the total benefit of the technology may be different from different perspectives; also the beneficiary, the decision maker, and the payer may be different stakeholders. Hence, a technology will need to fulfill the different success measures of these different stakeholders in order to be competitive and successful in the healthcare market.

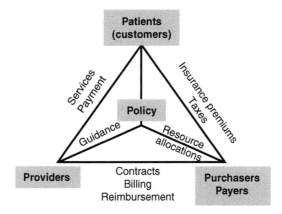

Figure 5.1 Stakeholders in healthcare. *Patients* (customers) consume healthcare services. They also finance the national healthcare system through various arrangements (e.g., taxation). Health policies are established at the national level by public *policymakers*. *Purchasers and payers* (public and private organizations) organize and pay for health services to be available for patients (customers). *Providers* (public and private, also third-sector organizations) provide health services to customers and are reimbursed through a variety of payment schemes.

As a result, the uptake of new technologies (especially those requiring organizational or legislative changes in order to be efficiently used) in the traditional healthcare domain is usually slow. This is further complicated by the fact that user concepts vary from one region to another, depending on the healthcare and welfare organization.[2] In addition, the concept of evidence-based medicine (EBM) is becoming popular in the healthcare domain. "EBM is the conscientious, explicit, and judicious use of current best evidence in making decisions about the care of individual patients."[7] The practice of EBM means integrating individual clinical expertise with the best available external clinical evidence from systematic research. Hence, the introduction of a new technology into the healthcare domain usually results in a slow evolutionary path with a series of clinical trials (Figure 5.2). On the contrary, wellness technologies are closer to individual consumers, and therefore the adoption of new technologies may be faster. For example, heart rate monitors for sports are not healthcare products but fitness products and therefore do not need approval from the authorities. As a result, the market penetration of these products may be significantly faster. However, all this depends on the usage models of the technology. The next section analyzes the most common usage models.

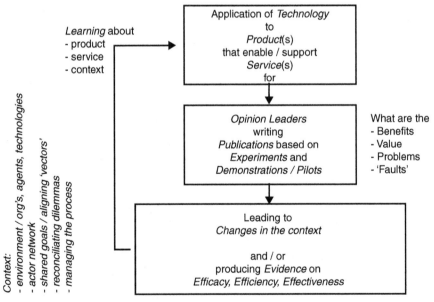

Figure 5.2 A schematic illustration of the process of introducing new technology into the healthcare market. A successful technology needs to demonstrate its efficacy (does it work? is it the right thing to do?), efficiency (is it being used right?), and effectiveness (does it make a difference in the larger context? is its use justified?). Usually this is achieved only through a series of evolutionary studies.

5.4 Usage models

Pervasive healthcare, especially mobile and personal health and wellness applications, typically involves several different stakeholders with varying needs. The domain may be divided into five categories based on the individuals' needs and characteristics:

1. Fitness and wellness (healthy individuals)
2. Risk management and prevention (individuals with a risk factor for a disease)
3. Disease management (patients with chronic illness)
4. Acute disease management and early discharge (patients with an acute illness)
5. Independent living (elderly and disabled individuals)

Figure 5.3 depicts how these models relate to each other in three dimensions. The first axis represents the degree of professional care versus self-care. For example, in the fitness domain, the individual is healthy and capable of taking care of himself or herself, whereas in acute disease management, the individual is a patient and very dependent on a clinician's support. The proactive-reactive axis refers to the nature and goals of care or health management. Proactive health management has the target of maintaining and promoting wellness and preventing illness, while a reactive approach means the treatment and management of an existing illness. The third axis represents the level of regulation involved in each usage model. For example, fitness applications are mostly considered consumer applications and are not subject to strict regulations.

In this section, we aim especially to describe the application domain, involved stakeholders, building blocks of a typical application, and key user requirements. This analysis is not intended to be fully comprehensive, but rather illustrative. Finally, we present some typical example applications from each usage model. Table 5.1 draws together the characteristics of

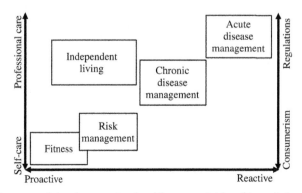

Figure 5.3 Usage models of pervasive healthcare outside of hospital conditions.

Table 5.1 Characteristics of the Usage Models

Usage Model	User's Participation	Involvement of Other People	Criticalness of Application	Need for Real-Time Data Transfer	Importance of Unobtrusive-ness	Importance of Ease of Use	Devices Owned By (Typically)	Stakeholders	Importance of Privacy	Role of EBM
Fitness	High	Low	Low	Low	High … moderate	Moderate	Individual	User	Moderate	Low
Risk management	High	Moderate	Low	Low	High	High	Individual	User (Care provider) Family Policy maker	High	Moderate
Chronic disease management	High … moderate	Moderate … high	Moderate … high	Low … moderate	High … moderate	High	Care provider or individual	User Care provider Payer Family Policy maker	High	High
Acute disease management	High … low	High	High	High	Moderate … low	Moderate … high	Care provider	User Care provider Payer Family Policy maker	High	High
Independent living	Low	High	High	High	Moderate … high	High	Care provider (or individual, family)	User Care provider Payer Family Policy maker	Moderate… high	High

different usage models, such as a list of the stakeholders involved, the degree of user participation, ownership of the technologies, and the role of EBM.

5.4.1 Fitness and wellness

5.4.1.1 Usage model of fitness and wellness technologies

Fitness and wellness applications are typically used by individuals in their exercising or sports hobbies to improve, control, document, or analyze their performances during or after the sports activity. Exercising and sports are typically carried out for fun and social reasons: to look good, to feel good, to improve fitness and health, or a combination of these. Hence, the driving force for engaging in fitness and sports is the experience—the health benefits are often secondary.

The fitness and wellness domain is basically a consumer business with a high degree of consumerism and a low degree of regulation by the authorities. The user makes the decision to purchase a product, pays for it, and uses it, usually independently. Therefore, technologies in this area emphasize design, branding, and marketing to appeal to potential users. The owner and consumer of the data captured by the system is the user, and therefore the system should be presented in a way that the average consumer finds valuable, understandable, and interesting. What this means in practice is highly dependent on the application and the users involved. Typically the focus of the application is on a single exercise or sports session, and potentially comparing performances of different sessions or with other users.[8–10] As the value of the data for a single session is high, the reliability of the application should be high as well to attract usage. On the other hand, the criticality of the data is low, and the loss of data does not lead to serious consequences.

The building blocks of the technologies for fitness and wellness include sensors, personal area networks for sensor communications, on-body devices for processing, and UI. These building blocks make up a stand-alone system, which is seldom compatible with other manufacturers' devices. Data collected by the system may be downloaded to an Internet site, a PC, a PDA, or a mobile phone for further analysis. Although only a single user is generally involved in the interpretation, there may be some peer user groups or personal coaches with whom the data are shared on a voluntary basis.

The critical success factors in this application domain are design, usability, and the ability to improve the user's experience (the "wow" effect). The user concept and the marketing approach are similar to other consumer applications rather than to those of health or medical technology. Thus, there is no need for proof of clinical benefits (e.g., EBM).

5.4.1.2 Fitness and wellness applications

There are a number of applications for fitness and wellness technologies. They include various sports instruments, such as pedometers,[11] heart rate monitors,[4] wrist-top computers for sports such as diving, golf, swimming, and cross-country running,[12] and armband devices for documenting sports.[13]

Suunto (Suunto Oy, Vantaa, Finland) manufactures sports instruments for measuring performance in different sports, such as diving, sailing, skiing, and golf.[12] The Suunto t6 is a wrist-top computer for recording exercise performance. The system consists of a heart rate transmitter belt for heart rate measurement, wrist-top computer with sensors (altimeter/barometer), and Suunto Training Manager for analyzing the data (Figure 5.4). The system may be augmented with peripheral observation devices (PODs) for measuring speed and distance during walking, running, or cycling. Communication between wireless measurement devices and the wrist-top device uses the 2.4 GHz ANT technology.[14]

The effect of training is analyzed offline with the Training Manager program to produce numerical and graphic performance feedback for users, as well as statistics such as respiratory rate, energy consumption, and EPOC (excess post-exercise oxygen consumption). EPOC provides information about the effect of the training session and helps to plan one's individual exercise program with the right intensity level. Suunto also offers a Web community where the users of Suunto sports instruments can share and compare exercise data with other users.[8]

The FRWD Outdoor Sports Computer (FRWD Technologies Ltd., Oulu, Finland) is designed for recording outdoor sports performance.[13] The system consists of the FRWD Recorder Unit, which is worn around the arm and gathers data during sports; the heart rate transmitter belt, which collects heart rate data; a mobile phone; the Mobile Player software for viewing performance graphs and numbers in real time during sports; and the FRWD Replayer PC software for offline analysis of the data (Figure 5.5).

Bike POD   Foot POD

Figure 5.4 The Suunto t6 wrist-top computer, Training Manager program, and wireless measurement PODs. (Adapted from Suunto Oy. With permission.)

Figure 5.5 The FRWD Outdoor Sports Computer consists of a Recorder Unit for recording sports performance, a mobile phone with the Mobile Player software for real-time viewing, and the Replayer PC software for offline analysis. (From FRWD Technologies Ltd. With permission.)

The FRWD Recorder Unit records heart rate, speed, distance, route, location, altitude, air pressure, time, and temperature. The unit connects wirelessly via Bluetooth to a Symbian Series 60 mobile phone for real-time analysis and feedback of the sports performance. The user may, for example, attach the mobile phone to a bicycle and see graphs and values of the measured and analyzed parameters (e.g., the physiological cost index), in real time.

The Replayer software may be used after sports to replay the route and altitude profile as a 2-D or 3-D presentation. The recorded data may be analyzed further with the software to extract interesting parameters such as total values for duration, distance, climb, altitude difference, calories, and efficiency; averages for speed, heart rate, temperature, and air pressure; peaks for speed, altitude, heart rate, etc.; and several others. Also comparing performance, or "racing" with oneself and other FRWD users, is enabled by the software—the performance files may be sent via e-mail or posted on Web sites. This feature also enables remote coaching.

5.4.2 Risk management and prevention

5.4.2.1 Usage model of risk management and prevention technologies

Chronic diseases such as diabetes mellitus or cardiovascular diseases develop gradually. The origin of a chronic disease is usually partially genetic and partially lifestyle related. Hence, the management of lifestyle factors, such as nutrition, exercise, and sleep, is important in preventing and managing these illnesses. Obesity is a major risk factor in many diseases. For example, it is, along with a sedentary lifestyle and an unhealthy diet, the most important risk factor of type 2 diabetes mellitus.[15] The future does not look promising, because more than 30 percent of Americans are obese (body mass index >30 kg/m^2), and a similar trend is seen in European countries with over 20 percent of the population obese in at least nine countries, including Greece, Germany, England, and Finland.[16,17] Another common health risk factor is poor sleep. According to the National Sleep Foundation, about two thirds of Americans suffer from sleep problems.[18] Poor sleep correlates with many health problems, for example, cardiovascular disease and mental problems, as well as an increased risk of accidents and falls.[19-21] The worldwide incidence of chronic diseases and the associated costs are increasing. According to the World Health Organization, for example, the total healthcare costs of a person with diabetes in the United States are two to three times higher than for people without the condition. The estimated cost of treating diabetes in the United States in 1997 was $44 billion.[22] Developing and using tools for managing the risks associated with a disease to prevent or delay its occurrence would bring great benefits to both society and the individual.

Individuals at risk for a disease due to lifestyle or genetic factors may use pervasive healthcare technologies to reduce or manage their risk factors. This may prevent them from becoming ill and increase the chances of detecting the disease early in its onset, which generally results in more successful treatment. Examples of risk management applications are weight management and lifestyle-based management of elevated blood pressure. Users of risk management and disease prevention technologies are generally healthy individuals with elevated risk factors. Compared to the fitness and wellness applications presented earlier, the main motivation for the users in this domain is concern for their health. Still, the users are independent and largely in charge of the decision making for purchasing and using technology. This results in a significant degree of consumerism, which influences the design of these technologies. However, health professionals may recommend technology and motivate users. Policymakers can potentially participate in funding the technology (either through partial or full coverage of the costs by health insurance or similar mechanisms) to improve preventive healthcare and reduce morbidity. Employers may also contribute to funding through an occupational health system. Hence, the healthcare authorities and regulations may play a role in successful risk management and prevention technologies. In the United States there is a push for flexible spending accounts (FSA) or medical savings accounts (MSA). This is part of a broader movement toward consumer-centric healthcare, in which consumers have more say on where to spend their healthcare dollars. In that sense, MSA and FSA may encourage adoption of these kinds of technologies.

A typical risk management application includes methods for the long-term monitoring of health-related parameters such as weight, blood pressure, blood glucose, exercise, and diet. The building blocks for such a system include a sensor or measurement device; short-range sensor (or device) communication to a local data gateway (e.g., a Bluetooth connection to a mobile phone); and a data processing, storage, and UI application running on a PDA, mobile phone, or computer. The data may be shared with a personal coach or a clinician (e.g., general practitioner or nurse), who participates as a healthcare coach. However, the main feedback is for the user in order to motivate and guide him or her in the management process; the user is the main consumer of the data. This is necessary not only for efficacy of management but also for economical reasons: for example, as illustrated above, obesity is such a pervasive problem in industrial countries that intimate personal coaching for each individual with this risk factor is not feasible. In turn, educating users about risk management is needed, and psychological models such as cognitive behavioral treatment (CBT) may be used to gain engagement.[23]

The usability requirements for risk prevention applications are higher than those for fitness applications. Ease of use is crucial and the wow effect is secondary to efficiency. Users are typically not interested in the technology itself, as is commonly the case with fitness technologies. Unobtrusiveness, comfort, and transparency of use are also important factors, as well as design. The reliability of the measurements needs to be high (e.g., monitor readings cannot drift) and the efficacy of the application should be demonstrated (EBM) in order to attract institutional support (i.e., participation of or recommendations from health professionals, compensation from health authorities or insurance companies, etc.). Risk management is a long-term task and therefore the value of a single measurement is low. Real-time feedback is seldom needed and communication may be asynchronous. Cost is also an important factor. Peer group support may be beneficial for risk management; the effect of this approach in weight management has been demonstrated by Weight Watchers.[24] Thus, peer support could also be integrated into pervasive risk-management tools.

5.4.2.2 Risk management and prevention technologies

There are many commercially available technologies for risk management and prevention. Many of these systems are for people who are at risk of widespread healthcare problems such as diabetes, high blood pressure, and obesity. The following two technologies are examples of systems that are designed to manage or reduce an individual's risk of health problems related to being overweight.

The bodybugg calorie management system (BodyMedia, Inc., Pittsburgh, PA, USA; Apex Fitness Group, Camarillo, CA, USA) is a Web-based system for weight management.[25,26] The bodybugg system consists of an armband device and a Web application (Figure 5.6). The bodybugg system uses SenseWear technology for measuring calorie consumption and physical activity. The armband contains several noninvasive sensors: a two-axis accelerometer, a heat flux sensor, a galvanic skin response sensor, and a temperature sensor. The data are wirelessly downloaded to the Web application, which analyzes and visualizes the data. The Web application provides the

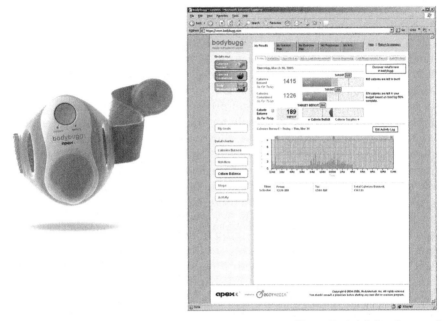

Figure 5.6 The bodybugg calorie management system. (From BodyMedia, Inc., and Apex Fitness Group. With permission.)

user with a personalized weight management plan: for example, the daily energy deficit needed to reach a weight goal in a certain time period. The user has the option to enter daily food intake in the Web application and to plan meals. The application provides feedback about energy expenditure measured with the armband (total expenditure and a minute-to-minute graph) and energy intake as well as detailed nutrition information based on user input.

The validity of the SenseWear technology in estimating energy expenditure has been investigated in several studies. Wadsworth et al.[27] found that the SenseWear armband accurately estimated energy expenditure in a protocol consisting of three consecutive fifteen-minute conditions of supine rest, walking, and supine rest. In some studies, however, the SenseWear armband significantly under- or overestimated energy expenditure during walking on a treadmill.[28,29] Jakicic et al.[30] suggest that these estimations could be improved by using exercise-specific algorithms for estimating the energy expenditure from the data.

The Polar WM42/WM41 (Polar Electro Oy, Kempele, Finland) is a heart rate monitor designed for weight management.[4] The system consists of a heart rate transmitter and a wrist unit (Figure 5.7). Heart rate data are wirelessly transmitted to the wrist unit via an inductive link. The Polar Weight Management Program in the wrist unit guides the user in weight management by calculating the recommended energy intake and expenditure. The application contains a weight diary for inputting weight measurements, an exercise diary for recording energy expenditure during exercise sessions, and a nutrition

Figure 5.7 Polar Weight Management products WM42/WM41, consisting of wrist units (shown in the figure) and heart rate transmitters. (From Polar Electro Oy. With permission.)

diary for recording daily energy intake. Weight and nutrition are manually input to the wrist unit, while energy expenditure during exercise is estimated based on heart rate. With this user information, the application calculates weekly exercise targets and the amount of calories that can be consumed daily. The targets are updated every week based on the user's weight management success. A Web service is also available for the users of Polar weight management products. The service provides a menu planner for controlling caloric intake (and plans and records meals) and an activity planner for planning exercises and writing notes about exercise sessions afterward. However, the efficacy both of the Polar solution and of the bodybugg system have not been evaluated in controlled weight management trials.

5.4.3 Chronic disease management

5.4.3.1 Usage model of chronic disease management technologies

The main difference between the chronic disease management domain and the two domains presented above is the presence of a diagnosed disease. The primary user is a patient and has a care provider (a secondary user), who is responsible for patient care. The most effective area for the applications of mHealth to disease management is supporting individuals with a chronic, possibly incurable, disease such as hypertension, diabetes, or asthma. Chronic diseases may manifest themselves when the patient is relatively young, require continuous management, and, if poorly treated, lead to serious complications. As such, the care of an individual with a chronic disease is often rather costly.

The course and status of a chronic condition are highly dependent on the individual's lifestyle and medication balance. Additionally, patients are generally still capable of managing the disease relatively independently at home.

Compliance to care is surprisingly low in many chronic diseases: for example about 50 percent of hypertensive patients do not take their medication as prescribed,[31] and in asthma, the estimated noncompliance rates range from 10 to 46 percent.[32] Poor compliance causes serious consequences; in bronchial asthma the consequences range from impairment of daily life to death,[33] and in diabetes the serious consequences include diabetic ketoacidosis (diabetic coma), cardiovascular disease and stroke, and lower-limb amputations, which could be prevented or delayed with diabetes therapy.[34]

The main rationale in the applications for chronic disease management is to improve the care balance and management of the disease. These applications are designed to allow individuals and clinicians to monitor changes in the disease, to guide self-management through lifestyle and medication, and to enable early intervention in an effort to prevent and shorten hospitalization episodes. Long-term monitoring may also help to understand the course of the disease better.[35] Hence, various different technologically based approaches have been developed to support the individual in taking a more active role in the disease management process and to improve patient communication with the clinician.

Typical building blocks of chronic disease management applications include a sensor or measurement device (e.g., to check blood pressure, blood glucose, peak expiratory flow, etc.), a short-range communication means to transfer the result to a local data gateway (e.g., a Bluetooth connection to a mobile phone), and a long-range communication means to transfer the data to an information system from where it can be accessed by a clinician or patient. There may also be a prediagnostic procedure at the individual's end of the system, and only relevant data are sent to the clinician. The patient, with or without the clinician, makes a decision on the subsequent course of action (e.g., changes in lifestyle or medication regime) based on the data. The reliability, privacy, and data integrity requirements of the disease management applications are high because care decisions are made based on the data. Because the focus is on long-term changes, the value of a single data point is relatively low, therefore asynchronous data transfer suffices. The loss of single data items is not critical. As a result, it is even feasible to use SMS (short message service) messaging to send such data.

This usage model has several different stakeholders with different roles: the purchase decision is made by a healthcare professional or administrator together with the individual; the payer may be a public or private healthcare organization, insurance company, or the individual patient; and both the healthcare professional and the individual are the users.[2] The efficacy of the applications usually needs to be convincingly demonstrated (EBM) and the interests of everyone involved have to be taken carefully into account. Although the users may be highly motivated to use the systems, they may not be accustomed to using these devices in their daily lives. Therefore the applications need to be easy to use and unobtrusive to promote long-term daily use. The cost may be weighed against the potential savings. Thus, high cost may not necessarily be a barrier if it can lead to large potential savings or other benefits. Integration into existing infrastructure (e.g., digital TV, mobile phone, etc.) would also be beneficial in order to make the applications more transparent.

5.4.3.2 Chronic disease management applications

There is a wide variety of technologies available for the management of chronic diseases. The selection of components, monitoring methods, and amount of explicit user interaction required of the technology is highly dependent on the type and severity of the disease being managed. The examples below illustrate this, with two different approaches to collecting patient parameters.

Philips Telemonitoring Services (Philips Medical Systems, Andover, MA, USA) are designed for remotely monitoring the specific parameters of patients with one or more chronic diseases.[36] The system is composed of a set of wireless measurement devices (personal scales, blood pressure and pulse monitor, pulse oximeter, ECG/rhythm strip recorders, and glucose meter) and a TeleStation home hub. TeleStation collects data automatically from the measurement devices using wireless communications. The data may also be input manually. TeleStation transmits the information to a clinician and presents the patient with health- or disease-related questions if the data cause concern. These AutoCheck Surveys are launched in response to unusual measurement results to obtain more information about the changed patient state. TeleStation enables a secure, two-way information flow between patients and remote clinicians (Figure 5.8).

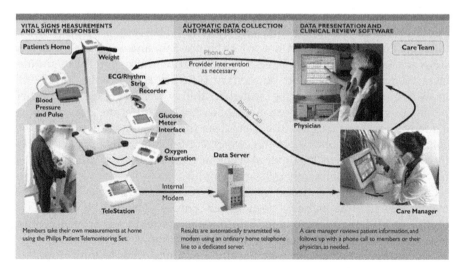

Figure 5.8 Philips Patient Telemonitoring Set consists of wireless measurement devices and a TeleStation home hub for two-way communication between patients and clinicians. (From Philips Medical Systems. With permission.)

The telemonitoring services system may be used for the management of congestive heart failure (CHF).[37] CHF is the leading cause of hospital admissions for seniors in the United States and the hospital readmission rate within six months is also high (about 44 percent). The treatment of CHF patients is very costly: accounting for 1 to 2 percent of the total healthcare budget in Europe and the United States. The treatment of CHF consists of a combination of drugs and lifestyle changes, and 20 percent of CHF-related hospitalizations are due to noncompliance with care.

The Trans-European Network Home Care Management System (TEN-HMS) trial studied the telemonitoring services system for home-based monitoring of CHF patients.[37] Four hundred and twenty-six CHF patients participated in the study, and they were randomized into three groups: the control group that received the usual care (UC) with specialist advice and a management plan; the first intervention group (nurse telephone support; NTS) that received patient care supplemented with monthly telephone contact from a nurse; and the second intervention group (home telemonitoring, or HTM) that consisted of nurse telephone support and self-measurements (twice per day) of weight, heart rate, blood pressure, and ECG that were automatically transmitted to the clinician. The trial period for the patients was a minimum of 240 days, followed by, on average, a 480-day follow-up period. Compared to the UC group, patients assigned to HTM or NTS had a significantly lower mortality rate. In addition, HTM reduced the duration of hospital admissions compared to NTS at the 240-day follow-up. However, no significant differences between the outcomes of NTS and HTM were observed at the 450-day follow-up. Therefore, the greatest contribution to improved care was achieved by rearranging the care process (NTS and HTM vs. UC).

The LifeShirt System (VivoMetrics, Ventura, CA, USA) is a wearable, noninvasive, ambulatory monitoring system.[38] The system consists of the LifeShirt Garment, LifeShirt Recorder (Figure 5.9), and VivoLogic analysis software. LifeShirt contains sensors for measuring more than thirty parameters related to cardiac and pulmonary function and activity. Peripheral measurement devices may be added to the system for measuring other parameters, such as blood pressure, blood oxygen saturation, EOG, periodic leg movement, and temperature. Data from the shirt and the peripherals are collected in the LifeShirt Recorder, which continuously encrypts and stores the data. The LifeShirt Recorder also contains a patient diary called Vivolog for recording symptoms, moods, and activities.

The parameters measured by the system are relevant to several chronic conditions, such as asthma, sleep apnea, chronic obstructive pulmonary disease (COPD), and coughing. The parameters extracted from LifeShirt measurements are well summarized by Wilhelm et al.[39] The system is currently not available for customers—only for clinical trials and research.

iMetrikus (iMetrikus, Inc., Carlsbad, CA, USA) is a healthcare technology company whose flagship product MediCompass provides an interactive Personal Health Record with secure, member-controlled access for storing

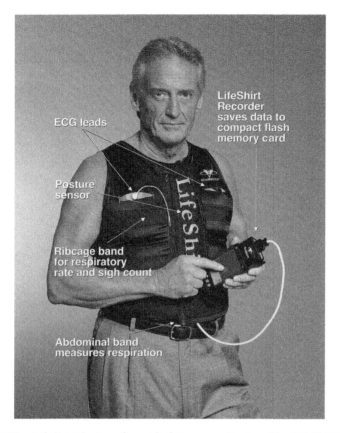

Figure 5.9 The LifeShirt System for ambulatory monitoring: The LifeShirt Garment and Recorder. (From VivoMetrics. With permission.)

and retrieving health data.[40] For those monitoring a chronic condition, MediCompass Condition Monitoring Modules facilitate information tracking and sharing by creating a connection between those with chronic concerns and their caregivers. The AirWatch (Figure 5.10) asthma monitoring device enables patients and healthcare professionals to share self-testing results between healthcare visits. AirWatch supports those monitoring asthma and COPD. It collects respiratory measurements (FEV-1, PEF) and tracks and trends patient status with respect to treatment plan. The uploaded data are easily and securely transmitted over a standard telephone line. The information is captured in the MediCompass data repository ready for the patient or clinician to access. MediCompass reporting and data analytic features provide clinicians the opportunity for early intervention to adjust a patient's regimen if necessary, before a health crisis occurs.

Lefkowich et al.[41] studied the accuracy and patient and clinician acceptance of the AirWatch system. The measurement results were found to be comparable to the conventional measurements of peak flow and FEV-1. The system was also well accepted by patients and clinicians. Furthermore,

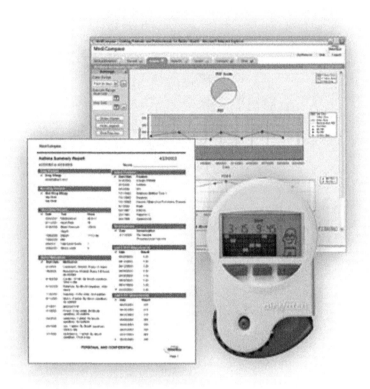

Figure 5.10 The AirWatch system and MediCompass personal health record. (From iMetricus, Inc. With permission.)

the system showed great promise for the outpatient management of asthma. Doherty et al.[42] studied the system for telemonitoring and found, albeit in a small sample of twenty-six patients, that the use of AirWatch significantly reduced emergency room visits and improved symptom scores and quality of life.

5.4.4   Acute disease management and early discharge

5.4.4.1   Usage model of acute disease management and early discharge technologies

The great majority of resources of the traditional healthcare system are used to dealing with acute diseases. In this domain, the patient has an acute illness, which is treated by the efforts of the healthcare providers. Pervasive technologies also offer promising possibilities to improve this area of healthcare.

The acute disease management usage model is characterized by the acute illness the patient has, the presence of the care provider who is in charge of treating the patient, and the time line of the management process, which can be temporally limited (as opposed to chronic disease management, which is an ongoing process). Examples of acute health conditions vary from sudden cardiac events to infectious diseases and to hospitalizations due to, for example, operations. A pervasive healthcare application in this domain may facilitate both the monitoring of the patient's condition outside of the hospital and the management of the patient's care (e.g., medication, rehabilitation), enabling early discharge or emergency response. Typically, the technology is given to the patient for a defined duration by his or her care provider and the patient is relatively passive in the process. The dependency on technology is high, sometimes critical; EBM and approvals from the authorities play essential roles. Technologies in this domain will not be used without strong proofs and validation studies.

The building blocks of acute disease management are fairly similar to those of chronic disease management. The system typically includes either a sensor or measurement device. The individual can use the system to take measurements of vital signs, such as an ECG, as a therapy device (e.g., for drug infusion), or both. As the data collected by the systems designed for acute disease management are used to make decisions that can have an extensive impact on the successful treatment of the patient, the criticality of the technology is significantly higher than in the case of chronic diseases. For example, asynchronous data transfer is typically not acceptable (e.g., using SMS messages) because no delay or loss of any data may be tolerated. The efficacy and correct functioning of the applications under all conditions need to be convincingly demonstrated. When using wireless technologies, for example, their functioning and coverage need to be guaranteed in all conditions.

The usage model is largely driven by the interests of healthcare providers and payers, who are in search of more efficient care with lower costs. Although patients do not play a large role in the selection of the technology, they also benefit from these systems through improved quality of life because of earlier discharge or faster response to emergency situations. The applications need to be acceptable for the patient users although, due to their medical condition, they may be ready to accept more obtrusive and effort-requiring solutions than in other categories presented earlier in this section. The compliance applications are typically used for a limited time by the same users (i.e., until the application is no longer needed), and they are typically owned by care providers. The cost of these technologies is generally high; however, they still offer a good alternative for cost-aware healthcare providers when compared to the costs of traditional ways of caring for the disease at hand.

5.4.4.2 *Applications of acute disease management and early discharge technologies*

The sensors, data analysis, and response capabilities of technology aimed at supporting the management of acute diseases vary considerably depending on the nature of the disease. The following example presents a system for management of cardiac problems.

SHL Telemedicine (SHL Telemedicine Ltd., Tel Aviv, Israel) offers remote diagnostic and monitoring systems.[43] The systems include devices for monitoring the heart, pulmonary function, blood pressure, and weight. SHL also provides medical call center services for the users of the products.

One of the specialized services is the cardiac emergency response system. Before subscribing to the service, the patients are interviewed extensively and a twelve-lead ECG is measured for baseline for future measurements. Outpatients can use the CardioBeeper twelve-lead or CardioPocket (Figure 5.11) one-lead ECG transmitter and LidoPen lidocaine auto-injector to manage cardiac problems. If the patient feels any symptoms, she or he may transmit an ECG to a dedicated monitoring center. The twelve leads of the ECG are transmitted sequentially, using a standard telephone line or a mobile phone. The transmission of twelve leads takes about sixteen seconds. The monitor center nurses, under physician supervision, diagnose the situation by comparing the transmitted ECG to previously stored baseline data for the same patient. They provide assistance to the user and if necessary, dispatch an ambulance. The service has more than 300,000 users

Figure 5.11 The CardioPocket is a leather wallet, which also serves as a one-lead (rhythm strip) ECG transmitter for diagnosing heart rhythm disturbances. In case of any symptoms, the user simply places the wallet against his or her chest and transmits the real-time ECG strip using an ordinary telephone line or mobile phone to a remote monitoring center for an immediate consultation. (From SHL Telemedicine Ltd. With permission.)

in Israel, the United States, and Europe (e.g., Italy and Germany). In a study conducted in Israel with 118 CHF patients, the use of the SHL system reduced hospitalization days by 66 percent when compared with conventional care.[44]

5.4.5   Independent living

5.4.5.1   Usage model of independent living technologies

The applications for independent living aim at helping elderly and disabled individuals to live securely in their own homes longer, thereby delaying institutionalization. A traditional application is a social alarm system, which is based on an alarm button that is worn on the wrist or as a pendant and is used for calling help in case of emergency.[45,46] A limitation of these devices is that they need to be worn at all times to be effective (i.e., they require high compliance from the individuals). However, a significant portion of the users (27 to 40 percent) do not wear these devices on the daily basis.[45]

New pervasive technologies embed sensors into traditional independent living systems, either into wearable devices or into the home. The sensors communicate wirelessly with a home unit, which processes sensor data and sends information and alarms to care personnel in remote locations. Many of the systems still incorporate manual alarm buttons. In addition, they are often capable of generating automatic alarms when the individuals cannot do it themselves, due to, for example, unconsciousness. Additional features, such as a hands-free phone, medication reminders, burglar and fire alarms, and access control (e.g., to prevent demented patients from wandering) may also be provided. Some systems also monitor compliance to make sure devices are used and worn as intended.

The monitored individuals are not the primary users of these applications, although they benefit from the systems through increased security (i.e., they use the systems implicitly). The primary (explicit) users are clinicians and family members. These users often make the decision to purchase an application or service. The main requirements for the successful implementation of independent living technology are proven efficiency, reliability, and low maintenance. Currently, the monitored individuals are typically very unaccustomed to technology, but on the other hand rely strongly on the correct functioning of the systems. The main requirements from the monitored individuals' point of view are unobtrusiveness and reliability.

5.4.5.2   Applications of independent living

Applications in the domain of independent living are primarily focused on the detection of events that require intervention from a clinician/caregiver or, often, an emergency response team. The following two examples highlight technology that helps to ensure safe, independent living while requiring minimal effort from the individual being monitored.

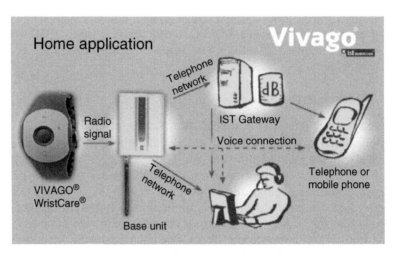

Figure 5.12 The IST Vivago WristCare for independent living. (From IST International Security Technology Oy. With permission.)

IST Vivago WristCare (IST International Security Technology Oy, Helsinki, Finland) is an intelligent social alarm system.[47] The system consists of a wrist unit, base station, and Vista alarm receiving and routing software (Figure 5.12). The wrist unit has a traditional manual alarm button that enables the individual to raise an alarm. It also has inbuilt sensors for measuring movements, skin conductivity, and temperature. The movement sensor is sensitive enough to record the muscle movements inside the wrist. The wrist unit transmits the sensor data to the base station using a low-power radio transmitter operating in the license-free 868 MHz ISM band within the range of a flat or a house. The wrist unit is battery-operated with a battery life of six to nine months.

The system also has adaptive automatic features, which enable calling for help when the individual is not capable (e.g., due to loss of consciousness). Alarms are generated based on unusual activity states (e.g., lack of movement) and unusual temperatures. Skin conductivity is used for detecting whether the wrist unit is worn; if the user does not wear the device, he or she may be reminded to do so. In addition, the system has self-diagnostic features for detecting, for example, when the battery power is low. The Vivago system may be extended with access control for demented patients. Activity data and any alarms generated by the system are transferred to defined recipients (call center, care personnel, relatives) via a public telephone line. The alarms may also be routed to a mobile phone. The recipient may call the user on the base station's hands-free phone to resolve the severity of the situation.

In addition to increased security, IST Vivago provides unique activity data, which is available online and can be collected for prolonged periods. The activity signal has been found to differ significantly between demented and nondemented elderly people[48] and to correlate with changes in their health status.[49] Furthermore, sleep/wake detection based on Vivago activity signals is as accurate as standard actigraphs[50] and correlates with sleep quality.[51] In addition to Finland, the system is used in Sweden, France, Germany, Great Britain, Ireland, and Japan.

Tunstall Group Ltd. (Yorkshire, U.K.) provides a wide range of personal and home reassurance telecare applications, which are used in many European countries, the United States, Canada, Australia, and New Zealand.[52] Tunstall products range from chronic and acute disease management to support of independent living in group and individual housing.

The Lifeline 4000+ home unit is a home hub designed to support independent living (Figure 5.13). The unit can register up to thirty-five wireless sensors, such as an infrared movement detector, smoke detector, fall detector, temperature sensor, bed and chair occupancy sensors, and gas detectors. Manual alarm devices, such as pendant buttons and pull cords, may also be included in the system. The wireless sensors and devices communicate with the home unit using the European 869 MHz social alarm frequency. Additional features include medication reminders, low-battery warnings, intruder alarms, and keyless door entry for care personnel in case of an alarm. The Lifeline 4000+ unit also has a hands-free phone that provides effortless communication between the user and the caregiver.

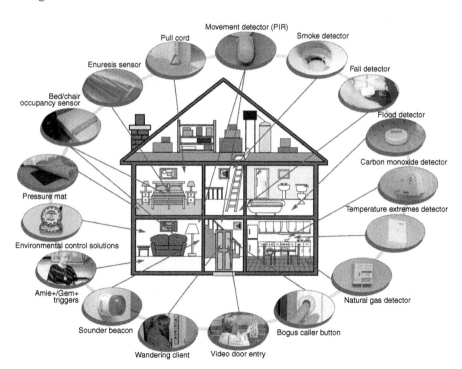

Figure 5.13 The Tunstall Lifeline 4000+ system. (From Tunstall Group Ltd. With permission.)

5.5   Discussion and conclusions

Pervasive mHealth technologies are designed to cover a wide range of user groups with drastically different needs. Many new emerging technologies are being rapidly adopted for use in pervasive health technologies and examples of possible uses are innumerable. For example, sensors are being increasingly embedded into garments, and new fibers with mechanical, optical, or electrical properties are under development.[35] These advances will allow increased usability, unobtrusiveness, and comfort for users. New low-power wireless sensing methods are also emerging. RFID technologies are especially promising for pervasive health technologies.[5] When sensors can reliably be integrated with RFID tags, they could become a powerful paradigm for health monitoring.[53] New wireless connectivity methods, such as the ZigBee/IEEE 802.15.4 standard,[54] will also create new opportunities for mobile monitoring in personal health and wellness systems. The transfer of data over wireless communication channels will become cheaper, faster, and ever more accessible. Mobile platforms will gain more computing power, memory, and advanced user-interfacing capabilities. These trends continue to contribute to the success of new mobile and personal health and wellness management systems and services.

There are two major groups of challenges in the development of successful mobile and personal health and wellness management systems. The first group focuses on nontechnical challenges. First, one needs to design technologies that enable new health and wellness services and simultaneously reengineer the service-provisioning framework to allow the new technology to be efficiently employed.[55] In other words, most of the application domains discussed in the previous section are based on providing a health or wellness *service* through the use of pervasive *technology*—and one without the other is not sufficient. As services usually need to be provided locally, successful developments of pervasive technology (which are often targeted for global markets) will require collaboration with local actors and the ability to tailor the technology to local needs. To keep this profitable requires the technology to support easy adoption by different service provisioning models and value chains. Second, in application domains related to health, EBM is increasingly applied. Hence, technologies and systems (including models for service provisioning) need to dependably demonstrate their effects and benefits in order to be adopted. Arranging studies for the efficacy of mobile and personal health management systems usually requires large interdisciplinary projects, possibly extending over several years, which is beyond reach for single entities in the domain. Hence, public-private partnerships and other such mechanisms are needed to overcome this challenge.

The second group of challenges focuses on technical challenges. For cost-efficient technical solutions in the health and wellness domain, generic platforms such as mobile phones, PDAs, and digital TVs should be employed. Clearly, a lot needs to be achieved in terms of standardization (of communication, interfacing, application platforms, operating systems, data contents, security, privacy solutions, etc.) to allow the safety and privacy demands of these applications to be implemented using these generic platforms. Furthermore, power economics seems to have become the main bottleneck for several applications designed to be used over long periods of time in daily life environments, especially in applications requiring wireless or wearable sensing and monitoring. Here, RFID-based technologies have interesting promise.[5] Usability and the cost of wearable monitoring solutions are still compromised. To be continuously used in daily life, intelligent biomedical clothing (IBC) should be cost-efficiently integrated into various different garments, because the people using the technology will no doubt continue to wear a large variety of clothes—including underwear—depending on the weather and their mood, activity, and social situation. This is clearly beyond reach within the next few years; therefore the focus of IBC will in the short term be on special activities, such as sports, or on professions, such as firefighters or soldiers.

   In this section we discussed different application domains and their usage models, stakeholders, and other key factors that affect mobile and personal health and wellness management systems. The focus has been on an analysis of the characteristics and requirements of the usage models, rather than on the technologies themselves. These concepts have been highlighted through the presentation of a few examples of technology applications for each domain. The examples we chose all represent applications of healthcare technologies, most of which are already commercially available and in real use. With this we wanted to emphasize that pervasive healthcare applications are not only a promise for the future improvement of healthcare, but also that they are already here and helping people to become and stay healthy. Hence, pervasive healthcare technologies are an evolution rather than a revolution. This evolution is continuing at a terrific rate and will undoubtedly result in new success stories benefiting patients, care providers, and society through improvements in health and better-managed costs. To achieve this, however, it is essential that we consider appropriate usage models and match the technology and services to the requirements set in each application domain.

References

1. Dishman, E. Inventing wellness systems for aging in place. *IEEE Computer*, 37, 34, 2004.
2. Kivisaari, S. et al. Health technology for the home: a revolution in the market concept through telecommunications. World Congress on Telemedicine for the Development of the Global Information Society for Health. Toulouse, 30 Nov.–1 Dec. 1995.
3. Korhonen, I. et al. IEEE/EMBS technical committee on wearable biomedical sensors and systems. Presented at Ambience '05, Tampere, Finland, 19–20 September 2005.

4. Polar Electro Web site, www.polar.fi. Accessed 29 March 2006.
5. Philipose, M. et al. Battery-free wireless identification and sensing. *IEEE Pervas. Comput.*, 4, 37, 2005.
6. *ISO 9241-11—Guidance on Usability*, 1998.
7. The Centre for Evidence-Based Medicine Web site, www.cebm.net. Accessed 29 March 2006.
8. Suunto Web site, suuntosports.com. Accessed 29 March 2006.
9. Polar outdoor guide Web site, www.polaroutdoorguide.com. Accessed 29 March 2006.
10. Motionbased Web site, www.motionbased.com. Accessed 29 March 2006.
11. Omron Healthcare Web site, www.omron-healthcare.com. Accessed 4 April 2006.
12. Suunto Oy Web site, www.suunto.com. Accessed 29 March 2006.
13. FRWD Technologies Ltd. Web site, www.frwd.fi. Accessed 29 March 2006.
14. ANT Technology Web site, www.thisisant.com. Accessed 4 April 2006.
15. World Health Organization. *Diabetes mellitus: report of a WHO study group*. Technical Report Series No. 727. Geneva, 1985.
16. National Public Health Institute Web site, www.ktl.fi/portal/english/osiot/ public_health_monitoring___promotion/monitoring___interventions/ nutrition_in_finland/nutrition_and_health/. Accessed 29 March 2006.
17. International Obesity Task Force. *EU platform on diet, physical activity and health, briefing paper*. 15 March 2005, Brussels. Available at: www.europa.eu.int/ comm/health/ph_determinants/life_style/nutrition/platform/ launch_en.htm. Accessed 29 March 2006.
18. National Sleep Foundation Web site, www.sleepfoundation.org. Accessed 29 March 2006.
19. Foley, D.J. et al. Sleep complaints among elderly persons: an epidemiologic study of three communities. *Sleep*, 18, 425, 1995.
20. Van Someren, E.J.W. Circadian and sleep disturbances in the elderly. *Exp. Gerontol.*, 35, 1229, 2000.
21. Brassington, G.S., King, A.C., and Bliwise, D.L. Sleep problems as a risk factor for falls in a sample of community-dwelling adults aged 64–99 years. *J. Am. Geriatr. Soc.*, 48, 1234, 2000.
22. World Health Organization, *Diabetes: the cost of diabetes*, Fact sheet no. 236. Available at: www.who.int/mediacentre/factsheets/fs236/en/. Accessed 29 March 2006.
23. Lamminmäki, E. et al. *Wellness diary for mobile phones*. Proc. of the 3rd European Medical & Biological Engineering Conference, EMBEC '05, Prague, Czech Republic, 20–25 November 2005.
24. Weight Watchers Web site, www.weightwatchers.com. Accessed 29 March 2006.
25. Bodybugg Web site, www.bodybugg.com. Accessed 29 March 2006.
26. BodyMedia Web site, www.bodymedia.com. Accessed 29 March 2006.
27. Wadsworth, D.D. et al. *A validation study of a continuous body-monitoring device: assessing energy expenditure at rest and during exercise*. Abstract presented at the annual meeting of the American College of Sports Medicine, Las Vegas, NV, 2005.
28. King, G.A. et al. Comparison of activity monitors to estimate energy cost of treadmill exercise. *Med. Sci. Sport. Exer.*, 26, 1244, 2004.

29. Fruin, M.L. and Rankin, J.W. Validity of a multi-sensor armband in estimating rest and exercise energy expenditure. *Med. Sci. Sport. Exer.*, 36, 1063, 2004.
30. Jakicic, J.M. et al. Evaluation of the SenseWear Pro Armband to assess energy expenditure during exercise. *Med. Sci. Sport. Exer.*, 36, 897, 2004.
31. Wang, P.S. et al. Noncompliance with antihypertensive medications. The impact of depressive symptoms and psychosocial factors. *J. Gen. Intern. Med.*, 17, 504, 2002.
32. Spector, S. Noncompliance with asthma therapy—are there solutions? *J. Asthma* 37, 381, 2000.
33. Siafakas, N.M. and Bouros, D. Consequences of poor compliance in chronic respiratory diseases. *Eur. Respir. J.*, 5, 134, 1992.
34. International Diabetes Federation Web site, www.idf.org. Accessed 29 March 2006.
35. Dittmar, A. et al. New concepts and technologies in home care and ambulatory monitoring, in *Wearable eHealth Systems for Personalised Health Management*, A. Lymberis and D. de Rossi, eds., IOS Press, 2004.
36. Philips Medical Systems Web site, www.medical.philips.com/main/products/ telemonitoring/. Accessed 29 March 2006.
37. Cleland, J.G.F. et al. Noninvasive home telemonitoring for patients with heart failure at high risk of recurrent admission and death. The Trans-European network-home-care management system (TEN-HMS) study. *J. Am. Coll. Cardiol.*, 45, 1654, 2005.
38. VivoMetrics Web site, www.vivometrics.com/site/system.html. Accessed 29 March 2006.
39. Wilhelm, F.H., Pfaltz, M.C., and Grossman, P. Continuous electronic data capture of physiology, behavior and experience in real life: towards ecological momentary assessment of emotion. *Interact. Comput.*, 18, 171, 2006.
40. iMetrikus Web site, www.imetrikus.com. Accessed 29 March 2006.
41. Lefkowich, D. III, Klimas, J.T., and Ross, R.N. Comparison of a new hand-held interactive airway monitoring system and a conventional peak expiratory flow meter in the management of asthma. *Am. J. Man. Care*, 2, 1227, 1996.
42. Doherty, G.B., Ross, R.N., and Ross, P.E. The effectiveness of an interactive electronic lung function monitoring in the total management of refractory asthma. *Dis. Manag. Health Out.*, 3, 89, 1998.
43. SHL TeleMedicine Web site, www.shl-telemedicine.com/. Accessed 29 March 2006.
44. Roth, A. et al. Telecardiology for patients with chronic heart failure: the "SHL" experience in Israel. *Int. J. Cardiol.*, 97, 49, 2004.
45. Porteus, J. and Brownsell, S.J. *Using telecare: exploring technologies for independent living for older people.* Anchor Trust, Oxford, 2000.
46. Särelä, A. et al. *IST Vivago—an intelligent social and remote wellness monitoring system for the elderly.* Proc. 4th Ann. IEEE Conf. on Inf. Tech. App. Biomed., 2003, 262.
47. IST International Security Technologies Web site, www.istsec.fi/ index.php?lang=eng. Accessed 29 March 2006.
48. Paavilainen, P. et al. Circadian activity rhythm in demented and non-demented nursing-home residents measured with telemetric activity. *J. Sleep Res.*, 14, 61, 2005.

49. Paavilainen, P., Korhonen, I., and Partinen, M. Telemetric activity monitoring as an indicator of long-term changes in health and well-being of older people. *Gerontechnology*, 4, 77, 2005.

50. Lötjönen, J. et al. Automatic sleep-wake and nap analysis with a new wrist-worn online activity monitoring device IST Vivago WristCare. *Sleep*, 26, 86, 2003.

51. Lamminmäki, E. et al. *Differences in light sleep and deep sleep measured with IST Vivago WristCare*. Proc. of the 3rd European Medical & Biological Engineering Conference, EMBEC '05, Prague, Czech Republic, 20–25 November 2005.

52. Tunstall Web site, www.tunstall.co.uk/. Accessed 29 March 2006.

53. MIMOSA project Web site, www.mimosa-fp6.com. Accessed 29 March 2006.

54. ZigBee Alliance Web site, www.zigbee.org. Accessed 29 March 2006.

55. Saranummi, N. et al. *A framework for developing distributed ICT applications for health*. Presented at Transdisciplinary Conference on Distributed Diagnosis and Home Healthcare (D_2H_2), Arlington, Virginia, 2–4 April 2006, 7.

chapter six

Sensors and wearable technologies for pervasive healthcare

Sungmee Park and Sundaresan Jayaraman
Georgia Institute of Technology, Atlanta, Georgia

Contents

6.1 Introduction 136
6.2 The healthcare challenge 137
    6.2.1 Meeting the challenge: The pervasive
          healthcare paradigm... 137
          6.2.1.1 Continuous monitoring—the patient view.............. 138
          6.2.1.2 Continuous monitoring—the healthcare
                  professional view... 139
          6.2.1.3 Continuous monitoring—the pharmaceutical
                  industry view 139
    6.2.2 Continuous monitoring—a way of life 139
6.3 Pervasive healthcare system: A patient-centric approach................. 140
    6.3.1 The life cycle of unit operations 140
    6.3.2 The enabling technologies 140
    6.3.3 Requirements of a pervasive healthcare system...................... 142
    6.3.4 Technology trends and pervasive information processing.... 143
          6.3.4.1 System- and information-related performance
                  requirements.. 144
6.4 Wearable biomedical systems for chronic care 145
    6.4.1 Classification of sensors... 145
    6.4.2 Ambulatory blood pressure monitoring (ABPM) 146
          6.4.2.1 Commercial ABPM systems 146
    6.4.3 Ambulatory cardiac monitoring: Holter monitors................. 148

6.4.4   Continuous glucose monitoring 149
            6.4.4.1   Continuous noninvasive glucose monitoring
                           by fluid extraction 150
            6.4.4.2   Continuous, nearly noninvasive glucose
                           monitoring by interstitial fluid harvesting 150
6.5   Integrated multiparameter biomedical
        monitoring system 151
        6.5.1   Textiles: The information infrastructure for
                    pervasive healthcare 152
        6.5.2   The Georgia Tech Wearable Motherboard (Smart Shirt) 152
                    6.5.2.1   The wearable motherboard architecture 153
                    6.5.2.2   Testing of the Smart Shirt 154
                    6.5.2.3   Launderability of the Smart Shirt 155
        6.5.3   The next generation: Fabric as a sensor 155
6.6   Looking ahead: Challenges and opportunities 156
References 158

6.1   Introduction

"Health is wealth" goes the age-old saying. It is as true today as it was when it was first coined. In fact, the economic vitality of a nation can be severely limited by the poorer health, premature deaths, and long-term disabilities of individuals without proper access to healthcare.[1] Consider the following facts pertaining to healthcare in the United States:

- Healthcare spending was $1.8 trillion in 2004. It is projected to reach $3.6 trillion in 2014, growing at an average annual rate of 7.1 percent from 2003 to 2014.[2]
- As a share of gross domestic product (GDP), health spending is projected to reach 18.7 percent by 2014, up from its 2003 level of 15.3 percent.[2]
- In the first half of 2005, 14.4 percent of the population was uninsured, which was not significantly lower than the 2004 estimate of 14.6 percent.[3]
- As of mid-year 2005, 5.7 percent of the population was unable to obtain needed medical care because of cost at some time during the past twelve months. The estimate is slightly higher than, but not significantly different from, the 2004 estimate of 5.5 percent.[3]
- In early 2005, 25.1 percent of adults twenty years and over were obese, which was similar to the 2004 estimate of 24.5 percent. Obesity is defined as having a body mass index (BMI) of 30 kg/m^2 or more.[3]
- 7.2 percent of adults over eighteen years old had never been diagnosed as having diabetes.[3]
- More than 1.7 million Americans die of a chronic disease each year, accounting for about 70 percent of all deaths in the United States.[4]

- Cardiovascular disease (including heart disease and stroke) alone is the leading cause of death in the United States, affecting over sixty million Americans and costing the nation more than $351 billion in direct and indirect healthcare costs per year.[4]
- Medical care for people with chronic diseases such as arthritis, asthma, cancer, diabetes, and heart disease accounts for more than 75 percent of the money spent as a nation on medical care.[4]
- The prolonged course of illness and disability from diseases such as arthritis, cancer, diabetes, heart disease, and stroke result in pain and suffering, poor quality of life, and disability for millions of Americans.[4]

Taken together, these facts represent a *silent* crisis that is dramatically affecting the quality of life for individuals—*silent*, because it doesn't have the violent and sudden devastation of a natural disaster or a military attack. Although seemingly less dramatic, the long-term social implications of lack of quality healthcare are as significant.

6.2 The healthcare challenge

In a landmark study, the Institute of Medicine concluded that "the U.S. healthcare delivery system does not provide consistent, high-quality medical care to all people."[5] Moreover, with universal access to information (e.g., through the Web), today's healthcare consumer is demanding more options and is taking more control in determining the course of healthcare. Therefore, the healthcare industry faces the following critical challenges:[6]

- Reducing healthcare costs while maintaining a high quality of care
- Providing access to care for as many people as possible
- Providing easy access to specialized professionals anywhere and anytime
- Shifting the focus of healthcare expenditures from *treatment* to *prevention* through wellness programs
- Controlling lengths of hospital stays and *decentralizing* the provision of healthcare
- Addressing the increase in the aging population and caring for chronically ill patients

The healthcare industry is facing a set of significant challenges on several fronts, namely, availability (or access), quality, and cost.

6.2.1 Meeting the challenge: The pervasive healthcare paradigm

According to Musich et al., seven major diseases accounted for 80 percent of deaths in the United States in 1990: heart disease, cancer, diabetes, arthritis, chronic bronchitis, influenza, and asthma.[7] For many of these health conditions, *early, systematic intervention* would be highly beneficial.

6.2.1.1   Continuous monitoring—the patient view

Let's take the example of an individual with Type 1 diabetes. Self-monitoring of glucose levels up to four times per day or more is important for preventing and detecting hypoglycemia (low sugar) and avoidance of hyperglycemia (high sugar). Moreover, regular monitoring of the sugar level enables the individual to adapt and respond appropriately by modifying the diet, treatment, or exercise regimen to maintain desired glucose levels and comply with the physician's recommendation. Thus, *continuous* monitoring of individuals, especially those with chronic conditions, is essential for the major facets of *prevention, detection, avoidance* and *compliance* associated with such a disease.

During the course of a typical day, an individual is likely to engage in a wide range of activities (e.g., working, shopping, exercising, etc.) in different physically and geographically distributed locations as shown in Figure 6.1, such as a hospital, gymnasium, theater, or restaurant. Regardless of the location or the activity, the individual must monitor the glucose level and respond appropriately and in a timely manner to prevent a potential fatality. This means the *point* of care represented traditionally by the hospital is now the *continuum* of care, spanning a wide range of locations as shown in the figure. Therefore, continuous monitoring is critical to achieve responsive care anytime, anywhere. Moreover, the individual (or the typical caregiver) is not necessarily a trained medical professional but an ordinary citizen. So the solution or technology for providing care anytime, anywhere, should be readily accessible to and usable by anyone.

Let us consider another scenario where a patient is recovering at home after heart surgery and is being monitored remotely by the hospital or a monitoring service. Such monitoring will help the patient feel more "secure" and will facilitate recuperation while simultaneously reducing the cost and time associated with recovery. For example, at the Montefiore Medical Center in New York with 140 heart-failure patients on its "Telescales" remote-monitoring program, the hospital has seen overall medical expenses decline by 18 percent compared with a group of patients who aren't being monitored.[8] A patient whose vital signs have been monitored continuously in the hospital feels that the "umbilical cord" is cut off when discharged from the hospital because typically there is no more real-time monitoring of the individual; this

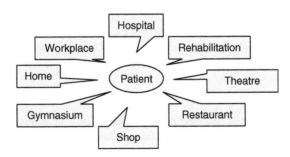

Figure 6.1 A typical day in the life of a patient.

causes a state of uncertainty in the patient's mind and the ensuing anxiety impedes the patient's recovery and recuperation. Because the daily cost of a hospital stay is typically greater than the daily cost of staying at home, the overall costs will also come down when the individual can go home sooner and be monitored remotely in real time. Moreover, in the event of an emergency, the doctor can be notified instantaneously. Thus, with *continuous* monitoring, the physician can administer the *right* treatment at the *right* time at the *right* cost, and indeed could save a life.[9]

6.2.1.2   *Continuous monitoring—the healthcare professional view*

The healthcare industry must meet the challenge of balancing cost containment with maintenance of desired patient outcomes. Consequently, healthcare professionals are trying to provide patient care more efficiently, and whenever possible, in the least expensive setting—be that an ICU (intensive care unit), a hospital general care unit, a skilled nursing facility, or an outpatient clinic. Even a patient's home is becoming the site of many types of care or monitoring that were once provided only by hospitals.[10] This has created a demand for portable, versatile medical devices that can be moved easily from the ICU all the way to a homecare setting.

The demand for high-tech medical devices is pushing the need for technologies that enable physicians to interact effectively and efficiently with *any* patient *anywhere* in the continuum of care *anytime*. In such an environment, the physical boundaries and distances that limit a doctor's healing area could potentially disappear and patients would have access to *any* doctor they desire. In fact, taken one step further, this approach could potentially lead to a network of specialty centers around the world, where each hospital could focus on a particular area of medicine rather than attempt to excel in *all* the specialties.

6.2.1.3   *Continuous monitoring—the pharmaceutical industry view*

Pharmaceutical companies developing new medicines and cures need to test them on subjects in as "natural" an environment as possible. Therefore, there is a need for *continuous* monitoring of test subjects as they go about their daily activities (see Figure 6.1) so that pharmaceutical companies can gain a better idea of the benefits and limitations of proposed treatment regimes.

6.2.2   *Continuous monitoring—a way of life*

In fact, *continuous* monitoring is quickly becoming a way of life. For instance, the OnStar Vehicle Diagnostics service from General Motors (GM) automatically performs hundreds of diagnostic checks on four key GM vehicle operating systems: the engine/transmission, antilock brakes, airbags, and OnStar communications system. Vehicles automatically send the results via electronic mail to owners each month. The report also provides maintenance reminders based on the vehicle's current odometer reading, remaining engine oil life, and other relevant vehicle ownership and OnStar subscription information.[11] If automobiles with finite price tags can be monitored for their "health" on a continuous basis, it is only appropriate that humans, whose lives are priceless, be continuously monitored for their care and well-being. Thus, the need for continuous monitoring in the continuum of care leads to the paradigm of pervasive healthcare: healthcare that is available to anyone, anytime, and anywhere with the potential to successfully enhance quality of life while reducing healthcare costs.

The remainder of this chapter is organized as follows: The principal modules constituting a patient-centric pervasive healthcare system are presented in Section 6.3 along with the main enabling technologies. Sensors, a key component of a pervasive healthcare system, are discussed in Section 6.4 with specific examples of wearable sensor systems for biomedical monitoring. In Section 6.5, the need for multiparameter sensing is established and a textile-based personalized health monitoring system (the Smart Shirt) is discussed. The challenges and opportunities for research in pervasive healthcare are presented in Section 6.6.

6.3 Pervasive healthcare system: A patient-centric approach

6.3.1 The life cycle of unit operations

Figure 6.2 shows the typical life cycle of unit operations associated with providing pervasive healthcare for an individual such as the diabetes patient discussed earlier. The first step is to *sense or observe* the patient's condition. The next unit operation is to *process* the sensory information to estimate the individual's vital signs such as heart rate, electrocardiogram (ECG), body temperature, pulse oximetry (SpO_2), and glucose level, among others. The next step is to *transmit* these vital signs to a doctor's office or a hospital. There the domain expert (e.g., physician) must *interpret* the signals and *diagnose* the condition of the patient. As shown in the figure, the next unit operation is to *respond* to the diagnosis by developing and administering a treatment. The final step is to *learn* from the experience so that future responses can build on the knowledge gained. Thus, a pervasive healthcare system must be designed to facilitate these unit operations in a cost- and time-effective manner.

6.3.2 The enabling technologies

A careful analysis of the unit operations identifies the key *enabling* technologies required for carrying them out to facilitate pervasive healthcare; as shown in Figure 6.3, these are:

- Sensors
- Signal processing system
- Communications system
- Decision support system

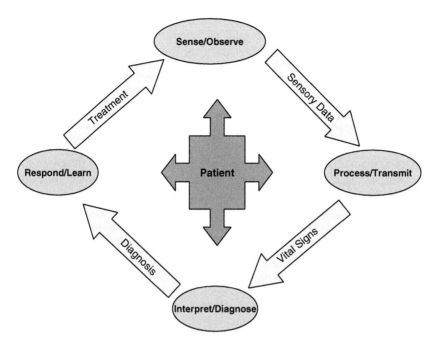

Figure 6.2 Life cycle of unit operations in pervasive healthcare.

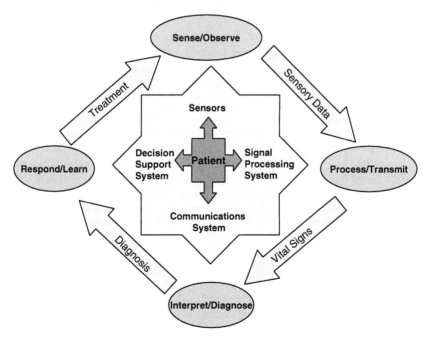

Figure 6.3 Key enabling technologies for pervasive healthcare.

The *sensors* will be responsible for gathering information from the individual and the environment (if appropriate) to provide contextual awareness. The *signal processing system* must then process the signals—biometric, video, environmental, etc.—and compute the appropriate parameters, including vital signs in terms of heart rate, body temperature, and ECG. The *communications system* must then transmit the computed data to the hospital or doctor's office (or a remote monitoring station) for storage and analysis. There, a knowledge-based *decision support system* can help the healthcare professionals interpret the information, diagnose the individual's condition, and develop an appropriate treatment to be administered in a timely manner thus completing the cycle. The treatment can be initiated by the individual based on the real-time recommendation from the healthcare professional or, if necessary, triggered automatically by the monitoring site (assuming the user is incapacitated or has previously authorized such automatic intervention). The specifics of each of these technologies and their relevant communications protocols, such as 802.11, Bluetooth, and ZigBee, are beyond the scope of this chapter and are not covered here.

Thus, these technologies—by enabling pervasive healthcare—can rapidly transform healthcare and the practice of medicine by improving the quality and safety of patient care as well as increasing the efficiency of healthcare providers.

6.3.3   Requirements of a pervasive healthcare system

First and foremost, the pervasive healthcare system must be *patient-centric,* meaning that it must be designed from the viewpoint of the patient, the ultimate beneficiary. Moreover, the resulting system must be *safe, effective,* and provide *timely, efficient,* and *equitable* care that does not vary in quality across the nation—aspects defined as key attributes by the Institute of Medicine in its "national statement of purpose" for a healthcare system.[5]

Figure 6.4 shows the key requirements from the patient's viewpoint. For example, *functionality* defines the functions to be carried out by the system or its intended purpose; in this case, it is the monitoring of a set of vital signs associated with the patient. *Usability* of the system is another critical requirement

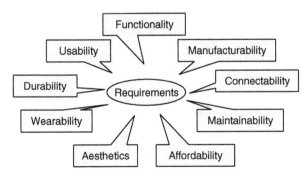

Figure 6.4 Key requirements: Patient view.

that significantly influences compliance, namely, whether the resulting system will indeed be used by the patient. If the system is not easy to use, the likelihood of its use will be greatly diminished thereby affecting compliance. The ability to integrate various sensors, devices, and processors—represented by *connectability* in the figure—determines the versatility and capability of the system. The other requirements shown in the figure include *wearability*, *maintainability*, *aesthetics*, *durability*, and *manufacturability*. Together, these requirements (and others that may be defined depending on the specific end-use application) must be considered in the design of a pervasive healthcare system. System performance- and information-related requirements are discussed in the following section.

6.3.4   *Technology trends and pervasive information processing*

Today, healthcare is at the threshold of a radical transformation as a result of the following technology trends:

- The significant advancements in computing and communications technologies are fundamentally changing when, where, and how individuals work and live. No longer confined to a workplace or chained to a static computing infrastructure *anyone* can process information *anytime* and *anywhere*, giving birth to the paradigm of *pervasive information processing* or "computing on the go."
- The advancements in, and convergence of, microelectronics, materials, optics, and biotechnologies, coupled with miniaturization, have led to the development of small, cost-effective intelligent sensors for a wide variety of applications.
- The transparency of the user interface coupled with the invisibility of "embedded" technology in the various devices and systems have contributed to the proliferation of sensors and sensor networks—from homes to outer space—and everywhere in between.

By effectively harnessing the benefits of these technological advancements, it is possible to create a pervasive healthcare system that will provide affordable healthcare to *anyone anywhere* at *anytime* and enhance the quality of life for everyone in the continuum of life from newborns to senior citizens. The challenge, however, lies in the development and realization of such a system.

In proposing the paradigm of pervasive or ubiquitous computing, Mark Weiser said, "The most profound technologies are those that disappear. They weave themselves into the fabric of everyday life until they are indistinguishable from it."[12] In Weiser's vision, computing and communications systems are totally integrated with the user. In a similar manner, the components or modules of the pervasive healthcare system must be tightly integrated with and unobtrusive to the patient so as to indeed "disappear." In short, the system must be *personalized*, be *mobile*, and facilitate *health information processing*, leading to the concept of personalized mobile health information processing, or PM-HIP for short.

6.3.4.1 *System- and information-related performance requirements*

The successful realization of a patient-centric pervasive healthcare system depends on rigorously meeting another set of performance requirements, namely parameters related to the system and information facets. While pervasive healthcare enables the healthcare professional to access patient records from anywhere at anytime, it also poses a set of significant challenges related to ensuring the *security*, *privacy*, and *confidentiality* of patient information (see Figure 6.5). Thus, in designing the pervasive healthcare system there is a critical need to balance easy access to patient information with the

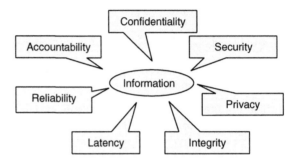

Figure 6.5 System- and information-related performance requirements.

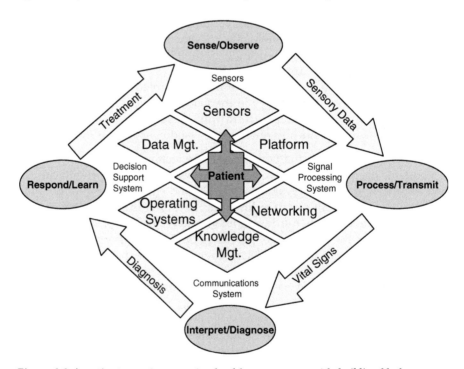

Figure 6.6 A patient-centric pervasive healthcare system with *building blocks*.

legal and ethical complications of inappropriate disclosures of such information. In this sensitive, information-rich domain, *accountability* becomes another important requirement as do data *latency* and data *integrity*. Any unnecessary delay in accessing or responding to the signals or erroneous data could mean the difference between life and death for the patient.

Thus, an overall vision of a patient-centric pervasive healthcare system emerges as shown in Figure 6.6. The underlying *building blocks* for realizing the system, such as sensors, platform, networking, operating systems, data management, and knowledge management, are also shown in the figure. For a pervasive healthcare system to succeed, all these factors must operate robustly and seamlessly.

We now discuss wearable biomedical monitoring systems for pervasive healthcare that address chronic diseases such as hypertension, heart disease, and diabetes.

6.4   Wearable biomedical systems for chronic care

A sensor is defined as "a device used to detect, locate, or quantify energy or matter, giving a signal for the detection of a physical or chemical property to which the device responds."[13] A biosensor has been defined "as an analytical device that incorporates a biologically active material in intimate contact with an appropriate transduction element for the purpose of detecting—reversible and selectively—the concentration or activity of chemical species in any type of sample."[14] Sensors transduce between the electrical, optical, thermal magnetic, mechanical, and chemical signal domains. For example, the glucose biosensor transduces from the chemical to the electrical domain. Thus, the sensor is at the heart of the wearable biomedical system because it is the component that captures data about the patient's condition (see also Figure 6.3).

6.4.1   Classification of sensors

At the top level, sensors can be classified as implantable, minimally invasive, or noninvasive. Implantable sensors require a hospital procedure to be put into place inside the body. Minimally invasive sensors penetrate the skin (subcutaneous) to obtain the signals. Noninvasive sensors are worn on the body. Sensors can also be classified as active or passive. Active sensors require power to operate, such as pulse oximetry sensors, while a temperature probe is an example of a passive sensor, which does not require a power source. Yet another way to view sensors is the mode in which the signals are transmitted for processing, as wired or wireless. In the former, the signals are transmitted over a physical conducting line to a processor. In wireless sensors, communication capabilities are built into the sensor, which transmits the signals wirelessly to a monitoring unit. Noninvasive sensors are typically used in systems for continuous monitoring because their use does not require extensive intervention from a healthcare professional. This is the type of sensor most commonly found in pervasive healthcare.

6.4.2   Ambulatory blood pressure monitoring (ABPM)

Hypertension, or high blood pressure, is a chronic condition affecting approximately fifty million people in the United States. However, there are different types of hypertension. For instance, 20 to 35 percent of patients may experience "white-coat hypertension," where they are hypertensive in the physician's office but normotensive at home.[15,16] Therefore, continuous monitoring of a patient's blood pressure is extremely important for the physician to determine which drug therapy regime will work best with each individual. According to the new Joint National Committee VII Guidelines,[17] ambulatory blood pressure monitoring (ABPM) is "warranted" for patients with white-coat hypertension. Additional studies have shown the following advantages of ABPM:[18–20]

- End-organ damage is more closely correlated with ambulatory blood pressure (ABP) than with clinic blood pressure (CBP) readings.
- ABP may be a better predictor of cardiovascular events and mortality than clinic blood pressure readings.
- Patients with hypertension whose nocturnal (sleep) blood pressure remains high (< 10 percent lower than daytime average) may have a worse prognosis.
- ABP provides a twenty-four-hour profile, allowing assessment of clinic effects, drug effects, work influence, etc.

6.4.2.1   Commercial ABPM systems

In an ABPM system, automatic measurements of blood pressure are obtained throughout a twenty-four- to forty-eight-hour period at specific intervals (say, every thirty minutes) or when triggered by the patient. The resulting stored data enables the physician to assess the blood pressure profile during the patient's typical day of activities. These systems indirectly measure blood pressure through auscultation (of Korotkoff's sounds) either with piezoelectric microphones, through oscillometric measurement of the vibratory signals associated with blood flow in the brachial artery, or through the combined use of both technologies.[16] Auscultatory devices record both systolic and diastolic pressures, whereas the oscillatory units record systolic and mean pressure and then calculate diastolic pressure through a variety of algorithms. Validation testing against mercury sphygmomanometry and intra-arterial measurement has confirmed the accuracy of these technologies.

In the typical ABPM machine, the blood pressure cuff is fitted around the upper arm and is fastened with Velcro. Figure 6.7 shows representative commercial ABPM systems from WelchAllyn and SunTech Medical, respectively. The monitor—typically worn on the belt and weighing around 250 to 300 grams—records the blood pressure at set intervals. The data are

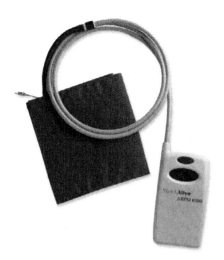 

Figure 6.7 Ambulatory blood pressure monitoring systems: (left) WelchAllyn ABPM 6100 (www.welchallyn.com); (right) SunTech Medical Oscar2 (http:// www.suntechmed.com).

downloaded in the physician's office and the software provides a complete analysis of the data over the study period. The physician can then provide the appropriate treatment to the patient.

Commercially available ABPM systems already provide pervasive healthcare and enhance the quality of life for individuals suffering from hypertension. In a study evaluating the potential value of ABPM for patients with established hypertension, a group of forty patients (thirty-three to sixty years old) that was being treated for hypertension was randomly selected from a general practice list and subjected to a single twenty-four-hour ABPM evaluation.[21] ABPM values were compared with CBP values obtained on the day of monitoring together with previous readings taken by the general practitioner. The study concluded that ABPM provided information over and above that obtained by CBP in a substantial proportion of patients, thus demonstrating the value of ABPM.

In another study, the objective was to describe patient satisfaction with ABPM performed in a primary care office in the United States, using modern ABPM technology.[22] There were 235 eligible respondents in the study. Three-fourths of the patients believed that undergoing the test was worthwhile considering the time and money involved, while most (90 percent) reported they thought the information provided by the test would be helpful to their physician in making treatment decisions, despite probable lack of insurance coverage, and appeared willing to experience some discomfort for the overall gain of the results obtained from undergoing the session. Patients reporting that their physician had clearly explained the benefit of undergoing the testing were more likely to report that they thought the results of the test would be more helpful in making treatment decisions. Few patients (20 percent) found that wearing the monitor was uncomfortable. Thus, ABPM has the potential to be valuable in pervasive healthcare.

6.4.3   Ambulatory cardiac monitoring: Holter monitors

As mentioned previously, cardiovascular disease is the leading cause of death in the United States. Holter monitoring provides a continuous recording of the heart's electrical activity (the heart rhythm) during the patient's typical day. The battery-operated monitor is usually worn for twenty-four hours to obtain the recording. Electrodes are placed on the patient's chest and attached to a small recording monitor that is carried in a pocket. Simultaneously, the patient keeps track of activities during the day. The stored data can be sent to the physician's office for analysis and treatment. Any irregular heart activity is correlated with the patient's activity at that time.

The Zymed DigiTrak Plus from Philips Medical Systems is a digital Holter monitor that is the size of a pager and weighs around 90 grams (Figure 6.8). It is typically worn on the belt or, in the case of young children, around the neck using a lanyard. The patient places five electrodes at designated locations on the body as shown in the figure. From the five EASI electrodes, three channels of ECG recording are made and stored in the recorder.

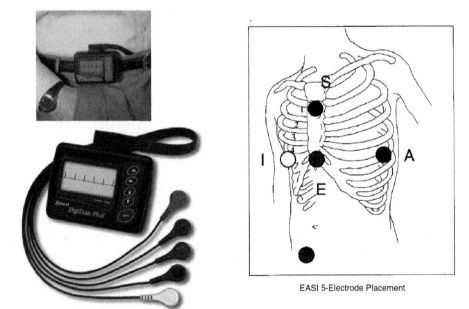

EASI 5-Electrode Placement

Figure 6.8 The Zymed Holter monitor from Philips Medical Systems showing electrode placements (from http://www.medical.philips.com/us/products/cardio graphy/products/holter/digitrak/).

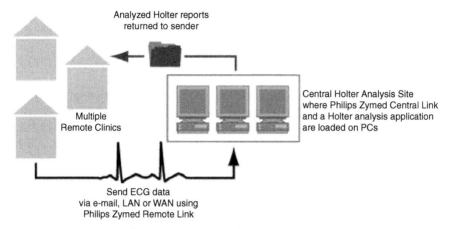

Figure 6.9 Pervasive monitoring using Zymed's Central Holter Analysis System (from http://www.medical.philips.com/us/products/cardiography/products/holter/central/).

The recorded data can be electronically sent to a central location (e.g., hospital or doctor's office) for analysis as shown in Figure 6.9. The results of the analysis can then be communicated back to the physician or patient for suitable intervention, thus facilitating pervasive healthcare for cardiac patients.

6.4.4   Continuous glucose monitoring

Diabetes mellitus is another chronic disease affecting nearly 7 percent of the population in the United States.[23] As stated earlier, proper disease management requires self-monitoring of the blood glucose level several times a day and taking necessary action—in terms of diet, exercise, and medication—based on the measured values. The typical procedure for glucose monitoring is invasive and it involves drawing a drop of blood from the patient (using a lancet), placing it on a test strip, and reading the glucose levels in the monitor. The process can be painful and prone to infection, especially because tests need to be repeated several times each day. This in turn affects compliance with glucose monitoring by patients. The traditional "finger-stick" measurement reveals the blood glucose value at a single moment or "snapshot" in time. An A1C test, which measures blood glucose control over a two- to three-month period, gives an average glucose value. Both can miss highs and lows in the user's blood glucose present at various times during the day—highs and lows that may lead to diabetes-related complications.[24]

Therefore, for successful continuous blood glucose monitoring of a diabetic patient, there is a need for less invasive procedures. Several techniques are under development. Noninvasive blood glucose monitoring generally involves either radiation or fluid extraction.[25] Nearly noninvasive blood glucose monitoring involves transcutaneous harvesting and measurement of interstitial fluid, which is the fluid surrounding every cell of the body.[26] We will now look at two systems that facilitate continuous monitoring of diabetic patients.

6.4.4.1   Continuous noninvasive glucose monitoring by fluid extraction

The GlucoWatch Biographer uses reverse iontophoresis (using an electric current to move a substance across body tissue) to monitor glycemia in diabetes patients.[27] The device is worn like a watch and the glucose readings are taken noninvasively by the AutoSensor (left in Figure 6.10). Induced by a low electric current, glucose is pulled through the skin and accumulated in two gel collection discs in the AutoSensor, which is a single-use component that snaps into the back of the watch and lasts for up to thirteen hours. The glucose reading is displayed on the screen as frequently as every ten minutes. A built-in alarm system alerts the patient when the reading exceeds the user-defined threshold level (for low or high glucose levels) or if there is a clear declining trend and the lower limit is likely to be reached in the next twenty minutes. However, the traditional finger-stick method must be used to calibrate the AutoSensor each time it is changed.

The GlucoWatch Analyzer software can be used to perform trend analysis by downloading the monitored data to a personal computer. This enables the patient to track patterns in glucose concentrations and to evaluate the effects of meals, exercise, insulin, and medication. Using this information, a healthcare professional can suitably fine-tune the diabetes treatment and thereby reduce the risk of long-term health complications from diabetes such as eye, nerve, kidney, and heart disease.

6.4.4.2   Continuous, nearly noninvasive glucose monitoring by interstitial fluid harvesting

The Guardian RT System from Medtronic provides real-time continuous monitoring of glucose values in subcutaneous interstitial fluid. A tiny sensor is inserted under the skin and is connected to a transmitter that is taped to

Figure 6.10 Continuous blood glucose monitoring systems: (left) G2 Biographer from GlucoWatch (www.glucowatch.com); (right) Guardian RT from Medtronic (www.medtronic.com).

the skin with a soft adhesive. The sensor can be worn for up to seventy-two hours. The transmitter uses RF (radio frequency) technology to wirelessly send glucose values automatically to the monitor (right in Figure 6.10). The monitor displays updated glucose readings every five minutes and alarms when values fall outside the user-defined threshold levels, for when the patient is hyperglycemic or hypoglycemic, respectively. The monitor stores up to twenty-one days of data; this data can be downloaded to the computer for detailed trend analysis. The Guardian Solutions software is similar to the GlucoWatch Analyzer software and enables the patient to better manage the disease in consultation with the healthcare professional.

We will now discuss cutting-edge technologies that enable *multiparameter* biomedical monitoring of mobile patients for personalized pervasive healthcare.

6.5   Integrated multiparameter biomedical monitoring system

While the systems described in Section 6.4 certainly facilitate continuous monitoring of patients, they are typically stand-alone systems for single parameters, such as for ABP, blood glucose, or cardiac monitoring. They are also distinct units and tend to be conspicuous on the individuals using them. If multiple parameters need to be monitored, the patient must wear multiple systems, which will be cumbersome and pose a whole host of challenges. Also, in the desired scenario of healthcare discussed in Section 6.3, these devices must *disappear* into the patient's lifestyle. Moreover, the patient, in essence, is an *information node* or *sensor* furnishing valuable real-time data and providing enhanced situational awareness to the health-care professional at the remote site. This data must be transformed into "knowledge" for diagnosis and treatment. A structured analysis of the sensing requirements for monitoring the individual leads to the following conclusions:[28]

1. Different *types* of sensors are needed to *simultaneously* monitor the various vital signs (e.g., heart rate, body temperature, pulse oximetry, blood glucose level).
2. Different *numbers* of sensors may be needed to obtain the signals to compute a single parameter (e.g., at least three sensors are required to compute the ECG).
3. The sensors need to be positioned in *different* locations on the body to acquire the proper signals (e.g., sensors for ECG).
4. Different *subsets* of sensors may be used at different times necessitating their easy attachment and removal, or *plug and play*.

In short, what is needed is the design and implementation of a "sensor network" on the individual to achieve the desired functionality of vital signs

monitoring for pervasive healthcare. Moreover, attributes such as functionality, modularity, and flexibility (such as plug-and-play capabilities) that are
required of the sensor network (Figure 6.4) suggest the choice of the motherboard paradigm. Just as special-purpose chips and processors can be
plugged into a computer motherboard to obtain the desired information
processing capability (e.g., high-end graphics), the motherboard paradigm
can provide an extremely versatile framework for the incorporation of sensing, monitoring, and information-processing devices.[10] However, because
the sensor network—in the form of the motherboard—needs to be on the
patient's body, it must be *wearable*. This leads to the paradigm of a "wearable
motherboard." The rest of this section presents an example of how a versatile,
networked pervasive healthcare system, including a wearable motherboard,
can be achieved.

6.5.1   Textiles: The information infrastructure for pervasive healthcare

Textiles are pervasive, can be personalized, and present a "universal" interface that is natural and easy to use.[10] They provide the ultimate flexibility
in system design by virtue of the broad range of fibers, yarns, fabrics, and
manufacturing techniques that can be deployed to create products for
desired end-use applications. Moreover, fabrics provide "variable" surface
areas (very small to very large) that may be needed for "hosting" varying
numbers of sensors and processors needed for creating sensor networks.
These sensor networks may be deployed on the individual or in the environment to enable pervasive healthcare. It is also possible to build in redun-
dancies for enhanced fault tolerance in the structure. Thus, textiles can serve
as an effective mobile information infrastructure that can be tailored to the
patient's specific requirements, making textiles an ideal platform for creating
a wearable motherboard.

6.5.2   The Georgia Tech Wearable Motherboard (Smart Shirt)

Research at Georgia Tech led to the realization of the world's first wearable
motherboard or "intelligent" garment.[29] Initially funded by the Defense
Advanced Research Projects Agency (DARPA) through the U.S. Navy, the
Georgia Tech Wearable Motherboard (GTWM), or Smart Shirt, uses optical
fibers to detect bullet wounds and uses special interconnected sensors to
monitor vital signs during combat conditions. The principal advantage of
the Smart Shirt is that it provides, for the first time, a very systematic and
personalized way of monitoring the vital signs of humans in an *unobtrusive*
manner.[30] A set of user requirements similar to those in Figures 6.4 and
6.5 was developed and used in the design of the Smart Shirt (the details
of the design methodology can be found in reference 31). Several versions
of the Smart Shirt have been produced; with each succeeding version, the

garment has been continually enhanced from all perspectives—functionality, capabilities, comfort, ease of use, and aesthetics.

6.5.2.1   The wearable motherboard architecture

Figure 6.11 shows the architecture of the wearable motherboard intended for medical applications. The comfort or base fabric (woven, knitted, nonwoven, etc.) provides the necessary physical infrastructure for the wearable motherboard. The base fabric is made from typical textile fibers (e.g., cotton, polyester, blends) where the choice of fibers is dictated by the intended application. The developed interconnection technology has been used to create a flexible and wearable framework to plug in sensors for monitoring a variety of vital signs including heart rate, respiration rate, ECG, body temperature, and SpO_2. In addition, voice can be recorded by plugging a microphone into the Smart Shirt.

These sensors can be positioned in desired locations on the body and will plug into the Smart Shirt. The motherboard or "plug-and-play" concept

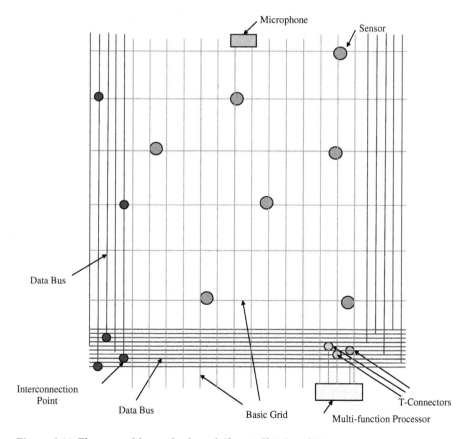

Figure 6.11 The wearable motherboard (Smart Shirt) architecture.

means other sensors (e.g., accelerometers to detect falls, carbon monoxide detection sensors) can be easily used with the structure. The flexible data bus integrated into the structure transmits the information from the suite of sensors to the multifunction processor known as the Smart Shirt Controller. This controller, in turn, processes the signals and transmits them wirelessly (using an appropriate communication protocol such as Bluetooth, 802.11b) to desired locations (e.g., doctor's office, hospital). The bus also carries information *to* the sensors (hence, the wearer) from external sources, thus making the Smart Shirt a valuable information infrastructure.[32]

6.5.2.2 Testing of the Smart Shirt

The vital signs monitoring capability of the Smart Shirt has been successfully tested in a variety of applications. The heart rate, respiration rate, ECG, and body temperature of individuals were measured using commercial off-the-shelf sensors that plugged into the Smart Shirt. Initial testing was done at Crawford Long Hospital in Atlanta, followed by another set of tests in the Department of Physiology at Emory University. An infant version of the Smart Shirt was subsequently tested in collaboration with the Egleston Hospital of Emory University School of Medicine.

The vital signs data were wirelessly transmitted to a personal computer or a medical ECG monitor (e.g., Nihon-Kohden hospital monitor). Figure 6.12 shows the display of key vital signs, including the ECG waveform. The garment is also comfortable and easy to wear and take off, similar to a typical undershirt. All these tests conclusively demonstrated the ability of the Smart Shirt to unobtrusively monitor the vital signs of individuals (from infants to

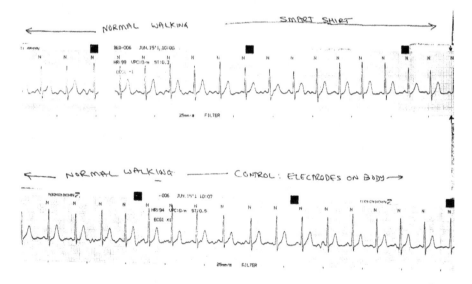

Figure 6.12 Wireless transmission of vital signs from the Smart Shirt. Top trace: from the Smart Shirt. Bottom trace: directly from the user.

adults) engaged in normal activities in an easy-to-use form with the convenience and familiarity associated with a garment.

6.5.2.3   Launderability of the Smart Shirt

The Smart Shirt successfully withstood a series of industry-standard launderability tests (washing and drying) typically carried out on textiles and apparel. In one such test, it was subjected to twenty wash/dry cycles; it functioned effectively after every wash/dry cycle, thus demonstrating the robustness of the wearable motherboard paradigm and embodying the principle of plug and play (the sensors and key electronic components are "unplugged" from the Smart Shirt prior to it being laundered).

The Smart Shirt facilitates multiparameter biomedical monitoring for mobile patients and enables personalized pervasive healthcare.

6.5.3   The next generation: Fabric as a sensor

Many sensors are not practical for long-term use and patient observation. For instance, the conventional ECG electrode has a conductive gel (silver-silver chloride) and an adhesive backing to affix the conductive lead to the patient's skin. Because the electrode sticks to the body, it is difficult to remove and can cause skin irritation when used continuously for extended periods of time. Therefore, for pervasive healthcare to realize its full potential, it is important to deploy sensors that do not have these shortcomings. The need to replace conventional ECG electrodes and eliminate the associated discomfort has led to the design and development of a fabric-based sensor and garment for obtaining the signals in monitoring vital signs.[33]

Figure 6.13 shows three different types of fabric-based sensor garments. In the first one, conductive materials (fibers/yarns) have been used to create a sensor (in the form of a stand-alone *patch*) that will replace the conventional ECG electrode for cardiac monitoring. This patch can be plugged into the lead wires of a Holter monitor (e.g., Zymed) or into the Smart Shirt discussed in Section 6.5.2. In another version, the conductive sensor has been integrated into the garment itself during manufacturing. The figure shows a knitted version of this next-generation sensing garment with integrated sensors in desired locations for obtaining ECG signals. It is important to have excellent electrical connectivity between the sensor and the body to ensure high fidelity signals. Therefore, to keep the sensors in place on the patient's body and ensure connectivity, spandex fiber is used as a form-fitting component in the sensing garment.

An infant version of the fabric-based Smart Shirt used for monitoring babies prone to SIDS (sudden infant death syndrome) is also shown in the figure. In fact, with this type of fabric-based sensing garment, it is possible to use the monitored vital signs to "close the loop" and provide a suitable response to the patient. For example, when the built-in sensor detects that a baby has stopped breathing, a small electrical pulse (shock) can be sent from a monitoring source to the child to stimulate breathing. By having

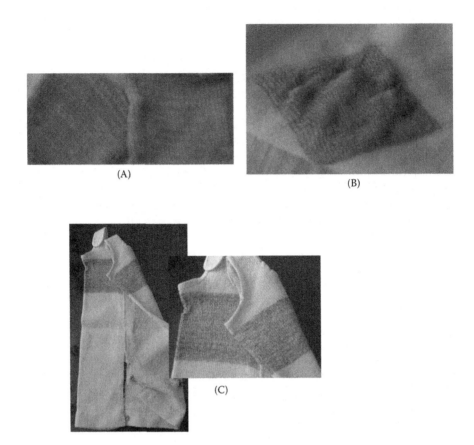

Figure 6.13 Fabric as a sensor for pervasive healthcare. (A) A pluggable stand-alone fabric-based sensor; (B) Smart Shirt with integrated intarsia knitted fabric-based sensor; (C) Infant Smart Shirt with fabric-based integrated sensors: multiple views.

a technology that is not only ubiquitous but also has the ability and intelligence to respond to the changes in the needs of the wearer, the quality of preventive care can be significantly enhanced, further reinforcing the paradigm that "investment in prevention is significantly less than the cost of treatment."

Thus, with the development of this new generation of fabric-based sensors (garments) such as the Smart Shirt, the textile-based pervasive healthcare system is a step closer to becoming a reality.

6.6   *Looking ahead: Challenges and opportunities*

Significant progress has been made in medical and information technologies over the last decade, resulting in the enhancement of the quality of life for everyone through better and increasingly pervasive healthcare. Change is the *only* constant today and the significant advancements in computing, communication, and biological systems offer exciting convergence opportunities that can be harnessed to meet the challenge of making pervasive healthcare a reality.

The pervasive healthcare system depicted in Figure 6.6 can be used as the basis for identifying these challenges and opportunities in each of the six *building blocks*: namely, sensors, data management, platform, operating systems, knowledge management, and networking. For pervasive healthcare to be truly successful, technology employing smart sensors with multifunction capability, integrated wireless communications capability, and low power consumption must be developed. The challenge is to balance functionality with the requirements of size and power consumption. Integration of these sensors into fabric-based networks that can be worn by patients or deployed in the environment points to the need for a standard or universal interface for these sensors.

In the typical pervasive healthcare scenario, multiple parameters (e.g., vital signs) will be continuously monitored. The challenge, and the opportunity, in the area of signal processing is to develop algorithms for multisensor data fusion that require minimal hardware so that the cost of the system can be minimized while keeping them "power-aware." Newer communications technologies (e.g., ZigBee) and their integration with smart sensors provide another active area of research. The success of pervasive healthcare depends on "closing the loop," that is, providing the right treatment at the right time to the patient without compromising the patient's other needs, such as privacy. The development of knowledge-based decision support systems that can facilitate such diagnosis and treatment presents ample opportunities for research. Of course, the challenge is to be able to "understand" and codify the physician's decision-making process. With the increasing use of computer-based physician order entry (CPOE) systems in hospitals, avenues are opening up for understanding the decision-making process and for documenting and assessing the impact of such decisions (treatments) on patients through electronic medical records (EMR). At a system level, issues related to the integration of the various enabling technologies shown in Figure 6.3 along with the requirements defined in Figure 6.4 and Figure 6.5 present interesting challenges and opportunities for research and development. Such a pervasive healthcare system will give rise to an integrated view of the patient as an information node. However, with it come significant amounts of data in various formats including text, graphics, audio, and video. For example, the medical record of an asthma patient intended for review by a specialist in a different location could, in addition to an x-ray image and quantitative data, contain an audio recording of the patient's breathing pattern. Tools and technologies for handling these multiple types and streams of data—while preserving the integrated view of the patient—must be developed to ensure their timely and effective utilization.

The ease with which personal data can be collected in real time using the various technologies described in this chapter will result in the creation of "knowledge banks" of human performance. This knowledge base can be used in clinical and pharmaceutical research and potentially lead to new treatments, drugs, and drug delivery systems. These benefits should be weighed in the context of potential invasion of personal privacy. Therefore, there is a critical need for a major initiative that brings together experts from the medical, insurance, and legal communities to address this important facet of advanced technologies so that society can harness the benefits from technological advancements and enhance the quality of life without sacrificing an individual's most prized possession: privacy.[9]

This chapter has shown how technology in general, and sensor technology in particular, is the catalyst that enables the transformation of today's healthcare to tomorrow's personalized pervasive healthcare.

References

1. "Insuring America's Health: Principles and Recommendations." Committee on Insuring Health, Institute of Medicine, Washington, DC, 2004.
2. http://www.cms.hhs.gov/statistics/nhe/projections-2004/proj2004.pdf. Last accessed November 20, 2005.
3. http://www.cdc.gov/nchs/data/nhis/earlyrelease/earlyrelease200509.pdf. Last accessed November 20, 2005.
4. http://www.cdc.gov/fmo/FY2006CDCCJ.pdf. Last accessed November 20, 2005.
5. "Crossing the Quality Chasm: A New Health System for the 21st Century." Committee on Quality of Healthcare in America, Institute of Medicine, Washington, DC, 2001.
6. Park, S., and Jayaraman, S. e-Health and Quality of Life: The Role of the Wearable Motherboard. In *Wearable eHealth Systems for Personalised Health Management*, ed. A. Lymberis and D. DeRossi (pp. 239–252). Amsterdam: IOS Press, 2004.
7. Musich, S.A., Burton, W.N., and Edington, D.W. Costs and Benefits of Prevention and Disease Management. *Disease Management & Health Outcomes*, March 1999, 5(3): 153–166.
8. Cherney, E. New Ways to Monitor Patients at Home—As Insurers Increasingly Cover "Telemedicine," Companies Launch Wave of Devices. *The Wall Street Journal*, April 18, 2006, p. D1.
9. Park, S., and Jayaraman, S. Quality of Life in the Internet Age: Role of the Georgia Tech Smart Shirt. *Atlanta Medicine*, Winter 2001, 74(4): 24–28.
10. Gopalsamy, C., Park, S., Rajamanickam, R., and Jayaraman, S. The Wearable Motherboard: The First Generation of Adaptive and Responsive Textile Structures (ARTS) for Medical Applications. *Journal of Virtual Reality*, 1999, 4:152–168.
11. http://www.onstar.com/us_english/jsp/ovd/index.jsp. Last accessed November 24, 2005.
12. Weiser, M. The Computer for the 21st Century. *Scientific American*, September 1991.
13. Kress-Rogers, E. Biosensors and Electronic Noses for Practical Applications. In *Handbook of Biosensors and Electronic Noses: Medicine, Food, and the Environment*, ed. E. Kress-Rogers (pp. 3–39). New York: CRC Press, 1997.
14. Arnold, M.A. and Meyerhoff, M.E. Recent Advances in the Development and Analytical Applications of Biosensing Probes. *CRC Critical Reviews in Analytical Chemistry*, 1988, 20: 149–196.

15. White, W.B. Ambulatory Blood-Pressure Monitoring in Clinical Practice. *New England Journal of Medicine*, 2003, 348(24): 2377–2378.
16. Marchiando, R.J. and Elston, M.P. Automated Ambulatory Blood Pressure Monitoring: Clinical Utility in The Family Practice Setting. *American Family Physician*, 2003, 67: 2343–2350.
17. Seventh Report of the Joint National Committee on Prevention, Detection, Evaluation, and Treatment of High Blood Pressure. *Journal of the American Medical Association*, May 21, 2003.
18. McGrawth, B.P. Ambulatory Blood Pressure Monitoring. *The Medical Journal of Australia*, 2002, 176(12): 588–592.
19. White, W.B. Ambulatory Blood Pressure as a Predictor of Target Organ Disease and Outcome in the Hypertensive Patient. *Blood Pressure Monitoring*, 1999, 4(3–4): 181–184.
20. O'Brien, E., Coats A., Owens, P., Petrie, J., Padfield, P.L., and Littler W.A. Use and Interpretation of Ambulatory Blood Pressure Monitoring: Recommendations of the British Hypertension Society. *British Medical Journal*, 2000, 320: 1128–1134.
21. Ernst, M.E. and Bergus, G.R. Favorable Patient Acceptance of Ambulatory Blood Pressure Monitoring in a Primary Care Setting in The United States: A Cross-Sectional Survey. *BMC Family Practice*, 2003, 4: 15.
22. Taylor, R.S., Stockman, J., Kernick, D., Reinhold, D., Shore, A.C., and Tooke, J.E. Ambulatory Blood Pressure Monitoring for Hypertension in General Practice. *Journal of the Royal Society of Medicine*, June 1998, 91(6): 301–304.
23. http://www.diabetes.org/diabetes-statistics/prevalence.jsp. Last accessed November 24, 2005.
24. The Diabetes Control and Complications Trial Research Group. *Journal of the American Medical Association*, 1997, 227(5): 374–375.
25. Klonoff, D.C. Noninvasive Blood Glucose Monitoring. *Diabetes Care*, 1997, 20: 433–437.
26. http://www.medihealthdme.com/education/diabetes_non.htm. Last accessed November 24, 2005.
27. Sieg, A., Guy, R.H., and Delgado-Charro, M.B. Noninvasive Glucose Monitoring by Reverse Iontophoresis In Vivo: Application of the Internal Standard Concept. *Clinical Chemistry*, 2004, 50: 1383–1390.
28. Park, S. and Jayaraman, S. *Sensor Networks and the i-Textiles Paradigm.* Proceedings of the Next Generation PC 2005 International Conference, Seoul, Korea, November 3–4, 2005, pp. 163–167.
29. *The Georgia Tech Wearable Motherboard: The Intelligent Garment for the 21st Century.* http://www.smartshirt.gatech.edu. Last accessed November 24, 2005.
30. Park, S., Gopalsamy, C., Rajamanickam, R., and Jayaraman, S. The Wearable Motherboard: An Information Infrastructure or Sensate Liner for Medical Applications. *Studies in Health Technology and Informatics*, 1999, 62: 252–258.
31. Rajamanickam, R., Park, S., and Jayaraman, S. A Structured Methodology for the Design and Development of Textile Structures in a Concurrent Engineering Environment. *Journal of the Textile Institute*, 1998, 89(3): 44–62.
32. Park, S. and Jayaraman, S. Enhancing the Quality of Life through Technology: The Role of Personalized Wearable Intelligent Information Infrastructure. *IEEE Engineering in Medicine and Biology*, May/June 2003: 41–48.
33. Jayaraman, S. and Park, S. Fabric-Based Sensor for Monitoring Vital Signs. US Patent 6,970,731, issued November 29, 2005.

chapter seven

Assistive technologies

Edmund LoPresti
AT Sciences, LLC, Pittsburgh, Pennsylvania

Contents

7.1   Introduction 161
7.2   Communication assistance 162
7.3   Mobility assistance... 165
7.4   Wayfinding 169
7.5   Cognitive assistance 172
7.6   Smart environments 175
7.7   Design considerations for assistive technology................................ 179
7.8   Conclusion... 184
References 185

7.1   Introduction

People with disabilities have long been early adopters of pervasive computing technologies, as these devices are well suited to helping people compensate for physical, sensory, communication, and cognitive limitations. This compensation provides increased educational, vocational, and recreational opportunities, thus improving quality of life. This chapter describes technologies for assisting people with various disabilities and demonstrates applications of pervasive computing, including mobile computing and smart environments.

More than any other user group, people with disabilities rely on assistive technologies for fundamental needs, such as movement, communication, and orientation in space and time. As such, this user group needs access to technology anywhere, anytime. Mobile computing is a key issue in the design of assistive technologies, and because of the promise of increased independence, people with disabilities are positioned to be early adopters of mobile computing. We will look at the use of mobile technologies for communication (augmentative and alternative communication systems), personal mobility (wheelchairs and walkers), and orientation in space and time (wayfinding and reminding technologies). We will also look at pervasive assistive technology incorporated into the built environment. Chapter 5 introduced the healthcare implications for smart home technology to support home-based monitoring, home-care infrastructures, and communication with medical professionals and family. In this chapter, we look at how smart home technology can assist people with mobility, sensory, and cognitive limitations. Pervasive computing technology can provide new ways to perform familiar tasks, such as turning on a lamp or responding to a doorbell, or provide novel forms of assistance for schedule maintenance and problem solving.

7.2   Communication assistance

One of the most fundamental human needs is to communicate: sharing with others our wants and needs, thoughts and feelings, hopes and dreams. For most of us, the primary form of communication is the spoken word. But many people are unable to speak or have speech that is incomprehensible to others. Approximately two million people in the United States alone are unable to speak adequately to meet their communication needs (ASHA 1991). The field of augmentative and alternative communication (AAC) investigates ways of providing people with alternatives to speech. Perhaps the most common form of AAC is gestures, including sign language. However, sign language requires a communication partner who also understands the language as well as the manual dexterity to form signs and the cognitive skills to learn the language. Other alternatives range from low-tech devices, such as an alphabet board on which people point out letters to spell words, to high-tech voice output communication aids (VOCAs).

A VOCA is a computerized device that accepts input from the user and produces speech, allowing the user to communicate with others (Figure 7.1). VOCAs must allow someone to communicate in all areas of their daily lives. This means VOCAs are typically mobile devices, although the definition of

Figure 7.1 Example of a VOCA (Vantage Plus, Prentke Romich Corporation, Wooster, OH) using Minspeak language encoding (Semantic Compaction, Pittsburgh, PA).

"mobile" ranges from handheld devices to bulky devices designed to be mounted on a wheelchair. Because people with communication impairments vary widely with regard to their physical, sensory, and cognitive skills, as well as their communication needs, VOCAs are available with a variety of configurations and capabilities. Key aspects of a VOCA include the platform, input methods, output methods, and processes for enhancing communication rate and accuracy.

A central feature of a VOCA is the platform. As previously mentioned, VOCAs are often fairly large, bulky devices. This allows them to have greater computing power; large batteries; large keys for selecting words, letters, or icons; and a rugged design. Because many VOCA users also use power wheelchairs, a large device can often be mounted to a wheelchair. However, for VOCA users who are not wheelchair users, a smaller design is desirable. Therefore, VOCA manufacturers offer some smaller devices that are more easily portable, although they may have a shorter battery life, a small user interface, or other drawbacks.

Another consideration with regard to the platform is whether a VOCA is a dedicated AAC device or a general purpose computer with AAC software. A general purpose computer allows a person to use a single device for both communication and other, more typical computer functions. A dedicated AAC device lacks these additional functions; however, it is easier to customize to the specific needs of people with communication impairments. There are also socioeconomic reasons for selecting a dedicated AAC device. In the United States, Medicare will pay for a VOCA if a user's clinicians can document medical need; however, Medicare will not pay for a computer so that a person can use AAC software.

The input method for a VOCA includes the selection set. The selection set refers to how words are encoded by the device and how this encoding scheme is presented to the user (Beukelman and Mirenda 1995, p. 9). For instance, words can be encoded by letters or by iconic symbols. Letters or spelled-out words may be more appropriate for some literate users while symbols can be useful for people who cannot read. Symbols can also support encoding schemes for rate enhancement. An encoding can also be one to one (e.g., an entire word spelled out on the device, or a symbol with a single meaning) or more complex (e.g., single letters allowing the person to spell out any word, or symbols that can be selected in series to construct different words). Although one-to-one encodings remove the cognitive load of having to spell out words or string together symbols, they also drastically reduce the user's available vocabulary to the number of words or symbols that can be displayed on the screen.

Aside from the types of items (letters, words, iconic symbols), characteristics of the selection set include the number of items and the size, spacing, and arrangement of items (Beukelman and Mirenda 1995, pp. 52–58). The items themselves may be on physical switches, similar to a computer keyboard, or may be virtual buttons on a touch screen. In order to make more items available, some devices feature a dynamic display: a series of screens, each with a different set of words or icons from which to choose. By navigating between screens, the user has a larger vocabulary of items available at the cost of remembering how to navigate to find the appropriate screen with the desired word. It is therefore important to have a logical and intuitive arrangement of screens.

The primary output of a VOCA is audible speech directed to the user's communication partner. Some devices use synthesized speech, while others use recorded speech (Beukelman and Mirenda 1995, pp. 73–78). Although recorded speech is more natural and easier to understand, it is highly memory-intensive and limits the user's available vocabulary to the available recorded words or phrases. Many devices also have some form of visual output, such as printing words in a screen at the top of the device as they are selected by the user. This primarily serves as feedback to the user but can also be shared with the communication partner for message clarification or for communication in quiet environments, such as a library, or noisy environments, where the device may not be audible. Some devices also include an internal printer or are able to connect with an external printer for written communication.

The user interfaces for VOCAs have been designed to be quick and easy to use, with minimal distraction so that the user and her or his conversation partner can focus on their interaction with each other, not on interaction with the device. The user must be able to get and hold someone's attention while completing a message. A key issue is speed as the user must be able to complete a message without losing the attention of his or her partner. Even with a patient partner, users want to be able to communicate quickly and efficiently.

A number of rate-enhancement methods have been developed for VOCAs. One is the use of abbreviations. For example, "adr" might expand to "address" or to the user's complete mailing address. A second technique is Minspeak, in which strings of meaningful icons represent words (Baker 1982; Figure 7.1). For example, a rainbow followed by an apple might mean "red" or an icon of a person walking followed by another symbol might be a verb. This technique allows common words to be represented by fewer icons than the number of letters required to spell the word, while still allowing many words to be accessed with a relatively small number of items in the selection set. For both letter encodings (such as abbreviations) and iconic encodings (such as Minspeak) the encoding scheme must be easy to learn and remember.

Another rate-enhancement technique is word prediction, where the device attempts to predict the word a person is spelling out based on the letters entered thus far, possibly with consideration of English word frequencies and previous words in the phrase. If the computer algorithm can make appropriate predictions, this technique can reduce the number of selections. However, this technique requires the user to divert his or her attention between spelling the word and reviewing the list of predicted words to see whether the desired word has been chosen. For some users, the delays introduced by these visual and cognitive tasks will exceed the time saved from typing.

In general, the success of a rate-enhancement technique must be judged by considering demands on the user's physical, cognitive, visual, timing, linguistic, and other abilities (Beukelman and Mirenda 1995, pp. 44–47). When designing pervasive devices, it is useful to predict time savings based on the average number of selections needed to produce a word, the average number of motor acts for each selection, and the average time for each motor act (Rosen and Goodenough-Trepagnier 1981), as well as the frequencies of different word lengths (Vanderheiden and Kelso 1987). In practice, it is also necessary to consider cognitive processing time involved in deciding which selections or acts are necessary (Light 1989). Short- and long-term use should also be considered. Strategies that require extensive learning and memorization will be slower at first but may be faster as learning proceeds. The best technique, and the best device, for an individual will depend on that person's balance of physical, sensory, and cognitive skills, as well as on his or her communication needs. Research continues to be needed on improved input methods and rate-enhancement techniques. Nonetheless, augmentative and alternative communication devices remain a leading example of mobile computing and make a crucial difference in people's everyday lives.

In the future, pervasive computing may also assist in communication for users of sign language. Some devices already exist to aid in lipreading by detecting those speech sounds that are not visible by observing lip movements (Cook and Hussey 1995, pp. 676–678). To aid sign language users in communicating with hearing individuals, researchers are working on wearable sign language recognition systems (Brashear et al. 2003). Computer recognition of sign language is complicated by the fact that sign language involves not only hand gestures but also arm position, head movements, and facial expressions, all of which may take place simultaneously.

The technologies discussed thus far have focused on the needs of people with speaking difficulties. However, pervasive computing has also had a major impact on technology for people with hearing impairments. Embedded computing provides better processing of sound for people with high-end hearing aids and cochlear implants (Goode, Rosenbaum, and Maniglia 1995; Arlinger et al. 1998). The improvement between analog amplification and digital auditory processing seems to be particularly significant for speech sounds (Arlinger et al. 1998), making this technology important to improved communication. Future directions include improved auditory processing schemes to translate the frequency spectrum of environmental sound, especially speech, into the available range for particular devices.

VOCAs and digital hearing aids share an application of portable computing to improve or enable social interaction. These technologies provide a challenging example of the need to balance computing power with portability, thus pushing the boundaries of computing hardware and software design.

7.3 Mobility assistance

Assistive technology increasingly incorporates computing in the area of mobility devices, such as wheelchairs, scooters, and walkers. Mobility devices would seem to be a natural platform for mobile computing. Indeed, as with automobiles, more high-end power wheelchairs have onboard computers to assist in smooth travel. Microcontrollers are used to tune characteristics such as maximum travel and turning velocity and acceleration and to filter the effects of a user's tremor or spastic movements (Cooper 1995). In addition to adapting to the user, onboard microcontrollers can adapt to the environment by filtering electrical noise or using dynamic control algorithms such as contour control, which allows the wheelchair to automatically follow a desired speed profile regardless of terrain or slope (Cooper 1995). Researchers are investigating software to measure a user's capabilities and automatically adapt the wheelchair's response (Ding, Cooper, and Spaeth 2004).

Power wheelchairs may be equipped with internal sensors that detect such signals as wheel speed and battery charge or detect internal problems (Cooper 1995). These sensors allow the system to improve performance and take appropriate action in the face of component failure. If a component fails while driving, the driver could be placed at extreme risk (e.g., stalling while crossing a street). It is crucial that the driver be able to remain in control of the wheelchair, albeit perhaps with reduced function, until reaching safety (Cooper 1995).

Mobile computing is also finding applications in mobility devices other than power wheelchairs. It is often desirable to use manual wheelchairs because they are more maneuverable, easier to transport, and require less maintenance than power wheelchairs. Power-assist manual wheelchairs allow the wheelchair user to propel the wheels in the same fashion as a standard manual wheelchair, but with a power assist chair the user receives assistance from motors built into the wheels' hubs and does not have to use as much strength (Cooper et al. 2001; Figure 7.2, top). The user's force in propelling the wheelchair is detected by sensors in the pushrims and used as the input signal to the motors. A computer translates the force input from the user so that the harder the user pushes, the more assistance is given. This represents a viable solution for people who are unable to generate sufficient propulsion force to use a manual wheelchair but who do not wish to use a fully powered wheelchair.

An emerging application for mobility technology and computing is the promotion of physical activity. Reduced physical activity can lead to functional decline, as well as cardiovascular disease and other health risks. However, exercise opportunities for manual wheelchair users are limited. Computer technology could aid in exercise programs for wheelchair users by providing motivation and feedback. One approach has been the Game-Wheels exercise system (O'Connor et al. 2002), which uses a set of instrumented rollers as a user interface to a video game system (Figure 7.2, bottom left). A user can propel his or her wheelchair on the rollers, like a walker on a treadmill, and the relative speed of the wheels provides an input signal to emulate a video game joystick. The video game in turn provides motivation to exercise by propelling the wheelchair.

Another application of onboard computer technology for mobility devices is navigation assistance. Smart wheelchairs and robotic walkers help

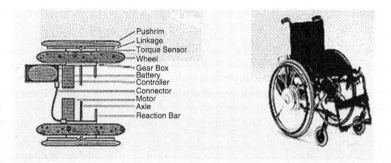
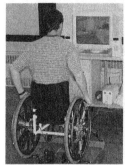

Figure 7.2 Applications of embedded computing in mobility devices. (Top) Push-rim-activated power-assist wheels (Yamaha, Japan; marketed in the United States as the Quickie Extender, Sunrise Medical, Carlsbad, CA). (Bottom left) GameWheels (University of Pittsburgh). (Bottom center) Obstacle-avoidance smart wheelchair utilizing PAPAW wheels (AT Sciences, Pittsburgh, PA). (Bottom right) PAMAID smart walker (Haptica Limited, Dublin, Ireland).

their users by warning of obstacles, by automatically acting to prevent collisions, or by helping the user to find his or her way around (Figure 7.2, bottom center). These features could be useful to people who have difficulty reliably controlling a wheelchair due to spastic movements or tremors (which could steer them into obstacles), difficulty seeing obstacles, or difficulty learning safe driving or navigation skills for their environment. The navigation routines used by smart wheelchairs often draw on existing mobile robotics techniques. However, additional requirements are important for a smart wheelchair, such as maintaining user comfort and acting in a consistent and intuitive manner (Levine et al. 1999).

Some smart wheelchairs and walkers have focused on obstacle avoidance (Levine et al. 1999; Miller and Slack 1995; Ulrich and Nourbakhsh 2000; Simpson et al. 2004). These systems have used a variety of sensor modalities to detect obstacles, including ultrasonic sonar, infrared range-finders, laser range-finders, and computer vision. Each of these sensors has strengths and weaknesses. By including sensors of different types, risk of missing an obstacle is reduced (Simpson et al. 2004).

In addition to obstacle avoidance, some wheelchairs and walkers provide autonomous navigation (Yoder, Baumgartner, and Skaar 1996; Nisbet et al. 1995). In these systems, the user indicates what room he or she would like to go to, and the wheelchair automatically determines a path and steers the wheelchair toward that location. These systems often depend on modifications to the environment, such as visible or magnetic tracks on the floor or visible markers on the walls, although some systems can be trained to navigate in unmodified environments by recognizing landmarks (Roefer and Lankenau 2000).

In addition to upright obstacles (e.g., walls, furniture, other people) it is important to detect drops (e.g., descending staircases or curbs) that could be even more dangerous. Drops can be difficult to detect with sufficient advance warning to stop the front wheels. One approach is to "learn" the appearance of the ground using computer vision and stop the chair when an unfamiliar floor appearance is detected (Ulrich and Nourbakhsh 2000). The user can move forward if it is simply a different surface (e.g., moving from tile to carpet) or look for an alternate route if it is a true obstacle or drop. Although this will keep the user from inadvertently going over a drop, this method assumes that the user is in a position, both mentally and visually, to ultimately differentiate between a safe and unsafe route. For other users, a system is needed that can autonomously make this decision.

Although most smart wheelchairs have been based on typical power wheelchairs, power-assist systems allow these features to be provided on manual wheelchair platforms as well. The Smart Power Assistance Module can detect obstacles near the wheelchair and modify the forces applied to each wheel's motor to avoid obstacles (Simpson et al. 2005; Figure 7.2, bottom center). This will stop the wheelchair and provide tactile feedback to the user by increasing the resistance of the wheels based on the proximity of obstacles. The user population for this system consists of individuals with both a visual impairment and a mobility impairment that makes it difficult or impossible to ambulate independently with a cane, guide dog, or other traditional mobility aid.

Mobile computing can also enable obstacle avoidance and navigation assistance in the context of a wheeled walker. Several navigational mobility aids based on walkers are currently in development. As with smart wheelchairs, some smart walkers simply provide obstacle avoidance (Lacey and MacNamara 2000; Figure 7.2, bottom right), while others can also assist with navigation and path planning (Dubowsky et al. 2000; Glover, Thrun, and Matthews 2004).

As microcontrollers and embedded computers become more powerful, they are able to provide an increasing level of intelligence for wheelchairs and other mobility devices. These improvements make wheelchairs more usable for all wheelchair users, in much the same way that intelligent automobiles provide desirable features for all drivers. At the same time, these features can make independent wheelchair use a possibility for a larger population, including people with visual or fine motor impairments who would otherwise depend on an attendant for mobility.

7.4 Wayfinding

A smart wheelchair is one example of a "wayfinding" device that helps people to navigate in their environment. Other devices are independent of a wheelchair or walker and are designed to address functional impairments in two areas: spatial orientation and wayfinding. Spatial orientation refers to the ability to establish and maintain an awareness of one's position in space relative to landmarks in the surrounding environment and relative to a particular destination. Wayfinding is the means by which a person employs spatial orientation to maintain a heading toward a destination regardless of the need to avoid or move around obstacles in his or her path (Ross and Blasch 2002). Therefore, a wayfinding device must have one or more means to locate someone in space and, based on this localization information, it must provide useful orientation cues to the user. A number of solutions have been attempted for both the sensing technology and the user interface.

To date, orientation and wayfinding aids that incorporate embedded computing have had difficulty competing with traditional orientation aids, such as canes and dog guides. Canes combine the sensing technology and the user interface, providing an extension to the user's own tactile sense in order to detect obstacles and environmental cues. The best sensing technology and user interfaces currently available are hard pressed to compete with the acute senses and responsiveness of a well-trained dog guide. Also, pervasive computing solutions cannot compete with dogs for companionship (although they are generally easier to clean up after). Nonetheless, pervasive computing is beginning to find a place in orientation and wayfinding for three reasons. First, sensing and mobile computing technologies are advancing to the point where they can provide real assistance without requiring the user to wear a lot of cumbersome equipment in a backpack. Second, technologies are emerging that can provide information that is unavailable to a cane or dog, such as the text of signs or absolute localization in unfamiliar locations.

Finally, secondary sensory or cognitive disabilities can increase the need for orientation and wayfinding assistance from a computer-based assistant (Ross and Blasch 2002). People with visual impairments often rely on other senses, such as hearing, touch, proprioception, and vestibular sense, that can become impaired with age or secondary disabilities. In addition, someone who has a cognitive impairment as well as a visual impairment may have difficulty learning traditional orientation and mobility skills. Learning these skills may also be difficult for older adults who acquire visual impairments (Lancioni et al. 1998). A high-tech solution has the potential to adapt to a user's needs, providing directions in a simple, straightforward manner and capitalizing on a person's available sensory capabilities. To succeed, such a high-tech solution must have reliable underlying sensor technology and a well-designed user interface.

A wayfinding system needs information about a person's current location, as well as objects in the environment. A variety of sensor options are presented in Table 7.1, together with their strengths and limitations. One

Table 7.1 Sensor Modalities for Localization or Obtaining Information about the Environment

Sensor Modality	Strengths	Weaknesses
Global positioning system (GPS)	Infrastructure is always present.	Low resolution (currently ten meters). May not be able to see three satellites in areas with tall buildings; will not work indoors.
RF triangulation	Infrastructure frequently present. If not, RF transmitters can be installed in an environment.	Limited resolution; noise.
RF or IR transmitters	High resolution. Also provide location-specific information (e.g., name of building).	Infrastructure cost to instrument environment. Requires a means to power beacons; or mobile device must power beacons, draining its own battery. Some devices require user to continually scan environment; potentially cumbersome.
Computer vision to identify landmarks	Has potential for high flexibility.	Computationally expensive. May require modifications to the environment to introduce recognizable landmarks.
Ultrasound, infrared	Obstacle detection.	No absolute localization.
Accelerometers, gyroscopes	Good resolution over short distances.	Prone to drift over time.

technology for obtaining absolute location information is a global positioning system (GPS). A GPS receiver determines latitude and longitude based on triangulation between three or more satellites. Because a GPS requires line of sight to these satellites, it is of limited use in urban areas with large buildings and cannot be used indoors. A more general technique is to use triangulation between three or more radio frequency (RF) transmitters (Friedman 1993). This technique can often take advantage of transmitters that are already in the environment, such as cellular phone towers for large distances and wireless access points for shorter distances (LaMarca et al. 2005; Kulyukin and Nicholson 2005).

Apart from the large-scale, context-insensitive location-finding methods such as GPS and RF, a system can determine its location by identifying objects in the environment. Some systems use short-range, wireless RF or infrared (IR) communication with transmitters in the environment, which transmit specific location information to the user (Ross and Blasch 2002). These methods can allow a location-finding system to pinpoint an exact location, such as a specific building entrance or street intersection. However, such systems will only work in instrumented environments and there can be a high infrastructure cost for systems that require landmarks or radio "beacons" to be distributed throughout the environment. Other systems have attempted to use computer vision to detect an object, symbol, or projected light pattern in the environment (Thrun et al. 2001).

In addition to determining location, a system that communicates with RF or IR beacons can provide information specific to that environment, such as the status of traffic signals at an intersection or room numbers in a building. This type of information can be provided in a variety of ways. For example, Talking Signs employ a coded beam of light to transmit information from the "sign" to a receiver held by the user (Brabyn and Brabyn 1983). In addition to information about landmarks, this is also important to gain information about obstacles in the environment. Ultrasound and IR sensors, such as those used in many smart wheelchair systems, have also been incorporated into handheld obstacle detectors for users with visual impairments (Cook and Hussey 1995, p. 655).

For tracking movement and orientation relative to a known starting location, it is possible to use accelerometers to monitor a person's movement (linear or angular). Such systems can provide good resolution over short distances but are prone to drift over time.

One could ameliorate the disadvantages of various sensing solutions listed in Table 7.1 by combining a variety of sensing technologies in a single system. For example, the Cyber Crumbs system includes a GPS, a digital compass, outdoor IR beacons at traffic intersections, wireless radio frequency identification (RFID) tags, and a system for "reading" IR patterns transmitted by indoor landmarks (Ross and Blasch 2002). This system allows for localization and orientation across outdoor and indoor environments.

Even a system with ideal sensor technology will not be of assistance if the user interface is not appropriate. The user interface must utilize a sensory modality that is available to the user, which is typically hearing or touch. It must then present information in a way that is meaningful to the user but not distracting from other tasks. For example, users with visual impairments often do not want their ears to be covered by headphones, as this tends to block environmental sounds needed for navigation (Ross and Blasch 2002). Therefore, any auditory information used to provide cues must be delivered so that it does not block sounds (e.g., using bone-conductive headphones) and does not distract from environmental sounds (e.g., providing auditory information no more often than necessary). The frequency of cues is also important because if cues are given too frequently, they could prevent the user from taking the initiative to search for the correct direction on his or her own (Lancioni et al. 1998).

Auditory information offers a great deal of flexibility in how information is encoded for the user. A system can use different tones to indicate the positions of landmarks or obstacles. One way to use a tone to provide orientation information is to play the tone in virtual space so that its position maps to the position of the object of interest. For example, when delivering the sound over a stereo channel, the virtual position of the tone can be controlled by the relative volume in the left and right stereo channels. Other methods include mapping the pitch, volume, or other aspects of the tone to the direction of the object relative to the user or the distance of the object from the user. Explicit verbal prompts can be used to describe objects in the environment or to provide navigation cues. Verbal prompts allow for more detailed information but may require more of the user's concentration.

Tactile cues can also convey navigation assistance or information about the environment. For example, vibration can be used to indicate the presence of an obstacle, with the strength or frequency of vibration indicating proximity. A tactile perception of movement (e.g., along a vibrating array) can be used to indicate a desired direction of travel. A Braille display can also be useful, although not all people with visual impairments read Braille.

In addition to supporting people with visual impairments, wayfinding technologies can be of use for those who have difficulty with navigation due to cognitive impairments (Lancioni et al. 1998). People with cognitive disabilities such as an intellectual disability, traumatic brain injury, or dementia may have difficulty performing tasks that require spatial planning, spatial memory, decision making, or mental representation (Liu, Guathier, and Gauthier 1991). As a result, this population can have trouble remembering how to get from place to place or adapting to a change in plan, such as a late bus. A wayfinding aid for people with cognitive disabilities might provide task-related information relevant to the person's location, as well as navigation assistance (Friedman 1993; Carmien 2002). People with cognitive disabilities may also need assistance with aspects of navigation other than the route, such as pre-trip planning (e.g., forgetting to gather necessary resources) and trip initiation (e.g., meeting the bus at the scheduled time) (Sohlberg et al. 2004). If a system is aware of a person's location, it can provide other context-specific assistance as described in Section 7.6.

Wayfinding technologies can assist people with visual or cognitive impairments in traveling independently, whether within a building or across a city. Work continues to be needed to develop flexible and reliable sensing methods to determine location and environmental context, as well as effective user interfaces.

7.5 Cognitive assistance

Assistive technologies can aid people with cognitive disabilities arising from a variety of diagnoses, including acquired brain injury (Wilson et al. 2001), intellectual disability/mental retardation (Davies, Stock, and Wehmeyer 2004), and dementia (Alm et al. 2004). Assistive technologies for cognition have been referred to as "cognitive prosthetics" (Cole 1999) and are broadly defined as "any computer-based system that has been designed for a specific individual to accomplish one or more designated tasks related to Activities of Daily Living (ADL), including work" (Lynch 2002). A number of studies have shown that cognitive assistive technologies can facilitate the performance of functional activities for adults having a variety of neurological impairments, thereby supporting independence in the home and community (Kime, Lamb, and Wilson 1995; Wilson et al. 2001; LoPresti, Mihailidis, and Kirsch 2004).

Cognitive impairments often affect the *executive functions*, which are cognitive skills associated with effective adaptation and accommodation to changing environmental demands. These include planning, task sequencing and prioritization, task switching, self-monitoring, problem solving, and self-initiation (Levine, Horstmann, and Kirsch 1992; Kapur, Glisky, and Wilson 2004). Disabilities can also affect *prospective memory*, or the ability to remember tasks that need to be performed, to remember the times associated with those tasks, and to remember to actually carry out the tasks at the appropriate times (Ellis 1996). The type and severity of impairment determines each individual's abilities, which can range from minor problems with complex navigation to an inability to complete simpler activities of daily living, such as handwashing.

As a result of these functional impairments, people with cognitive disabilities are particularly suited to take advantage of portable organizational aids originally designed for the mainstream population, such as voice recorders and handheld computers (Garland 2004). Some people need additional features, such as a simplified interface, more intelligent scheduling assistance, or task guidance (LoPresti et al. 2004). Once integrated into assistive memory aids, these features can benefit the mainstream population as well.

It is desirable for cognitive assistive devices to be portable and, ideally, pervasive. People with cognitive disabilities who rely on cognitive assistive technology as an external memory want to always have this resource available (Inglis et al. 2004). People with cognitive disabilities can have great difficulty learning even one novel technology interface and therefore it is important that their interface to this external memory be consistent across environments (Garland 2004). Users of this type of technology should be able to carry a consistent device with them or have the ability to access a consistent interface across distributed hardware devices, so that it is available to them at work, at home, or in the community.

Most people with or without cognitive disabilities have occasional difficulties with managing daily schedules. Therefore, many mainstream technologies exist to meet this need, ranging from low-tech solutions such as calendars and written lists to voice recorders and scheduling software. Low-tech solutions can be useful for some people with cognitive impairments as these solutions are inexpensive, easy to use, and have no social stigma that might otherwise be attached to "rehabilitation" devices (LoPresti et al. 2004; Garland 2004). However, these solutions are limited with regard to the amount of information that can be stored and how information can be presented to the user. Furthermore, written lists and calendars provide no cues to the user as to when he or she needs to perform a task. For individuals with deficits in self-initiation, a device that can call the person's attention to itself (i.e., an auditory alarm) will be more likely to facilitate activity performance (Kime et al. 1995). Therefore, it is often desirable to have a single, easily portable device that provides both an external cue and relevant information.

Electronic memory aids are commercially available for the mainstream population. These include digital voice recorders, scheduling and reminder software for laptop or palmtop computers, or portable interfaces that download information from a desktop computer (e.g., the ability to download the day's schedule to a digital wristwatch). These devices may be more readily available than devices designed for people with disabilities, may be more socially acceptable, and may have more established technical service. However, they are too complex for some people with cognitive disabilities. Devices have been designed specifically for individuals with cognitive disabilities, providing features such as customized user interfaces and the ability to easily contact caregivers. Some offer more elaborate features, such as automatically generating and revising a user's schedule in response to unexpected events (Levinson 1997).

Some design choices will serve people with a variety of cognitive impairments. Uncluttered visual displays will reduce distractions for people with attention problems. Providing information in nontext formats (e.g., graphics, video, audio) will support users who cannot read. Minimizing the number and complexity of decision-making points can make a device easier to learn and to use. Presenting information sequentially, rather than all on one screen, reduces the amount of information the user must process at one time (Wehmeyer 1998).

There are some trends within groups of people with the same type of cognitive impairment. For instance, people with acquired brain injuries may recover cognitive skills over time, people with a dementia such as Alzheimer's disease tend to decrease in cognitive ability over time, while people with developmental disabilities, such as mental retardation, are relatively stable. These different prognoses must be taken into account when designing technology for different groups. However, there is still variation in the capabilities of people who are diagnosed with the same type and severity of impairment. As such, customization is typically necessary. This is not only to adapt a device to a particular individual's needs but also to continually adapt as the user's capabilities change (Cole 1999). In addition to adapting a device to an individual's strengths and weaknesses, it is also desirable for customization to reflect other factors such as user priorities and the environment where the activity is performed (home, community, school, or work).

Although automated reminding may increase a user's independence, it is important to recognize the limitations of such a device. It is sometimes important to facilitate communication with a human caregiver or support person when a device is unable to provide assistance. One project, ISAAC, combined cognitive orthotic software with a cellular phone, a digital camera, and a GPS satellite navigation receiver to provide support center staff with information about the client's needs (Jonsson and Svensk 1995). Another project, the MemoJog, involves an interactive memory aid that communicates via a mobile phone (Inglis et al. 2004). This enables communication with the caregivers' computer system so that, for example, an alarm can be raised if critical messages are not acknowledged by the user. In addition to calling on human support in an emergency, it is important to facilitate support during the setup and review of a person's schedule. Often, a user's therapist or family member will program the user's schedule into the system. Some systems allow this to be done remotely and are also capable of logging data for the caregiver to review at a later date (Thöne-Otto, Walther, and Schulze 2003; Jinks, Kellinger, and Garber 2004).

Most existing memory aids are designed to present scheduled, one-step tasks (e.g., "at 5:00, cook dinner"). However, many tasks are not limited to one step. For example, cooking dinner involves a number of subtasks related to preparing kitchen utensils and following a recipe. A person may need assistance with such multistep activities due to problems remembering the steps in the task, problems remembering the proper order of steps in a sequence, or other difficulties (Davies et al. 2004). Therefore, people may also benefit from a task guidance system that is able to provide messages in sequence. One such system, COGORTH (COGnitive ORTHotic), was designed to guide through multistep tasks (Kirsch et al. 1987). COGORTH checked a user's performance for errors, provided error-correcting or help procedures, managed interruptions of a task when a higher priority task must be completed, and managed a user's environment through control of electric appliances, telephone, and audio signals. The user could receive prompts through a video display or audio messages and could respond by clicking buttons on the computer display. An inappropriate response from the user, or lack of response within a certain amount of time, would cause COGORTH to conclude that assistance was required. COGORTH was evaluated in a series of efficacy studies, which suggested an improved adherence to schedules for people from a wide range of patient types and cognitive disabilities (Kirsch et al. 1987).

Many people experience cognitive impairments due to congenital or acquired disabilities. These individuals can often benefit from technologies that have been developed to help all of us overcome our cognitive limitations, from Post-its to personal digital assistants. People with cognitive disabilities are likely to benefit even more if these cognitive assistive technologies have specialized interfaces, support communication with caregivers as well as automated prompts, and provide step-by-step guidance as well as high-level reminders.

7.6   Smart environments

The assistive technologies discussed thus far have been mobile devices that a user can carry anywhere (or that carry the person), whether they are handheld devices, embedded in wheelchairs, or wearable computers. People with disabilities can also take advantage of stationary devices that have computing power embedded in the environment, creating smart homes, smart workplaces, or other smart environments. Chapter 5 presented the potential for pervasive computing in the home to support home-based medical care through home-based monitoring, home-care infrastructures, support for patient-doctor-nurse collaboration at a distance, and support for peer-to-peer social awareness. In this section we focus on the use of pervasive computing in the home to support increased independence and quality of life. These uses include cognitive, physical, and sensory supports.

For individuals with cognitive disabilities, smart environments can provide context-sensitive reminders. Sensors in the environment can gather information about a person's progress in daily tasks and allow the system to provide appropriate reminders in an intelligent fashion based on a person's activities, available resources, and social context. Smart environments can also provide warnings of environmental hazards and can potentially take action to prevent hazardous situations.

There are a number of motivations for context-sensitive reminding. First, knowledge of context can allow a device to be more specific and relevant in its prompts and cues. If a device is aware of the user's location, for example, it could give reminders relevant to that location. Information about the user's environment might also provide cues to the device on what reminders might be important (e.g., handwashing if the person is in the washroom) or unnecessary (e.g., a reminder to go to the cafeteria if the person is already there). Social cues might allow the device to know when a reminder would be inappropriate, such as when the user is talking with another person and might not want to be interrupted.

Second, traditional memory aids are limited to reminding the user about items on their daily schedule. However, users might wish to initiate tasks on their own and still receive assistance to complete those tasks. A context-sensitive reminding system could potentially infer a person's desired task based on observed activities and provide appropriate reminders for the user-selected task (Pollack et al. 2003; Philipose et al. 2004). This would help satisfy the desire expressed by some users for a memory aid "being responsive to you but not controlling you" (Inglis et al. 2004).

Third, traditional memory aids give the same reminders at all times, whether a particular reminder is needed or not. This can be frustrating for an individual who is sometimes able to remember tasks independently. For example, consider a young man with memory impairment following traumatic brain injury whose memory aid contains his daily schedule. Initially, he needs all the reminders in order to make it to meals, attend therapy sessions, and be at other appointments in a timely fashion. As time goes on, however, he begins to recover some memory capabilities. He is still prone to forget appointments, so the memory aid is still programmed with all daily activities. As he is now able to remember the majority of his appointments, most of the time the device will remind him of something he has already done, leading to frustration and, potentially, device abandonment. If the device could determine what tasks the user has already done or is success-fully completing, then the device could give only those reminders that are needed, reducing frustration while continuing to support task completion.

Similarly, a context-sensitive reminder could support the method of vanishing cues (Friedman 1993). This training technique provides as much cue information as patients need to make a correct response and then grad-ually withdraws it across learning trials (Glisky, Schacter, and Tulving 1986). A context-sensitive memory aid could apply this technique automatically, only providing those cues that are truly needed and otherwise allowing the user to independently perform their daily tasks. A therapist could monitor the user's progress and adjust the maximum or minimum support provided by the device, or the speed of response of the device, in order to promote greater independence on the part of the user. Unfortunately, a person's abilities may decrease instead of increase. In this instance as well, a con-text-sensitive system can detect difficulties a person is having and adapt as the user's abilities change over time (Adlam and Orpwood 2003).

Finally, the sensors used by a context-sensitive reminder system could be used not only to determine when the resident might need reminders but also to detect situations and respond autonomously. For example, if the user experienced a fall, the system could automatically alert his or her caregiver or the system could automatically turn off an oven that was left on.

An example of a system that provides context-sensitive assistance is the COACH (Cognitive Orthosis for Assisting aCtivities in the Home), an adapt-able device to help people with dementia complete handwashing with less dependence on a caregiver. The COACH used artificial neural networks, plan recognition, and a video camera to automatically monitor progress and provide prerecorded verbal prompts. Prompts are only given when the user needed assistance. The system has been able to adapt its cueing strategies according to a user's preferences and past performance (Mihailidis, Barbenel, and Fernie 2004).

Context-sensitive reminding systems can be based outside the home as well. Some systems have been based in vocational settings (Friedman 1993; LoPresti, Friedman, and Hages 1997). One system was capable of determin-ing a person's location and tracking his or her progress in vocational tasks. The computer provided voice prompts only as needed to help the user maintain his or her schedule. Continued difficulty adhering to the schedule would cause the computer to automatically call for human assistance (Fried-man 1993). Using a similar device, two subjects with intellectual disability were able to maintain levels of productivity comparable to those obtained when a human job coach closely guided each subject (LoPresti et al. 1997).

In addition to providing assistance at home or in the workplace, it is desirable for context-aware prompts to be available while a person is out in the community. Thus, instrumented smart environments might be aug-mented by a mobile device that takes advantage of ubiquitous sensor data (e.g., a GPS). For example, the Opportunity Knocks project uses multiple technologies (cell phone, GPS receiver, and wireless modem) to learn a user's

typical daily routines (including locations of activities), monitor for variations in the individual's typical day-to-day activities, and then
decide whether a prompt is necessary if the routine is unexpectedly changed
(Patterson et al. 2004).

A smart environment might also incorporate home automation equipment to assist people with physical and sensory disabilities. In assistive
technology parlance, home automation systems are often referred to as
"environmental controls" or "electronic aids to daily living." People who
are unable to easily control standard interfaces (light switches, door knobs,
remote controls) often utilize environmental control units. Such environmental controls provide access to multiple appliances from a central interface.

Home automation systems allow multiple home appliances to be operated through a computer or general-purpose remote control. The system can
control standard appliances, such as a lamp, thermostat, oven, and washing
machine, and can also emulate existing remote controls for devices such as
stereos or televisions. It can be extended to control traditionally nonelec-
tronic appliances by adding automatic door openers or curtain controls.
These features offer convenience to mainstream populations, but they can
offer independence for people with physical impairments. A person with
limited mobility or fine motor control can operate multiple devices through
a single, consistent control interface or even delegate control to a computer
by establishing a schedule.

The main strength of pervasive computer technology in this application
is the ability to separate a device from its control interface. Many objects in
the home have specific interfaces, such as door knobs and light switches.
Through a home automation system, all devices in the system can be controlled by pressing buttons. For people who simply lack the finger or hand
strength to use door knobs or light switches, a standard remote control or
computer interface might be sufficient. Others can use a remote control with
larger buttons, an alternative computer interface, or voice control, thus wid-
ening the circle of people who can access home appliances.

In some cases, the home automation interface can be integrated with
other assistive technology. Some power wheelchairs and augmentative and
alternative communication devices have environmental control options, so
that an individual can use the same interface for controlling household
appliances as they already use for the wheelchair or communication aid
(Cook and Hussey 1995, p. 602). For people with limited physical abilities,
who may have highly customized interfaces, this can increase the person's
ability to access multiple devices (Angelo and Trefler 1998). Otherwise, a
person with limited movement may need a caregiver to switch input devices
when the person wishes to change tasks. Such integrated controls do have
some drawbacks. For instance, there is a risk that if a person's wheelchair
fails, they will also lose access to their home appliances. Also, integrated
controls may not be desirable for people with more fine motor control, who
can benefit more from input devices that are customized for a particular
device.

In addition to converting a device's input method to suit a user's abilities, computing technology allows for converting a device's output. Many appliances assume that the user is able to see, in order to read LED displays, dials, or other indicators. Some appliances can be fitted to provide auditory feedback options or, in some cases, Braille displays. As more appliances become available that are designed for use in smart homes, it will become easier to determine the device's status and translate the information into whatever sensory modality works best for a particular individual.

Similarly, people with hearing impairments need access to auditory information in their environment such as doorbells, smoke alarms, or a child's cry (Cook and Hussey 1995, p. 681). There are devices that provide visual or tactile cues to replace such audible information. Some devices simply replace an auditory alarm, such as wiring the "doorbell" to display a flashing signal rather than an auditory "bell." Others actually detect sounds, for example, monitoring sound in a room and flashing a light in response to any loud sound. Some such devices are designed to a special purpose and respond only to a specific frequency, such as that of a smoke alarm. Others have microphones in specific locations, so that ambient noise will interfere much less with sound detection from a specific object or area.

In order to fully achieve the potential of home automation for people with disabilities, there is a need for communication standards between devices in the home, as well as between the home and the outside world (Adlam and Orpwood 2003). Currently there is no such standard between devices from different manufacturers. Ideally, proposed standards such as the V2 specification (Zimmerman et al. 2002) will become commonplace for smart appliances. However, it is likely that some companies will not voluntarily design to such standards, while many low-tech and older devices will not meet the standard. In such situations, software could be designed to act as an intermediary between various devices.

Smart environments can serve people with a variety of disabilities. For people with cognitive disabilities, environmental sensors can enable context-aware reminding. For people with physical impairments, home automation systems allow people to operate home appliances from an accessible interface. For people with sensory impairments, information can be translated into a usable sensory modality. While current systems provide a substantial benefit, there is room for improvement in developing usable, intelligent human–computer interfaces and efficient machine–machine interfaces.

7.7 Design considerations for assistive technology

A common theme in assistive technology is the need for an appropriate user interface. People with disabilities have a wide range of physical, sensory, and cognitive limitations and often have a combination of different types of impairments (LoPresti et al. 2004; Kapur et al. 2004). A list of commonly impaired abilities is presented in Table 7.2, along with some of the implications of

Table 7.2 Possible Impairments and Their Implications for Input, Output, and Interaction Design

Ability	Implications of Impairment	Design Considerations
Fine motor control	If fine motor control is impaired, a person may not be able to select small buttons.	Use larger buttons (may require using fewer buttons or increasing the overall size of a device) or allow an alternative input mode such as speech or scanning.
Speaking	If a person is unable to speak, voice recognition will not be an option (although a system trained to the individual may in some cases be able to recognize his or her speech patterns, even if the speech is unclear to humans).	Provide an alternative option (e.g., have a keypad available as a backup to speech recognition).
Vision	May have difficulty reading text, viewing images, or getting feedback that relies on color or other visual cues.	For low vision, consider a larger display or change the appearance of text and images (high contrast, different color combinations). For users who are blind, support alternative output (auditory or tactile feedback).
Hearing	Will result in difficulty receiving auditory prompts (e.g., speech, alarms, auditory encoding of information).	Provide multimodal feedback (a combination of visual, auditory, and/or tactile). For spoken feedback, provide for closed captioning.
Oral comprehension	Some people can hear, but have difficulty with comprehension due to problems such as inability to process the speed of spoken language.	An abstract auditory alarm will still be useful, but alternatives to standard speech output. An option to slow down the speech playback will work for some users. For others, visual display of information will be needed.
Tactile sense	A person with an impaired sense of touch will have trouble using a device with small buttons, and won't benefit from vibrating or other tactile alarms.	Provide or support visual or auditory alternatives.

Ability	Implications of Impairment	Design Considerations
Reading	A person may have difficulty reading due to visual problems or a learning disability.	Provide nontext visual cues (photos, icons, animations, etc.) or speech output.
Speed of response	A person might be slow to respond because he or she needs extra time to move or to plan a response.	Response times for devices should be customizable to allow for slow responses.
Learning	A person who has difficulty learning may not benefit from a device with a complex user interface, although it may offer desirable features.	Allow a user interface to be simplified; for example, by having the option to hide infrequently used toolbars and other user interface elements.
Attention	A person who is easily distracted will need a simple, clean user interface.	Design for simplicity wherever possible (which tends to be a good design for all users); or allow the interface to be simplified through a customization process. Avoid distracting animations, or allow them to be turned off.
Planning	A person may have difficulty determining the correct series of steps to accomplish a goal.	Support planning (by presenting options in a clear and helpful way) or remove the need to plan (by only presenting one option).
Problem solving	A person may have difficulty recovering if something goes wrong (either a fault with the device or a mistake on the user's part).	Devices should, as much as possible, be failsafe and/or provide simple instructions to help the person recover in the case of a failure.
Self-initiation	A person may not independently initiate tasks (e.g., forgetting to check a to-do list regularly).	Devices should call themselves to the person's attention at the appropriate times if the user has difficulty initiating activities.
Flexibility	Some users will have difficulty adapting to changing interfaces.	Provide a consistent, predictable user experience.

impairments for the input methods, output methods, and interaction designs supported by a device. It is important to consider cognitive skills as well as physical abilities. For example, a memory aid that provides appropriate alarms but requires a great deal of planning and problem solving from the user will be useful for some clients but will only exacerbate difficulties for others. Because of the variety of abilities and limitations, a device needs to be customizable, both in terms of choosing the right device for the individual and recognizing the individual's specific needs. Another possibility is to support an interface to existing assistive technologies. An example of this approach is associating a text description with graphics on a Web site. The text can then be accessed by auditory Web browsers that are already available for people with visual impairments.

Often there are different input methods available for a particular device. For communication aids, the most common selection technique is for the user to press buttons (physical switches or virtual buttons on a touch screen), although many users employ a mouse, joystick, or head-operated mouse emulator. Power wheelchair users may control their chairs through a traditional joystick, through head-operated controls, or controls operated by a single finger. Some users may have no effective means to directly select from an array of buttons (e.g., from a keyboard or remote control) and may instead use "scanning" (Beukelman and Mirenda 1995, pp. 62–65; Cook and Hussey 1995, p. 590). In scanning, the user interface device is a single switch. The device highlights items on the screen one at a time and the user presses the switch when the desired item is selected. In many scanning techniques the user first selects a region of the screen (e.g., row, column, quadrant), then scans through items in that region. Because of the need to wait for the desired item to be highlighted, scanning is generally a very slow input method. Switches used in a scanning interface may be controlled by large movements (e.g., of the shoulder), small movements (e.g., blinking), or breath (via pneumatic switches). Ongoing research is developing new ways for computer input, both for mainstream use and for the field of assistive technology. In the future, a device might use computer vision or sensors embedded in clothing to learn and recognize arbitrary gestures made by the user. Other researchers are exploring use of brain activity signals, recorded from the scalp or from implanted electrodes (Moore 2003).

A device's output can be visual or auditory and can range from abstract (color coding or simple tones) to complex (written text or synthesized speech). Some users may need a tactile interface, such as a vibrating alarm or a refreshable Braille display (Cook and Hussey 1995, pp. 649–655). A single device might have multiple forms of output. For example, a reminder device might have a visual display that shows upcoming events, an auditory or tactile alarm for when an event comes due, visual or auditory information conveying the current message, and auditory feedback when the user presses buttons. Each of these forms of output must be customized for the individual.

Another common theme is the need for technology to be either invisible or aesthetically pleasing. Many people will not accept smart home technology unless the sensors and instrumentation blend into the home environment (Forlizzi, DiSalvo, and Gemperle 2004). Similarly, an augmentative communication device should mediate a conversation without becoming the focus of the user's or conversation partner's attention. Many people look forward to the day when assistive devices can be embedded invisibly in clothing or built into an environment. However, just as some people wear eyeglasses and others prefer contact lenses, people vary with regard to how much technology they are willing to have visible or even want visible. Some people want no visible assistive technology as they feel it may brand them as "disabled" and invite ridicule or pity (Seelman 2005). Others prefer to provide cues. For example, by carrying a cane a person might benefit from assistance or at least extra patience from the people around him or her. In any case the appearance of technology, as well as its function, is a factor in its long-term acceptability by users with disabilities.

The user's response to a device will also be affected by the user's overall attitude toward technology. As in the general population, some people will find technology fun and exciting, some will approach it as potentially useful, and some will be predisposed to think they are incapable of controlling technology or that a high-tech solution is incapable of serving their needs.

While many of the technologies described in this chapter are designed exclusively for the needs of people with disabilities, it is also possible to address the needs of this population through universal design. The universal design approach aims to develop technologies that can meet the needs of a diverse population. Providing mainstream technologies that are accessible removes the stigma that can be associated with dedicated assistive technologies, and it is often cheaper because these devices have a larger market (Seelman 2005). However, it is impractical to design a device that will meet the needs of all people all the time. Therefore, some dedicated assistive technologies will probably always be needed.

The best way to ensure that a device or system will work for people with disabilities is to include members of the target user population as early and as often as possible in the design and evaluation of the device, utilizing user-centered or participatory design approaches. In user-centered design, technology development is guided by frequent interactions with representatives of the user population to discuss general needs and possible features and to review prototypes. In participatory design, members of the user population are continually involved as members of the design team and suggest features and point out possible difficulties with the design (Cole et al. 1994).

The person with a disability is not the only stakeholder. The stakeholders will include not only the end user but also caregivers, rehabilitation professionals, and insurance or other third-party payers. To be successful, a device must be usable and attractive to each of these populations. For example, if there is an interface for a caregiver or rehabilitation professional to customize the system for an individual client, this interface must be easy to use and not cause undue stress for a caregiver. The device must also, ultimately, be affordable; whether the end user is paying for it him- or herself or whether it will be paid for through public or private insurance or other financial support.

Assistive technology design must take into account the users' range of physical, sensory, and cognitive abilities as well as psychosocial factors related to how the user wants to be seen and how the user perceives technology. Technology design should also consider all the people involved in using, purchasing, or servicing a product.

7.8 Conclusion

People with disabilities have long been pioneers in the field of wearable and portable computing, incorporating in every aspect of their lives computers that assist with communication, mobility, and control of their environment. Sometimes they use technologies that were designed with their needs in mind, while other times they make use of systems designed for the mainstream population. These devices have had a tremendous impact on the independence, community involvement, health, and quality of life for people with disabilities.

However, there is still room for improvement. Developers and researchers continue to explore ways to make user interfaces more intuitive and adaptive to the needs of diverse users. Power issues remain a challenge, with the batteries representing a large portion of the weight of power wheelchairs and many other devices; nonetheless battery life presents a limitation to people's independence. On a more positive note, new and emerging technologies continually present opportunities to serve users in entirely new ways. Advances in technology can provide more enhancements to people's ability to function in various settings, bringing increased quality of life to millions of people.

Computer technology has been a tremendous boon to people with disabilities, in part because it allows for transforming the information landscape. People with visual disabilities can hear text instead of reading it; people with motor impairments can create text using a switch operated by eye blinks when they cannot hold a pen or type on a keyboard. With a computer, context-sensitive help is only a click away. Pervasive computing can effect this same transformation in the physical environment. People who cannot see or hear the physical environment can get an interpretation, in their preferred sensory modality, of where they are and who or what is nearby. Someone who cannot physically operate a lamp switch or oven dial need only speak or blink to control the environment. Context-sensitive help will be available for everything from the expected arrival time of a bus to how to sort laundry to the name of relatives at the family reunion. Achieving this potential will require substantial work in sensors, user interface, networking, and power management among other technical issues, not to mention the social and market factors involved in getting this technology to the people who need it most. However, early examples are beginning to appear. As more emerge they will further level the playing field: by augmenting the abilities and adapting to the limitations of all users, and in the process improving quality of life for millions of people with disabilities.

References

Adlam T.D. and Orpwood, R.D. (2003). Technology, Autonomy and Cognitive Disability. *Proc. UbiComp*, www.healthcare.pervasive.dk/ubicomp2003/papers/Final_Papers/12.pdf.

Alm, N., Astell, A., Ellis, M., Dye, R., Gowans, G., and Campbell, J. (2004). A Cognitive Prosthesis and Communication Support for People with Dementia. *Neuropsychological Rehabilitation* 14(1/2):117–134.

American Speech-Language-Hearing Association (ASHA). (1991). Report: Augmentative and Alternative Communication. *ASHA* 33(Suppl. 5):9–12.

Angelo, J. and Trefler, E. (1998). A Survey of Persons Who use Integrated Control Devices. *Assistive Technology* 10:77–83.

Arlinger, S., Billermark, E., O'Berg, M., Lunner, T., and Hellgren, J. (1998). Clinical Trial of a Digital Hearing Aid. *Scandinavian Audiology* 27(1):51–61.

Baker, B. (1982). Minspeak: A Semantic Compaction System That Makes Self-Expression Easier for Communicatively Disabled Individuals. *Byte* 7:186–202.

Beukelman, D.R. and Mirenda, P. (1995). *Augmentative and Alternative Communication: Management of Severe Communication Disorders in Children and Adults*. Baltimore: Paul H. Brooks.

Brabyn, L.A. and Brabyn, J.A. (1983). An Evaluation of "Talking Signs" for the Blind. *Human Factors* 25(1):49–53.

Brashear, H., Starner, T., Lukowicz, P., and Junker, H. (2003). Using Multiple Sensors for Mobile Sign Language Recognition. *Proc. 7th IEEE International Symposium on Wearable Computers*.

Carmien, S. (2002). MAPS: PDA Scaffolding for Independence for Persons with Cognitive Impairments. *2002 Human Computer Interaction Consortium*.

Cole, E. (1999). Cognitive Prosthetics: An Overview to a Method of Treatment. *NeuroRehabilitation*, 12, 39–51.

Cole, E., Dehdashti, P., Petti, L., and Angert, M. (1993). Design Parameters and Outcomes for Cognitive Prosthetic Software with Brain Injury Patients. *Proceedings of the Rehabilitation Engineering Society of North America (RESNA)* (pp. 426–428). Arlington, VA.

Cook, A.M. and Hussey, S.M. (1995). *Assistive Technologies: Principles and Practice*. St. Louis: Mosby Press.

Cooper, R.A. (1995). Intelligent Control of Power Wheelchairs. *IEEE Engineering in Medicine and Biology Magazine* 15(4):423–431.

Cooper, R., Fitzgerald, S., Boninger, M., Prins, K., Rentschler, A., Arva, J., et al. (2001). Evaluation of a Pushrim-Activated Power-Assisted Wheelchair. *Archives of Physical Medicine and Rehabilitation* 82(5):702–708.

Davies, D.K., Stock S.E., and Wehmeyer M.L. (2004). A Palmtop Computer-Based Intelligent Aid for Individuals with Intellectual Disabilities to Increase Independent Decision Making. *Research and Practice for Persons with Severe Disabilities*, 28(4):182–193.

Ding, D., Cooper, R.A., and Spaeth, D. (2004). Isometric Joystick Tuning Interface and Assessment. *Proc. RESNA*.

Dubowsky, S., Genot, F., Godding, S., Kozono, H., Skwersky, A., Yu, H., and Yu, L.S. (2000). PAMM—A Robotic Aid to the Elderly for Mobility Assistance and Monitoring: A "Helping Hand" for the Elderly. *IEEE International Conference on Robotics and Automation.*

Ellis, J. (1996). Prospective Memory or the Realization of Delayed Intentions: A Conceptual Framework for Research. In M. Brandimonte, G.O. Einstein, and M.A. McDaniel (Eds.), *Prospective Memory: Theory and Applications* (pp. 1–22). Mahwah, NJ: Lawrence Erlbaum.

Forlizzi, J., DiSalvo, C., and Gemperle, F. (2004). Assistive Robotics and an Ecology of Elders Living Independently in Their Homes. *Journal of HCI Special Issue on Human-Robot Interaction,* 19(1/2): 25–59.

Friedman, M. (1993). A Wearable Computer That Gives Context-Sensitive Verbal Guidance to People with Memory or Attention Impairments. *Proc. RESNA,* pp. 199–201.

Garland, D. (2004). Considerations in the Selection and Use of Technology with People Who Have Cognitive Deficits Following Acquired Brain Injury. *Neuropsychological Rehabilitation* 14(1/2): 61–76.

Glisky, E.L., Schacter, D.L., and Tulving, E. (1986). Learning and Retention of Computer-Related Vocabulary in Amnesic Patients: Method of Vanishing Cues. *Journal of Clinical and Experimental Neuropsychology* 8: 292–312.

Glover, J., Thrun, S., and Matthews, J. (2004). Learning User Models of Mobility-Related Activities through Instrumented Walking Aids. *Proc. IEEE International Conference on Robotics and Automation.*

Goode, R.L., Rosenbaum, M.L., and Maniglia, A.J. (1995). The History and Development of the Implantable Hearing Aid. *Otolaryngologic Clinics of North America* 28(1): 1–16.

Inglis, E.A., Symkowiak, A., Gregor, P., Newell, A.F., Hine, N., Wilson, B.A., Evans, J., and Shah, P. (2004). Usable Technology? Challenges in Designing a Memory Aid With Current Electronic Devices. *Neuropsychological Rehabilitation* 14(1/2):77–88.

Jinks, A., Kellinger, F., and Garber, H.J. (2004). PC/Web-Based Task Reminder for Brain-Injured Adults. *Proc. RESNA.*

Jonsson, B., and Svensk, A. (1995). ISAAC—A Personal Digital Assistant for the Differently Abled. *Proceedings of the 2nd TIDE Congress* (pp. 356–361). Paris, France.

Kapur, N., Glisky, E.L., and Wilson, B.A. (2004). Technological Memory Aids for People with Memory Deficits. *Neuropsychological Rehabilitation* 14(1/2): 41–60.

Kime, S., Lamb, D., and Wilson, B. (1995). Use of a Comprehensive Program of External Cueing to Enhance Procedural Memory in a Patient with Dense Amnesia. *Brain Injury,* 10, 17–25.

Kirsch, N.L., Levine, S.P., Fallon-Kreuger, M., and Jaros, L. (1987). The Microcomputer as an "Orthotic" Device for Patients with Cognitive Deficits. *Journal of Head Trauma Rehabilitation,* 2(4), 77–86.

Kulyukin, V. and Nicholson, J. (2005). Wireless Localization Indoors with Wi-Fi Access Points. *Proceedings of the 2005 Rehabilitation Engineering and Assistive Technology Society of North America (RESNA) Conference,* Atlanta, Georgia.

Lacey, G. and MacNamara, S. (2000). User Involvement in the Design and Evaluation of a Smart Mobility Aid. *Journal of Rehabilitation Research and Development* 37: 709–723.

LaMarca, A., Chawathe, Y., Consolvo, S., Hightower, J., Smith, I., Scott, J., Sohn, T., Howard, J., Hughes, J., Potter, F., Tabert, J., Powledge, P., Borriello, H., and Schilit, B. (2005). Place Lab: Device Positioning Using Radio Beacons in the Wild. *Pervasive 2005.*

Lancioni, G.E., O'Reilly, M.F., Oliva, D., and Bracalente, S. (1998). Guiding a Person with Blindness and Intellectual Disability in Indoor Travel with Fewer Auditory Cues. *Journal of Visual Impairment and Blindness* 92(9): 609–614.

Levine S.P., Bell, D., Jaros, L., Simpson, R., Koren, Y., Borenstein, J. (1999). The NavChair Assistive Wheelchair Navigation System. *IEEE Transactions on Rehabilitation Engineering* 7(4): 443–451.

Levine, S.P., Horstmann, H.M., and Kirsch, N.L. (1992). Performance Considerations for People with Cognitive Impairments in Accessing Assistive Technologies. *Journal of Head Trauma Rehabilitation* 7(3), 46–58.

Levinson, R. (1997). PEAT: The Planning and Execution Assistant and Training System. *Journal of Head Trauma Rehabilitation* 12(2).

Light J. (1989). Toward a Definition of Communication Competence for Individuals Using Augmentative and Alternative Communication Systems. *Augmentative and Alternative Communication* 5: 137–144.

Liu, L., Guathier, L., and Gauthier, S. (1991). Spatial Disorientation in Persons with Early Senile Dementia of the Alzheimer Type. *American Journal of Occupational Therapy* 45: 67–74.

LoPresti, E.F., Friedman, M.B., and Hages, D. (1997). Electronic Vocational Aid for People with Cognitive Disabilities. *Proc. RESNA,* pp. 514–526.

LoPresti, E.F., Mihailidis, A., and Kirsch N. (2004). Technology for Cognitive Rehabilitation and Compensation: State of the Art. *Neuropsychological Rehabilitation* 14(1/2):5–39.

Lynch, W. (2002). Historical Review of Computer-Assisted Cognitive Retraining. *Journal of Head Trauma Rehabilitation* 17(5): 446–457.

Mihailidis, A., Barbenel, J.C., and Fernie, G. (2004). The Efficacy of an Intelligent Cognitive Orthosis to Facilitate Handwashing by Persons with Moderate to Severe Dementia. *Neuropsychological Rehabilitation* 14(1/2): 135–172.

Miller, D.P. and Slack, M.G. (1995). Design and Testing of a Low-Cost Robotic Wheelchair Prototype. *Autonomous Robots* 2(1): 77–88.

Moore, M. (2003). Real-World Applications for Brain-Computer Interface Technology. *IEEE Transactions on Neural Systems and Rehabilitation Engineering* 11(2):162–165.

Nisbet P., Craig, J., Odor, P., Aitken, S. (1995). "Smart" Wheelchairs for Mobility Training. *Technology and Disability* 5:49–62.

O'Connor, T.J., Fitzgerald, S.G., Cooper, R.A., Thorman, T.A., Boninger, M.L. (2002). Kinetic and Physiological Analysis of the GAME Wheels System. *Journal of Rehabilitation Research and Development* 39(6): 627–634.

Patterson, D.J., Liao, L., Gajos, K., Collier, M., Livic, N., Olson, K., Wang, S., Fox, D., and Kautz, H. (2004). Opportunity Knocks: A System to Provide Cognitive Assistance with Transportation Services. *Proc. UbiComp,* 433–450.

Philipose, M., Fishkin, K.P., Perkowitz, M., Patterson, D.J., Fox, D., Kautz, H., and Hähnel, D. (2004). Inferring Activities from Interactions with Objects. *IEEE Pervasive Computing* 3(4): 50–57.

Pollack, M.E., Brown, L., Colbry, D., McCarthy, C.E., Orosz, C., Peintner, B., Ramakrishnan, S., and Tsamardinos, I. (2003). Autominder: An Intelligent Cognitive Orthotic System for People with Memory Impairment. *Robotics and Autonomous Systems* 44: 273–282.

Roefer, T. and Lankenau, A. (2000). Architecture and Applications of the Bremen Autonomous Wheelchair. *Information Sciences* 126(1): 1–20.

Rosen, M. and Goodenough-Trepagnier, C. (1981). Factors Affecting Communication Rate in Non-Vocal Communication Systems. *Proc. RESNA*, pp. 194–195.

Ross, D.A. and Blasch, B.B. (2002). Development of a Wearable Computer Orientation System. *Personal and Ubiquitous Computing* 6(1): 49–63.

Seelman, K.D. (2005). Universal Design and Orphan Technology: Do We Need Both? *Disability Studies Quarterly* 25(3).

Simpson, R., LoPresti, E., Hayashi, S., Nourbakhsh, I., and Miller, D. (2004). The Smart Wheelchair Component System. *Journal of Rehabilitation Research and Development* 41(3B): 429–442.

Simpson, R.C., LoPresti, E.F., Hayashi, S., Guo, S., Ding, D., Ammer, W., Sharma, V., and Cooper, R. (2005). A Prototype Power Assist Wheelchair That Provides for Obstacle Detection and Avoidance for Those with Visual Impairments. *Journal of NeuroEngineering and Rehabilitation* 2: 30.

Sohlberg, M.M., Fickas, S., Hung, P., and Lemoncello, R. (2004). Community Navigation Profiles for Six Individuals with Severe Cognitive Impairments. Presentation at the annual meeting of the National Academy of Neuropsychology, Seattle, WA.

Thöne-Otto, A.I.T., Walther, K., and Schulze, H. (2003). MEMOS—Evaluation of an Interactive Electronic Memory Aid for Brain-Injured Patients. *Journal of the International Neuropsychological Society* 9(4): 583.

Thrun, S., Fox, D., Burgard, W., and Dellaert, F. (2001). Robust Monte Carlo Localization for Mobile Robots. *Artificial Intelligence* 128(1–2).

Ulrich, I. and Nourbakhsh, I. (2000). Appearance-Based Obstacle Detection with Monocular Color Vision. *Proceedings of AAAI 2000*, 866–871.

Vanderheiden, G. and Kelso, D. (1987). Comparative Analysis of Fixed-Vocabulary Communication Acceleration Techniques. *Augmentative and Alternative Communication* 3: 196–206.

Wehmeyer, M.L. (1998). National Survey of the Use of Assistive Technology by Adults with Mental Retardation. *Mental Retardation* 36(1): 44–51.

Wilson, B.A., Emslie, H.C, Quirk, K., and Evans, J.J. (2001). Reducing Everyday Memory and Planning Problems by Means of a Paging System: A Randomised Control Crossover Study. *Journal of Neurology, Neurosurgery, and Psychiatry* 70(4): 477–482.

Yoder J.D., Baumgartner, E.T., and Skaar, S.B. (1996). Initial Results in the Development of a Guidance System for a Powered Wheelchair. *IEEE Transactions on Rehabilitation Engineering* 4(3): 143–151.

Zimmermann, G., Vanderheiden, G.C., and Gilman, A. (2002). Prototype Implementations for a Universal Remote Console Specification. *CHI 2002 Conference on Human Factors in Computing*, 510–511.

Design and development of pervasive healthcare technologies

chapter eight

Human factors and usability of healthcare systems

Andre Kushniruk and Elizabeth Borycki
University of Victoria, Victoria, British Columbia, Canada

Contents

8.1 Introduction 192
8.2 Background: Human–computer interaction (HCI) in
healthcare and pervasive computing 193
8.3 HCI and the system development life cycle 195
8.4 Usability engineering methods for the analysis
and improvement of pervasive computing applications
in healthcare 197
    8.4.1 Usability testing 198
    8.4.2 Usability inspection 200
    8.4.3 Modeling of workflow 202
8.5 Examples of HCI approaches to analyzing pervasive
applications in healthcare 203
    8.5.1 Use of PDAs by physicians 203
    8.5.2 Use of Web-based applications by chronically
    ill healthcare consumers 203
    8.5.3 Use and usability of smartcard technology 204
    8.5.4 Pervasive computing and the elderly—toward
    a framework for analysis of pervasive
    applications in healthcare 205
    8.5.5 Monitoring health status 207
    8.5.6 Promoting safety 208

8.5.7   Aiding cognition..209
8.5.8   Promoting functional independence..210
8.6   Considerations for new and mixed modalities
        of interaction in pervasive healthcare...210
8.7   Discussion and conclusions 212
References 213

8.1   Introduction

Pervasive computing promises to change the way that healthcare is carried out by providing access to patient and medical information "anytime and anywhere." The possibilities for integrating pervasive systems into the daily work practices of healthcare workers depend on the convergence of information technologies into a networked computing environment that is always on and available in unobtrusive ways. However, today's emerging pervasive computing technologies face serious challenges with regard to how such an innovative technology can actually be effectively and efficiently integrated into the day-to-day living and cognitive and physical activities of humans. Thus, we may consider pervasive computing within two dimensions: the technological dimension and the human dimension. Advances in technological aspects of pervasive technologies are moving ahead at a rapid pace including the widespread use of wireless networks, wearable computing, remote monitoring, and surveillance technologies.[1] Despite the incredible potential of this model of computing there are a number of critical barriers and issues related to human–computer interaction that will need to be addressed to achieve its full potential, particularly in complex areas such as healthcare. In this chapter we deal with what we believe to be one of the greatest current challenges to the introduction of pervasive computing in healthcare: the human factor. The technological advances that are occurring in areas such as networking technology, mobile computing, and wearable computing need to be followed more intensively with work in the analysis of human–computer interaction. We will introduce usability engineering approaches to the design, development, and deployment of pervasive computing in healthcare settings. These approaches reflect research that is currently being undertaken in a number of areas including cognitive science, human–computer interaction, and the ethnographic study of workflow. We analyze how we can make new models of computing actually pervade day-to-day work practices in complex healthcare settings. Specifically, we focus on an examination of how pervasive computing can be integrated into healthcare practices in such a way that users can maximize the benefits of new computing capabilities.

Topics discussed in this chapter include requirements for the engineering of new applications and use of scientific methodologies for the study of human factors—"the unique area of engineering that tailors the design of technology to people, rather than expecting people to adapt to technology,"[2] (p. 1). This includes discussion of designing unobtrusive and effective system–user interactions, design and testing of pervasive applications in geriatrics, application of usability engineering methods, and user testing issues. A range of applications where pervasive computing has begun to appear in healthcare will be used to illustrate human-centered methodological approaches to the design and evaluation of pervasive computing. This will include discussion of integration of computerized patient record systems and databases with new pervasive applications; the relation of pervasive computing to telemedicine applications; the need for integrating standards; and usability considerations in the use of mobile devices and Internet-based medical applications. We then discuss the importance of an improved understanding and consideration of human cognition and communication modalities including speech, handwriting, head-mounted displays, and combined approaches. Furthermore, we argue that principled examination of how healthcare workers interact with each other and the world around them is needed for the development of effective and useful pervasive computing applications. The chapter ends with a brief description of a human factors laboratory in development to examine the intersection between human–computer interaction and pervasive computing in healthcare.

8.2 Background: Human–computer interaction (HCI) in healthcare and pervasive computing

In the emerging field of health informatics, a wide range of innovations has appeared, ranging from advances in patient monitoring to the computerized patient record (CPR) system—an electronic repository that allows healthcare workers to enter patient information into a database electronically and retrieve needed patient data and relevant medical information about patients at point of care. These types of systems promise to revolutionize healthcare and form the basis for further advances in healthcare information processing and care, including automated alerts and reminders to physicians and automated invocation of rules and guidelines that interact with patient data to provide advice to healthcare workers. However, perhaps in no other field of research have issues related to human–computer interaction come more to the fore when attempting to disseminate such innovative information technologies. Despite repeated efforts at multiple levels, the widespread use of integrated healthcare information systems, including the CPR, has remained elusive.[3] Deployment of such systems has faced a variety of problems including delayed implementation times, difficulty in integrating new systems with existing systems, and even the difficulty of new information technologies that introduce error into the healthcare system if not designed and deployed carefully.[4]

It has been argued that issues of HCI (i.e., issues involving the study of the design, implementation, and evaluation of computers for human use) may be the most serious barrier to successful implementation of innovative healthcare information technologies in general.[5] In this chapter we take a broad perspective on HCI that encompasses three levels: level 1—the level of the individual user interacting with a system; level 2—the level of the user interacting with an information system in order to carry out real work tasks; and finally level 3—the organizational level where the interaction with an information system is considered in the context of its impact and effect on the organization as a whole. This characterization of the use of new information technologies in healthcare builds on a three-level model of HCI,[6] which provides a useful framework for considering the complex problem of understanding how to best design, test, and deploy innovative healthcare information technologies such as pervasive computing (see Figure 8.1).

Using this model we can consider problems in the acceptance of new technology at each of the three levels. For example, the goal of successful adoption of a CPR system may fail at level 1 if the design of the computer screens and instructions are such that users cannot easily learn how to use the system to enter patient data. Even if a system is designed to work well at level 1, problems may occur once the system is inserted into the complex day-to-day activities and workflow of healthcare work practices, which may involve a variety of team members, contexts, environments, levels of urgency, and complexity of tasks. However, careful analysis and adjustments made to provide effective systems at level 2 do not guarantee uptake and acceptance of a new healthcare information technology, because the effect and impact of deploying such a technology at the organizational level (i.e., level 3) may be an issue. For example, in the context of the CPR, privacy and confidentiality issues at an organizational or political level may restrict the deployment of this technology within an organization such as a hospital. Nowhere in healthcare may careful consideration of each of these three levels of HCI be more germane than in consideration of pervasive computing, as pervasive computing spans all the levels of individual users of systems. Through the application of this new technology in complex work roles and activities, issues may emerge with the increased possibilities for widespread access and dissemination of patient information, along with organizational concerns regarding privacy and confidentiality.

Another perspective when considering pervasive computing in healthcare relates to the extent of human interaction with the system. We can consider

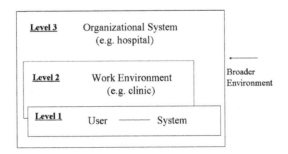

Figure 8.1 A three-level model of HCI in healthcare.

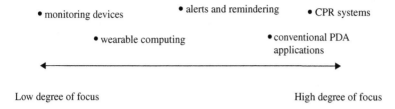

Figure 8.2 Degree of user focus/awareness of system in user interactions in healthcare.

user interaction with pervasive information technologies in healthcare along a continuum, from applications that require continual attention from the user to applications where the technology is "invisible" or interacts to a very limited extent with the user. For example, the user interfaces of many hand-held applications (see the right side of the continuum in Figure 8.2) typically represent an extension of conventional desktop user interfaces to mobile applications. With these types of applications, users must focus their attention on the user interface (e.g., to enter medical values into a personal digital assistant [PDA]) when the device is in use and must explicitly insert its use into their work activities. The introduction of a new technology must be understood in terms of how it changes the work activity of the user. To this end, many aspects of human factors from the study of conventional user interfaces are applicable. However, many applications of pervasive computing, including remote monitoring devices and wearable computing, are designed to be used ubiquitously *while* the user carries on work activities (i.e., without switching the focus of attention to interacting with the technology). This is presented in the left-hand side of the continuum shown in Figure 8.2. The implications of this new type of user-system interaction include the following: (1) interaction of the system with the environment is through a variety of modes and involves context awareness by the system; (2) the system may need to be operated with minimal cognitive awareness and effort on the part of the user, which is achieved through the system's context awareness and ability to automatically adapt to new user situations; and (3) a wide range of tasks may be performed by the system without human–system interaction.[7] The implications of these key aspects of many types of pervasive applications will be discussed throughout this chapter.

8.3   HCI and the system development life cycle

In addition to considering the HCI issues described above, to practically design and successfully deploy complex healthcare innovations such as pervasive computing, we must consider the development of the technology along a continuum from initial project planning through analysis and design to eventual system deployment. A useful framework for considering the various methods for improving HCI in pervasive systems (described later in this chapter) is the system development life cycle (SDLC). The SDLC defines a set of stages for system development. The traditional SDLC is characterized by the following phases: (1) planning, (2) analysis, (3) design, (4) implementation, and (5) maintenance and support. Although the exact nature of the SDLC employed in developing an information system may vary in terms of the relationship between these phases and their scheduling, these phases represent activities that are important for developing information systems. One implication of considering HCI aspects of pervasive computing along the SDLC is that the analysis approach taken will depend on where in the life cycle the project is at, as illustrated in Figure 8.3. For example, during the initial phases of development of pervasive applications, a variety of methods may be employed for prototyping different user interface designs by having users respond to and comment on mock-ups of display screens (e.g., for mobile applications) and applying methods of usability testing, as described in the next section. Later in the development cycle, usability testing may involve an analysis of representative users interacting with a system in artificial or realistic scenarios (also described in the next section). Summative evaluations, where nearly completed software products are beta tested, occur late in the SDLC. In health informatics this type of testing of systems has typically involved clinical controlled trials, where one group of subjects (e.g., physicians) receive a new information technology, such as a new handheld application, and another group serves as a control group (i.e., continues to use whichever method they were using before). Such studies attempt to isolate the effect of a system on healthcare practice, but they have been criticized for their inability to provide designers of complex healthcare systems with concrete and detailed information related to HCI, which might lead to dramatic improvements in their design.[8] The remainder of this chapter discusses methods for analyzing HCI that can be employed during the formative evaluation of systems in order to provide iterative input into device and system design and refinement throughout the SDLC.

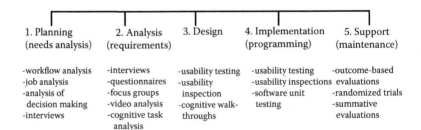

Figure 8.3 HCI approaches to system evaluation throughout the system's development life cycle (SDLC).

8.4 Usability engineering methods for the analysis and improvement of pervasive computing applications in healthcare

Usability engineering is a rapidly emerging area in the field of HCI and has provided a set of methodologies for the analysis of complex human interactions with computer-based systems. In this section we describe some of the main methods that are employed for gaining insight into many detailed aspects of HCI in the study of pervasive computing applications. These approaches can be considered along a continuum from experimental laboratory-based studies (presented as the left side of Figure 8.4) to the study of use of devices in naturalistic real-world settings (the right side of Figure 8.4). There is a category of study of HCI that falls between the two categories and involves the use of realistic simulations of real settings and contexts, which mimic those a user would expect to encounter. For example, a laboratory study of a handheld application for entering medical prescriptions might involve subjects coming to a usability laboratory and having their interactions with the application recorded as they respond to artificial medical cases. They might also be asked to verbalize their thoughts as they complete their tasks. A simulation-based study of the same application might involve subjects interacting with a "simulated patient" (i.e., a trained research collaborator playing the role of a patient), while the subject uses the application. A naturalistic study of the same application might involve remote logging and tracking of user interactions with a device as the subjects carry out actual day-to-day activities in a medical clinic. It should be noted that an in-depth analysis of HCI in healthcare may involve a laboratory study of user interaction with a device or application, which leads to testing under simulated conditions and then, finally, to testing in naturalistic settings. To analyze HCI effectively, all aspects of user interaction with the application or device are recorded so that they can be interpreted. This is often done by several methods including videotaping, recording interactions at the device level, and user feedback during each stage of trials. This information is critical for conducting usability tests.

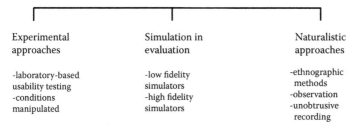

Experimental approaches	Simulation in evaluation	Naturalistic approaches
-laboratory-based usability testing -conditions manipulated	-low fidelity simulators -high fidelity simulators	-ethnographic methods -observation -unobtrusive recording

Figure 8.4 A continuum of approaches to evaluating healthcare information systems.

8.4.1   Usability testing

Usability of systems refers to the degree to which they are useful, effective, efficient, and enjoyable.[9] One of the most powerful methods for understanding and analyzing the usability of pervasive technologies is through usability testing.[10] Usability testing refers to the evaluation of information systems by conducting an in-depth analysis of user interactions with the system as users carry out tasks the system or device was designed to support, under either artificial or realistic conditions. Typically this involves video recording the interactions of users with the pervasive device, especially when the testing occurs in a simulated environment. Elements that may be captured are screens of a computer application, logs of the system's behavior, and the physical and verbal behavior of subjects as they interact with the new device and others in their work environment. Under artificial conditions, subjects may be asked to "think aloud" while interacting with a device or carrying out a task that involves the use of pervasive technology. The amount and type of data captured is highly dependent on the type of trial (i.e., it is generally easier to capture more data in a simulated environment versus a real one). Any resulting audio and video recordings of the interaction can then be analyzed posttrial using methods involving the coding and classification of user problems.

Usability testing is based on cognitive task analysis. Cognitive task analysis emerged from the fields of cognitive science and psychology and involves a detailed analysis of humans as they carry out complex reasoning and decision-making tasks.[11] Usability testing typically includes the following stages:[12]

Stage 1—Identification of evaluation objectives. The objectives at this stage may range from testing a prototype design with potential users to obtain feedback on preliminary design issues to testing of near-completed applications to assess reliability and safety.

Stage 2—Selection of test subjects and computer applications. Subjects selected for study should represent the intended end users of the pervasive application. For example, subjects with a particular cognitive deficit should be recruited for studying a system designed to provide support to such populations.

Stage 3—Selection of representative experimental tasks. The selection of appropriate and realistic tasks for subjects to carry out is essential to ensure the validity of human-factors testing. An example would be the selection of realistic clinical tasks to study physicians' use of a PDA application in an emergency ward.

Stage 4—Selection of an evaluation environment. The evaluation environment may range from a laboratory setting (e.g., a usability test site with built-in monitoring equipment, one-way mirrors, etc.) to real conditions (e.g., monitoring use of pervasive technologies within an operating room).

Stage 5—Data collection of user (i.e., subject) interactions with the pervasive technology while carrying out representative tasks. This is the stage where subjects carry out tasks using the technology under study. Data collection can include video recording overt, physical user actions; audio recording verbalizations of subjects; screen recording of computer-based devices; and logging or tracking of system responses and processing.

Stage 6—Analysis of the process data. The data collected in stage 5 are analyzed using a range of methods. This includes coding of transcripts of video and audio interactions to identify issues in the user interaction, problems with human–computer interaction (e.g., problems locating information, navigating through information resources, etc.), and problems with system-sourced activities (e.g., inappropriate context switching by a pervasive application).

Stage 7—Interpretation of findings for system redesign and refinement. The ultimate purpose is to provide feedback to the designers of new technologies about how to improve the design. The cycle of prototype testing, feedback, and redesign may occur several times before the designers and potential users are satisfied with the performance of the device.

The following example illustrates the usability testing stages outlined above. For example, in a study with the objective of assessing the utility and usability of a mobile application for alerting physicians about patient emergencies through a PDA device (stage 1), physicians might be asked to carry out routine daily activities while the experimenters observe (and video record) the physicians before, during, and after simulating an emergency patient alarm using the PDA (i.e., representative subjects carrying out representative tasks using a representative device; stages 2 and 3). The video recordings of user interactions can then be analyzed to assess the efficacy and usability of the device in realistic situations in a laboratory environment (stage 4). Subjects can also be asked about their experience with the system immediately after interaction with the device has taken place, using a method called cued recall. In cued recall, subjects are asked to recount their thoughts about the device while they are shown video recordings of their interactions. The investigators may have specific queries for the subjects as to their thoughts and actions at key points in their interactions with the system or device under study (stage 5). Subsequent to this, researchers can analyze video and audio recordings (stage 6) and incorporate their findings into the iterative design of the system (stage 7).

Usability testing can be easily adapted to analyze pervasive healthcare information technologies and systems. For example, this could range from usability testing involving video recording physicians or nurses carrying out tasks while using a PDA to simulations involving elderly subjects undertaking daily activities while wearing portable monitoring devices. This approach has been employed in the study of devices ranging from mobile handheld applications to analysis of the usability and usefulness of automated alerts and reminders for physicians in simulated emergency situations.[4,13]

8.4.2 Usability inspection

A second methodology adapted from HCI in healthcare that has emerged from a cognitive task analysis approach is known as usability inspection.[14] Usability inspection involves a usability analyst or inspector stepping through or "walking through" use of an interface or system in the context of some real task or activity. For example, an approach known as the cognitive walkthrough involves the analyst (or a team of analysts) stepping through the activities that use a new, and possibly pervasive, technology while recording goals, actions, system responses, and potential problems. Another example is a methodology known as heuristic evaluation. This methodology focuses on identifying violations of principles of human-factors design when a system is used to carry out a task.

Jacob Nielsen[10] has outlined a set of principles or rules to consider when conducting such analysis, including the following:

1. Visibility of system status—The state of the system's processing should be visible to users of a system when they so desire that information.
2. Matching the system to the real world—Real-world language and conventions should be used in user interfaces.
3. User control and freedom—Users should feel like they are in control.
4. Consistency and standards—The user interface and system operations should be consistent.
5. Error prevention—Designers should design interfaces to prevent errors.
6. Minimize memory load—Systems should support recognition (e.g., using menus) rather than recall.
7. Flexibility and efficiency of use—Systems should allow for customization and adaptability.
8. Aesthetic and minimalist design—Often the simplest and most minimal designs are the best.
9. Appropriate error handling—Help users recognize, diagnose, and recover from errors.
10. Help and documentation—Help should be available to users when needed.

These principles can be extended when considering pervasive computing in healthcare. Based on our current work with pervasive devices, we propose the following heuristics or guidelines for the evaluation of pervasive applications:

1. Unobtrusiveness—Direct interaction of a user with a pervasive system should be limited to only the parts of the task where such interaction is necessary (i.e., allowance for visibility when required).
2. Privacy and security—Use of a pervasive device must not violate privacy and security restrictions under normal conditions of use.

3. Emergency override capability—Under exceptional conditions, security and access restrictions may need to be overridden. Such exceptional cases need to be identified and logged for subsequent audit.
4. Appropriate context awareness—Pervasive systems must be able to track the context of use and respond to differing contexts in an appropriate manner.
5. Failure backup—Failure of a pervasive device or its supporting network should be made apparent to the user through some form of notification.
6. Alternative modes—Allowances should be made so that alternative modes of user interaction are available during system failure and recovery periods.
7. Information and alerts prioritization—The system should appropriately prioritize and display alerting or reminder information. This information should be displayed only at essential points in user workflow to avoid cognitive overload.
8. User controllability—The user should assume manual control of the system when needed in the absence of traditional interface cues (e.g., a physician can shut off decision support functionality in situations where he does not feel it is appropriate).
9. Appropriate modality for interactions—The type of system–user interface should be fitted to the task and user's needs.
10. Consistency across modalities—The user should be able to switch methods of interaction and easily recognize where common information is located.
11. Seamless modal switching—The user should be able to switch devices without having to re-input any information.

Applying appropriate design principles and guidelines in the development of effective pervasive applications is difficult, as such applications may often demand seemingly conflicting requirements (e.g., the transparency of operation for ubiquitous applications under normal operating conditions, while providing visibility of operation under failure or certain emergency contexts). The ability to disengage from an interaction with a pervasive system or application may be important at times. This would require users to know where they were in an interaction with the system, which may be difficult if the system is designed to be highly unobtrusive with limited explicit and visible interaction with users. Furthermore, the varied contexts of use of systems in healthcare (e.g., differing situations of task complexity, urgency of medical situations, and differing clinical environments) may make it extremely difficult to develop systems that can appropriately adapt to differing conditions of use. Guidelines such as those presented above can be applied in a principled manner by using a cognitive task analysis approach. For example, analysts may step through the use of a pervasive device, recording violations of any of the above guidelines during testing. In addition, the same guidelines can be used to analyze data collected from the study of subjects interacting with pervasive systems under artificial conditions, simulations, or naturalistic settings. These types of techniques essentially form a category basis for coding and quantifying problems observed by analysts and investigators when reviewing video data of user interactions.

8.4.3   Modeling of workflow

As discussed so far in this chapter, for pervasive applications to be effective they must embody context sensitivity that allows for the system to adapt automatically to changing contexts and situations. The effectiveness of this is directly related to how well the system is integrated with and operates within an appropriate workflow (i.e., the steps in a process from its start to its end) from the perspective of the users.[15] A variety of new methods have appeared for analyzing workflow and the effects of systems on user interactions and task goals. For example, Cysneiros and Kushniruk[16] describe the application of an ontology of nonfunctional software requirements that can be used to help analysts and decision makers reason about issues such as the effects of introducing pervasive devices (e.g., wireless handheld devices) on nonfunctional requirements such as privacy and usability. Furthermore, application of techniques from the area of workflow modeling can more accurately characterize and describe changes in workflow before and after the introduction of devices and healthcare applications (e.g., wireless access to patient records). Along these lines, accurate analysis of the contexts where the pervasive system will be used is essential to develop mechanisms for appropriate automatic switching of contexts that are supported by a system. Recent advances in workflow modeling and requirements gathering using nonstandard approaches offer hope in improving this critical aspect of pervasive systems, particularly through the appropriate mapping of system responses to situational contexts in the healthcare domain. For example, Jorgensen and Bossen[17] describe the application of extended unified modeling language (UML) through case modeling to obtain requirements and refine the design of a pervasive patient record system. Using their approach, three levels of analysis are suggested, beginning with prose descriptions of work processes and proposed computer support, followed by development of formal executable models of the work processes, and completed with the development of testable animations that represent the proposed solutions. Using such an approach, system requirements can be interactively examined by stakeholders (e.g., designers and end users) in the context of potential work processes. Such analysis may be essential for development of pervasive applications as they are often intended to be tightly integrated into complex work activities.

8.5   Examples of HCI approaches to analyzing pervasive applications in healthcare

This section presents a number of projects that examine human–computer aspects of pervasive healthcare systems, with a focus on how the principles and methods described in the previous sections can be practically applied in design and deployment of pervasive systems.

8.5.1   Use of PDAs by physicians

A recent research study examined the relationship between usability problems and prescription errors when physicians used a mobile PDA-based application to write prescriptions and interact with a remote CPR system.[4] A simulation approach was chosen for studying interactions of physicians with the PDA device when carrying out a set of realistic tasks (e.g., interviewing patients and ordering medications). The study design involved full video and audio recording of the screens of the PDA and of the subject's interaction with both the device and with a simulated patient, while the subject carried out tasks that the device was designed to support. To analyze the resulting data, transcribed audiotapes of the physician subjects' verbalizations were linked to the corresponding video sequences of their interactions with the handheld device. Results indicate that specific categories of usability problems can be identified with considerable accuracy early in the design of a device. Examples of these include problems due to limited screen size and lack of display capability in order for users to have the opportunity to view a summary of medications entered prior to submitting a medication request. The occurrence of usability problems in user interactions with the device was noted under one of the following categories: display visibility problems, navigational problems, input problems, printing problems, and speed problems. By coding the data in this manner, as well as by coding actual medication errors during the simulation sessions, it was possible to estimate the extent and probability of an identified usability problem predicting an actual medication error *prior* to finalizing the design of the application.[4,13] Work such as this promises to ensure that pervasive applications actually improve healthcare efficiency and safety, rather than inadvertently increasing error, as has been shown in recent studies of use of automated medication entry applications.[18]

8.5.2   Use of Web-based applications by chronically ill healthcare consumers

Pervasive applications in healthcare using application platforms such as the Internet and innovative devices such as smartcards (described in Section 8.5.3) will likely make patient data accessible not only to healthcare professionals at various and mobile locations but also to a wide range of stakeholders including patients themselves. As described earlier, CPR systems constitute large repositories of patient information (e.g., vital signs, medical images, and patient demographics). In a related line of research, we studied the extension of a Web-based CPR system, which allowed thousands of physicians at Columbia Presbyterian Hospital in New York to access the records of their patients from various locations via the Internet.[19] The project aimed to extend access of the records not only to physicians but also to patients.[20] Security issues were paramount in designing an application that would allow for an architecture that is open enough to keep the system versatile, while limiting access to appropriate users in appropriate contexts. Use of the resultant system was evaluated over a thirty-six-month period and involved thirteen subjects in a process of iterative development and refinement; the study provided details about patient and physician system access such as review of patient data, access to educational resources, and decision support—all activities were automatically recorded into a log file. The results of the study indicated that allowing such access could have a positive impact on face-to-face doctor–patient encounters while keeping security risks at a minimum. The technique to remotely analyze such applications resulted in the design of an application known as the Virtual Usability Laboratory (VUL).[21] The VUL is designed to unobtrusively monitor users of Web-based applications remotely while also providing the ability to automatically query users after they interact with the application being studied (e.g., querying users after they interact with a remote CPR system). Data from remotely monitoring a large number of users can then be integrated within a centralized database connecting usability data, such as browsing patterns, system invocations, and user interactions, with medical and demographic data related to those users. Successful and widespread use of pervasive technologies within healthcare will require close integration with the growing number of CPR systems being installed at various institutions. Additionally, monitoring and tracking users will be needed to assess how to design and refine usable and useful applications. Effective pervasive applications in healthcare need to interface seamlessly with interoperable databases and CPR applications. This goal is still somewhat elusive because there is now a plethora of different CPR systems in use in healthcare, with individual hospitals often having a number of different CPR systems. However, the movement toward adopting messaging and medical vocabulary standards (e.g., using HL-7 standards for the exchange and integration of electronic information) will allow for increased integration of hospitalwide applications and clinical information systems with innovative mobile-based applications.

8.5.3 Use and usability of smartcard technology

Another exciting example of a pervasive application in healthcare is the integration of smart data and smartcard technologies with CPR systems. The use of smartcards (card-sized devices with a microprocessor that contain patient-specific information and that can be carried around by patients) has been adopted in a number of countries.[22] To date, such applications have been limited to storage of a relatively small amount of basic demographic and medical data about patients. Data on these mobile devices can be integrated with applications running clinical guidelines and rules, resulting in alerts or reminders to users (e.g., physicians or patients) via a remote device. Wearable electronic bracelets and radio frequency identification devices (RFIDs) can be used for patient identification and can contain patient-specific information about drug allergies, ensuring that appropriate medications are given to patients.

The usability of such devices in real hospital settings has been problematic for a number of reasons. In particular, issues in appropriately updating the remote source of patient data as well as security issues, such as privacy and the need to detect unauthorized access to such patient-specific information, remain problematic. In general, developing interfaces with pervasive devices allows for unfettered access by authorized users while preventing unauthorized access has remained one of the more problematic issues to be resolved before such devices become common. In a recent article, Coiera and Clarke[23] argue for the development of a framework for considering "e-consent" to guide access to electronic health information. Their model distinguishes between a general consent model (where blanket consent is given to healthcare organizations to access the patient's data, and exceptions to general access are specified) and a general denial model at the other extreme (where access is denied by default, unless specified otherwise). Application of usability engineering approaches, such as simulation techniques described above to test such devices in a range of realistic situations and scenarios, will be necessary to ensure that such systems actually support workflow and security concerns and do not hinder patient care in real hospital settings.

Thus, although potential integration and remote access to institutional CPR systems are goals of a number of research projects, they need to await advances in both wireless networking and the resolution of privacy and confidentiality issues. In one recent study conducted in Canada, the factors influencing the adoption of smartcards were investigated after tracking use of the cards by 299 professionals and 7,248 patients, who used smartcards for a year.[24] From interviews with both patients and physicians, results strongly suggested that users were very sensitive to the need for clear benefits of use of the technology in order to adopt it widely. From the physicians' perspective the relative advantage of the system was linked directly to how complete the recorded information is, while the factor most likely to influence adoption of the system by patients was how much ownership the patients felt they had over their own health data.

8.5.4   Pervasive computing and the elderly—toward a framework for analysis of pervasive applications in healthcare

Another area in healthcare where pervasive applications currently holds considerable promise is in the support of care for the elderly and impaired populations. In recent years there has been a shift in the provision of patient care. Improvements in the quality of healthcare information and communication technologies (ICTs), medical care, treatments, and the rise of consumerism have reduced the need for hospitalization, shifting the burden of care from the hospital to the community and from the health professional to the individual healthcare consumer. Such advances in the provision of care have led to improvements in patient quality of life, increased cognitive and functional independence, better patient monitoring, self-management of and control over disease, and reduced health service utilization (i.e., the number of hospital, physician, and home care visits and hospital length of stay).[25] Nowhere has this trend been more evident than among the elderly (i.e., those over sixty-five). Increasingly, elderly healthcare consumers are living better, more independently, and staying at home longer (without the need for institutionalization). In addition to these trends, elderly persons have become the most significant and growing group of computer users, using new technologies to learn about, monitor, and manage their health. It is this interest in the potential assistive nature of new technologies among elderly consumers that has led to the growth, development, design, and evaluation of a wide range of pervasive computing technologies for the elderly. In recent years, significant interest has emerged among software and hardware developers and elderly consumers about the potential health and independence benefits associated with the use of pervasive computing technologies in the home by an older adult.

As we age we experience significant physiologic changes. Systems within the body do not function as well as they had when we were young. Furthermore, as we age, a combination of genetic and lifestyle factors causes us to be more prone to developing acute and chronic conditions that affect our cognitive and physical functioning, therefore impeding our ability to live independently or with our families at home. For example, the ability to perform activities of daily living (ADLs) (e.g., bathing oneself) and instrumental activities of daily living (IADLs) (e.g., grocery shopping) will often determine if an older adult is able to manage at home or requires some form of assisted living or institutionalized living arrangement. Software and hardware designers and developers are exploring the value of pervasive computing technologies in supporting older adults' changes in cognition and function. More specifically, four key areas of research have emerged: (1) promoting safety, (2) aiding cognition, (3) improving social interaction, and (4) promoting functional independence. These areas of research overlap in the domains of aging and pervasive computing technologies including the remote monitoring of health status. The evaluation of HCI aspects of such systems can be understood in terms of several critical aspects. Figure 8.5 illustrates a framework reflecting the critical aspects that need to be considered when analyzing the impact of pervasive technologies designed to support the elderly and impaired.

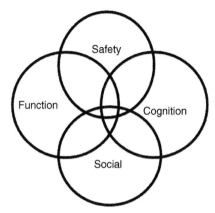

Figure 8.5 A framework for considering HCI aspects of pervasive systems for the elderly and impaired.

8.5.5 *Monitoring health status*

A wide range of research projects is currently underway to explore the value of using pervasive computing technologies to monitor patient vital signs (i.e., heartbeat, respiration, blood pressure, and temperature). In recent work, pervasive computing technologies are being used in a nursing home to monitor residents' vital signs.[26] It is believed that the ongoing monitoring of vital signs will help healthcare professionals to identify healthcare crises before they occur. In this case, the technology would help the health professional observe trends in bodily functioning over the course of a month, week, day, or minute by minute. If changes in bodily function can be identified by changes in monitored vital signs, it is possible that early medical attention could be sought before a critical event, such as a stroke or heart attack, occurred. Another benefit of using such devices is a potential increase in the efficiency and effectiveness of health professionals. Automating the collection and recording of vital signs would mean that health professionals such as nurses would need to spend less time manually collecting and documenting information, an activity that has been found to take up a significant amount of time. This would allow health professionals to spend more time reviewing and evaluating patient health data and conducting other forms of assessment, resulting in better patient care. The benefits of using such technologies in the clinical setting are significant, but the implementation of such technology highlights the importance of usability for both health professionals and patients. From a patient perspective, pervasive technologies that transmit information about a patient's health status need to be lightweight, easy to wear, and of a sufficiently small size that they do not impair mobility or independence.[7] For health professionals, new knowledge and skills need to be developed to help them relate information about physiologic function to critical health events to patients—an area of research that is being

studied increasingly. Decision-support systems that help health professionals to review and understand data will be critical, as such sheer volumes of
data that could be logged automatically could lead to cognitive overload
and subsequently to increased, rather than decreased, work for the health
professional. Pervasive computing technologies could help health professionals spend less time on administrative activities and more time assessing
and managing the care and treatment of patients. However, care must be
taken in technology design to ensure that users are not overwhelmed with
information.

8.5.6   Promoting safety

Pervasive computing applications have been cited to improve patient safety.
The aim of many of these applications is to prevent injury arising from
physical or cognitive deficits while promoting independence. For example,
pervasive computing technologies (e.g., tracking systems) are being used in
acute and long-term care settings to prevent cognitively impaired patients
from wandering away.[27] These technologies can be broken down into those
that are used for the purposes of tracking patients and those that monitor
patient movements and, based on these movements or identified behaviors,
signal health professionals that the patient may have or has already injured
themselves. These technologies are a significant advance over traditional
methods of caring for patients. For example, current systems used to care for
cognitively impaired patients who wander limit patient independence,
restricting the patients' movements (such as the use of locked units), or use
alarms that can unintentionally startle or frighten patients. New tracking
systems allow patients to wear bracelets or badges that provide information
to a central monitoring system about where they are located in a facility at
any point in time. Although still in development, another pervasive computing technology that holds great promise toward increasing patient safety is
the falls monitor.[28] Currently, clinicians are limited in their ability to prevent
a fall or help a person who has just fallen. Clinical attempts to prevent falls
have resulted in the development of pen-and-paper scales that measure an
individual's level of risk and the implementation of strategies that can be
used to reduce the likelihood of a severe fall. These include injury-proofing
the home (e.g., removing mats) and developing clothing (e.g., hip pads) that
cushions individuals when they do fall to hopefully prevent fractures. Pervasive computing technology could move us beyond these limited
approaches to try and avoid falls, providing us with information about the
types of activities and movements that lead to falls. In addition to this, current
systems have successfully warned health professionals or patients of the
potential for experiencing a fall, encouraging individuals to change their
behavior to reduce the risk. Pervasive computing technologies have also
been used to successfully identify the movements of an individual who has
fallen.[28] Deploying such technology in the home could help an older adult
make decisions about the types of behaviors he or she undertakes that may

place him or her at risk for falling. Additionally, an automatic fall detection system could get immediate assistance by notifying caregivers or health professionals if the individual did experience a fall. This would effectively reduce complications caused by the "long lie" experienced by people if they are severely injured by a fall and have to wait for someone to find them. These systems are currently in development and their use will depend on the ability of the system to recognize impending or actual falls accurately, without generating an unacceptable level of false positives (i.e., false alarms).[29] Presently, such systems have some ability to detect some types of falls and not others. More research is needed to develop a comprehensive set of fall-related behaviors and then construct a system that is effective in detecting the potential for a fall as well as a fall itself. Implementation of this type of pervasive technology may be difficult as results from a focus group suggest that potential users perceive these types of systems as infringing on their privacy.[28] Furthermore, there are concerns about who would have access to the monitoring data. Finally, there is a need to create usable interfaces that allow an older adult who has sensory-motor, cognitive, and functional impairments to interact with these systems.

8.5.7   Aiding cognition

More recently, pervasive computing technology advances have been created to act as assistive devices and help the cognitively impaired. A large focus in this area is the development of reminder systems. Reminder systems have taken various forms: from systems that are used by those individuals who have mild impairments (e.g., medication reminder systems for people who forget to take their medication) to those used by people with significant impairments (e.g., a cueing device to help someone remember the steps involved in washing, cooking, or dressing). One such system provides the individual with a visual reminder of the last task completed.[30] For example, if the individual had just finished brushing her teeth, a picture would appear providing her with a reminder that she had brushed her teeth and another picture would be provided cuing her to wash her face. Although this is an interesting approach toward helping older adults with cognitive problems, it remains unclear if such systems could be successfully employed by the cognitively impaired adult as a form of cueing or as a reminder. For example, could a cognitively impaired individual learn to use such a system or would it go ignored or unused because the system was too difficult to learn or perceived as a form of cognitive overload? Also, would such a device lead to cognitive overload during the actual task? Many questions remain unanswered and need to be addressed. The application of usability approaches to these problems may go a long way to shedding light on such issues. As this is a particularly vulnerable user group, careful user testing will be needed to evaluate the success of the device in assisting the user, along with the use of a minimum number of simple user interfaces to match the cognitive limitations of the individual. In the case of a progressive condition such as Alzheimer's disease, the device should automatically adapt over time as the user's cognitive impairment worsens. Current work along these lines could include approaches based on usability testing and simulations of actual situations with representative subjects with varying levels of cognitive deficit.

8.5.8   *Promoting functional independence*

One of the most promising areas where pervasive computing technologies can be used is in promotion functional independence. Here, the environment can adapt to the functional deficits of an individual who has compromised sensory motor function or has musculoskeletal disorders that have limited his or her ability to perform ADLs. In such cases, the individual is now able to control his or her environment using voice recognition.[31] The environment can also be contextually aware and can adapt to the individual's movements and needs as he or she moves from room to room, making proactive decisions about what the individual will want and responding appropriately.[32] For example, a wheelchair-bound individual can now move freely throughout his home without having to open doors or turn on lights, as these are automatically and preemptively controlled by the system. Such systems hold much promise to help individuals who have functional impairments to live longer in their own homes. Further development of these technologies is needed, as number of user-interaction issues remain to be fully explored. Some of these include (a) how a balance will be achieved between the "system's perceived importance of the interaction and the user's perception of the appropriateness of the interaction and level of intrusiveness of the event,"[33] (b) what constitutes the context appropriateness of the system's interaction with the user and how it will be determined, (c) how the user will control and judge the current state of the interaction (e.g., can the user cut short or disengage the system when his or her needs are fulfilled? can he or she easily move to another activity when he or she no longer needs to address the first activity? how does the user engage and disengage the system if there are dependencies between devices that have been networked?), (d) how the user judges functionalities that are being used for the first time, (e) how the user switches between different modes while engaged in an interaction sequence,[33] and (f) whether the individual wants a seamless interface with his or her home and how this affects quality of life.

8.6   *Considerations for new and mixed modalities of interaction in pervasive healthcare*

One of the greatest challenges in the design of usable and useful pervasive applications is the improved integration and appropriate application of varied modalities that are afforded by such systems. As argued earlier in this chapter, technological advances may have outpaced our understanding of the cognitive and perceptual aspects involved in the use of varied interface modalities. Healthcare has provided a unique and complex testbed for studying some of the more exciting technological advances in user interfaces.[34,35] Work in healthcare has also highlighted some of the major issues and barriers related to HCI that have restricted the adoption of pervasive technologies to date. The complex conditions that surround healthcare professionals (e.g., emergency rooms, operating rooms, the patient bedside, etc.) necessitate flexible modes of interaction with healthcare information systems. In particular, the issue of data entry by health professionals has been one of the biggest barriers to direct use of both conventional and pervasive applications. Although improvements have taken place in areas related to pen technology, handwriting recognition, PDAs, and wireless connections,[35] applications that require entry by pen or stylus are limited in the context of pervasive applications. Speech recognition solutions for data input into pervasive devices and for entering spoken commands to such systems are attractive, hands-off alternatives.[36,37] However, given the environmental considerations involved in many clinical activities (e.g., physician activities in a noisy clinical environment or emergency room), the use of speech input may be limited. Other than use in a private office setting, privacy concerns further limit the use of speech in many clinical contexts. Other types of interfaces, including head-mounted displays (HMDs), provide possible alternatives for appropriate healthcare applications.

Given the array of input and output options, work along the lines of developing user interfaces that allow for switching among different modalities of interaction is becoming an increasingly recognized area in design of user interactions with pervasive computing.[33] Advances in understanding how to integrate input modality into pervasive applications (e.g., by employing advanced workflow modeling techniques) will likely lead to improved models of interaction that allow for adaptive mixed modalities of interaction based on the contexts encountered by the healthcare professional. Along these lines, research from the area of adaptive user interfaces is extremely relevant to the issues that arise from user interaction with pervasive systems.[38,39] Adaptive technologies use intelligent approaches for modeling users, contexts, and user needs to help support a number of functionalities, including a number of different dialogue modes and the switching of modes in a way that is easy for the user to understand, learn, and interact with. Work in this area is conducted along several lines, including adaptive content selection by systems, adaptive navigation support, and adaptive presentation of information to users;[39] for example, providing physicians with particular medication guidelines specific to the hospital site where they are currently located. According to Perry et al.,[33] the central issues for developing adaptive approaches to pervasive systems include allowing for smooth transitions between modes and the ease of learning how to use different modes by users. In addition, issues related to context sensitivity include designing systems with location-appropriate interactions (i.e., systems that provide users with appropriate interactions given a particular location), activity-appropriate interactions (i.e., systems that provide users with appropriate interactions given a specific task or activity), and multiuser interactions, as well as appropriate assessment by the system for interaction (e.g., automated assessment of what interactions should take place and in what order or priority).

The difficulties of designing context-sensitive adaptive systems have been described elsewhere in this chapter but include the following: (1) the need for users to have a sense of control, (2) consistency across modalities as much as is possible through use of common semantics, and (3) appropriate switching between modes based on user needs, tasks, and locations. Complicating the development of such systems is the fact that target populations may be extremely varied (e.g., cognitively impaired older adults have varying levels and types of impairments), making the design of a "one size fits all" system for a particular class of users inappropriate. The development of effective pervasive systems in healthcare depends on the use of appropriate methods for guiding the design and testing of particular applications.

8.7 Discussion and conclusions

We have presented a number of different frameworks and methodological approaches for considering human factors of pervasive computing systems in healthcare. In addition, the application of a number of HCI methods to pervasive computing in healthcare has been discussed. We argued throughout that the success of these systems ultimately depends on how well they support and augment real workflow and human activities. A number of the issues related to human–computer interaction and pervasive computing system design are currently being explored in usability laboratories, such as the one we established for the analysis of new healthcare innovations. This has included study of systems ranging from CPR systems to PDA applications designed to provide context-sensitive access and presentation of information to a range of users including healthcare providers, patients, and laypeople. This has involved taking a "portable" approach to studying complex interactions with pervasive systems with recording equipment that can be taken into the actual settings where pervasive systems would be used. Work such as the projects described above has been focused on understanding how advanced HCI methodologies and techniques can be applied to improve the usability and usefulness of a range of healthcare systems including emerging pervasive applications.

The devices, systems, and concepts presented here highlight some applications of pervasive healthcare. Indeed, new pervasive healthcare technologies are already helping patients live longer and more independently. Some systems, such as the use of remote telemetry monitoring via telephone by nurses, have been developed to enable health professionals to remotely monitor patients, allowing patients to recuperate in the comfort of their homes without compromising their health and safety. No one particular area of pervasive computing in healthcare has experienced more growth than that of technologies aimed to assist with the unique sensory, cognitive, and functional concerns of the elderly and disabled.[40] A shift has taken place from providing patient care in the hospital to providing increasing levels of complex patient care in the community. With this shift a greater emphasis has been placed on individual patients and their families and health professionals. However, issues related to providing effective, robust, and usable applications that truly support users are likely to be the greatest barrier to the widespread adoption of pervasive healthcare applications. To overcome this challenge, careful design and extensive usability testing of new applications are required to ensure that applications meet user needs.

From the perspective of HCI, the complexity of providing information and support to both health professionals and patients on an "anytime and anywhere" basis is enormous. Only the use of principled theoretical and methodological frameworks from the field of HCI will meet this challenge. As discussed, developing system context sensitivity, adaptive system capabilities, and appropriate models of user interaction is essential for successful deployment of new pervasive technologies. The critical barriers from the human-factor dimension of healthcare systems need to be addressed from an interdisciplinary and flexible perspective. This includes developing methods of workflow analysis and new approaches to requirements data gathering, which will be used in conjunction with usability testing to lead to improved system designs and effective user interactions.

References

1. Hansmann, U. et al. *Pervasive Computing*, 2nd ed. New York: Springer, 2003.
2. Vincente, K. *The Human Factor*. New York: Routledge, 2004.
3. Kahone, I.S. Computer-based patient records. In J. van Bemmel and A. McCray, eds., *Yearbook of Medical Informatics*. Stuttgart: Schattauer, 1998.
4. Kushniruk, A.W. et al. Technology-induced error and usability: The relationship between usability problems and prescription errors when using a handheld application. *Int. J. Med. Inf.*, 74, 519, 2005.
5. ACM SIGCHI. *Curriculum for Human–Computer Interaction*. ACM Special Interest Group on Computer–Human Interaction Curriculum Development Group. New York, 1992.
6. Eason, K.D. Ergonomic perspectives on advances in human–computer interaction. *Ergonomics*, 34, 721, 1991.
7. Lukowicz, P. et al. Wearable systems for healthcare applications. *Meth. Inf. Med*, 43, 232, 2004.
8. Kushniruk, A.W. Evaluation in the design of health information systems: Applications of approaches emerging from usability engineering. *Comp. Bio. Med.*, 32, 141, 2002.
9. Preece, J., Rogers, Y., and Sharp, H. *Interaction Design: Beyond Human–Computer Interaction*. New York: John Wiley & Sons, 2002.
10. Nielsen, J. *Usability Engineering*. New York: Academic Press, 1993.
11. Gordon, S.E. and Gill, R.T. Cognitive task analysis. In *Naturalistic Decision Making*, C.E. Zsambok and G. Klein, eds. Mahwah, New Jersey: Lawrence Erlbaum Associates, 1997.

12. Kushniruk, A.W. and Patel, V.L. Cognitive and usability engineering approaches to the evaluation of clinical information systems. *J. Bio. Inf.*, 37, 56, 2004.
13. Kushniruk, A. et al. The relationship of usability to medical error: An evaluation of errors associated with usability problems in the use of a handheld application for prescribing medications. *In Proc. MedInfo 2004*, 1073, 2004.
14. Nielsen, J. and Mack, R. *Usability Inspection Methods.* New York: John Wiley & Sons, 1994.
15. Haag, S. et al. *Management Information Systems.* 2nd ed. New York: McGraw-Hill, 2004.
16. Cysneiros, L. and Kushniruk, A.W. Bringing usability to the early stages of software development. In *Proc. 11th IEEE Int. Requirements Engineering Conference*, 2003, 359.
17. Jorgensen, J.B. and Bossen, C. Executable use cases: Requirements for a pervasive health care system. *IEEE Software*, 2, 34, 2004.
18. Borycki, E.M. and Kushniruk, A.W. Identifying and preventing technology-induced error using simulations: Application of usability engineering techniques. *Healthcare Quart.*, 99, 2005.
19. Cimino, J.J., Socrates, S.A., and Clayton, P.D. Internet as clinical information system: Application development using the world wide web, *JAMIA*, 2, 273, 1995.
20. Cimino, J.J., Patel, V.L., and Kushniruk, A.W. The patient clinical information system (PatCIS): Technical solutions for and experience with giving patients access to their electronic medical records, *Int. J. Med. Inf.*, 68, 113, 2002.
21. Kushniruk, A.W. and Ho, F. The virtual usability laboratory: Evaluating web-based health systems. *Proc. e-Health*, 2005.
22. Kohler, C.O. et al. *Health Cards '95.* Amsterdam: IOS Press, 1995.
23. Coiera, E. and Clarke, R. e-Consent: The design and implementation of consumer consent mechanisms in an electronic environment. *JAMIA*, 11, 129, 2004.
24. Aubert, B.A. and Hamel, G. Adoption of smart cards in the medical sector: The Canadian experience. In *Yearbook of Medical Informatics*, R. Haux and C. Kulikowski, eds. Stuttgart: Schattauer, 2003.
25. Shortliffe, E.H. et al. *Medical Informatics: Computer Applications in Health Care and Biomedicine.* 2nd ed. New York: Springer, 2001.
26. Maiolo, C. et. al. Home telemonitoring for patients with severe respiratory illness: The Italian experience. *J. of Telemed. Telecare*, 9, 67, 2003.
27. Hauptmann, A.G. et al. Automated analysis of nursing home observations. *Pervasive Comp.*, 2, 15, 2004.
28. Sixsmith, A., and Johnson, N. A smart sensor to detect the falls of the elderly. *Pervasive Comp.*, 2, 42, 2004.
29. Lee, T. and Mihailidis, A. An intelligent emergency response system: Preliminary development and testing of automated fall detection. *J. Telemed. Telecare*, 11, 194, 2005.
30. Dishman, E. Inventing wellness systems for aging in place. *Comp. IEEE*, 5, 2004, 34.
31. Stanford, V. Using pervasive computing to deliver elder care. *Pervasive Comp.*, 1, 10, 2002.
32. Helal, S. et al. Enabling location-aware pervasive computing applications for the elderly. In *Proc. of the First IEEE International Conference on Pervasive Computing and Communications*, 2003.

33. Perry, M. et al. Multimodal and ubiquitous computing systems: Supporting independent-living older users. *IEEE Trans. Inf. Tech. Biomed.*, 8, 224, 2004.
34. Patel, V.L. and Kushniruk, A.W. Interface design for health care environments: The role of cognitive science. In *Proc. of the AMIA 98 Annual Symposium*, C. Chute, ed., 29, 1998.
35. Tang, P.C. and Patel, V.L. Major issues in user interface design for health professional workstations: Summary and recommendations. *Int. J. Bio-med. Comp.*, 34, 139, 1994.
36. Lucas, B. VoiceXML for Web-based distributed conversational applications. *Comm. ACM*, 43, 53, 2000.
37. Shneiderman, B. The limits of speech recognition. *Comm. ACM*, 43, 63, 2000.
38. Kantorowitz, E. and Sudarsky, O. The adaptable user interface. *Comm. ACM*, 32, 1352, 1989.
39. Brusilovsky, P. and Maybury, M.T. From adaptive hypermedia to the adaptive Web. *Comm. ACM*, 45, 31, 2002.
40. Kushniruk, A.W. Technology, health care and the elderly: Where are we headed? *Perspec. J. Ger. Nurs.*, 27, 10, 2004.

chapter nine

Routes and requirements for realizing pervasive medical devices

Michael P. Craven
University of Nottingham, United Kingdom

Contents

9.1 Introduction 218
9.2 Health technology assessment 219
    9.2.1 The evidence-based environment 219
    9.2.2 Reimbursement issues 220
    9.2.3 Health technology and the role of HTA bodies 221
9.3 Medical devices and regulation 221
    9.3.1 Definitions, principles, and classes 221
        9.3.1.1 Definitions 222
        9.3.1.2 Essential principles 223
        9.3.1.3 Device classes 225
    9.3.2 Regulatory matters 228
        9.3.2.1 Conformity 228
        9.3.2.2 Standards 229
        9.3.2.3 Postmarket surveillance, vigilance, and adverse incident reporting 230
        9.3.2.4 Data issues 231
    9.3.3 Clinical studies 232
        9.3.3.1 Ethics 233
        9.3.3.2 Advances in clinical trials for devices 234

9.4   Routes for medical devices innovation...235
      9.4.1   New product development processes.......................................235
      9.4.2   Special considerations for pervasive medical device
              development...237
9.5   Conclusions...237
Acknowledgments..238
References 238

9.1   Introduction

Pervasive computing brings together information and communications technology (ICT) through a wide variety of computing devices. Although most medical devices are not computing devices per se, more are coming to rely on ICT whether by means of logging and telemetry functions (e.g., portable or home healthcare devices) or through their deployment in networked hospital environments. Furthermore, with the trend toward electronic patient records, it is likely that data from all manner of medical equipment will be transmitted through healthcare ICT systems. As a result of this convergence of technologies, pervasive healthcare design engineers need to have a good understanding of the regulatory environment in which medical device development and deployment is placed and an awareness of the evidence-based and cost-sensitive processes of healthcare technology assessment. This is on top of adherence to the increasingly sector-specific engineering methods and standards that guide electrical and computing equipment product designers in order to ensure safety-critical quality devices and systems.

This chapter aims to introduce the routes and requirements for realizing medical devices, enabling the pervasive healthcare engineer to enter the minds of the regulators and assessors and to better understand the industrial processes. In the first section, the concept of health technology assessment is introduced, which is the framework that judges the effectiveness and value of a device. The second section presents the regulations pertaining to medical devices in some detail including definitions and classifications, standards and quality systems, and the key aspects of conformity processes in Europe and North America. A brief overview of clinical studies includes ethics approval and some recent advances in clinical trials that are particularly suited to medical devices. Also covered in this section are deployment and data issues, the former to ensure continued monitoring of the device and the latter arising from the link between the predominantly independent worlds of devices and computer networks that are coming together in pervasive healthcare. The third section examines the product design process for medical devices that is being developed by the Multidisciplinary Assessment of Technology Centre for Healthcare (MATCH) research program. The concluding section reiterates the multidisciplinary environment of the pervasive healthcare innovator and points to sources of information and advice.

9.2   Health technology assessment

9.2.1   The evidence-based environment

Design engineers should already be familiar with the concepts of product life cycle, quality control standards and regulations, and the need to take a user-centered approach. For successful healthcare technology development, however, we must include two additional concepts: first, *evidence-based practice*, which began in clinical medicine but is now commonplace across healthcare including areas such as information management and facility design; and second, an understanding of *reimbursement* processes for healthcare technologies. For the purposes of this chapter these two concepts are tied together into the process known as health technology assessment (HTA).

Evidence-based practice is described by Booth as follows:[1]

> Evidence-based medicine (EBM) is "the conscientious, explicit, and judicious use of current best evidence in making decisions about the care of individual patients. The practice of evidence-based medicine means integrating individual clinical expertise with the best available external clinical evidence from systematic research" (Sackett et al., 1996). Evidence-based practice is, by implication, the systematic application of rigorous scientific methods to the evaluation of the effectiveness of health care interventions. This can be broadened to include such considerations as appropriateness, clinical decision-making, economic evaluation, health technology assessment, outcomes measurement and risk management.

The entrance of evidence-based practice in the context of healthcare technologies means not only that a new design should show efficacy in relation to risk for a specific therapeutic procedure by the gathering of appropriate clinical evidence (which may be mandatory, depending on the device class), but also that it must prove its clinical effectiveness with respect to existing alternatives in terms of both efficacy and cost, according to the healthcare allocation policy of the payer (e.g., equitable provision, or otherwise, of a treatment). Therefore evidence-based practice as related to healthcare technologies can be regarded as the antithesis of the technology-push mentality found in the marketing of consumer electronics.

The implication for design engineers is that knowledge of its future assessment should feed into the development process of a device or system at as early a stage as possible, so that the appropriate design choices are made. As Baxter shows, early investment in the front end of the design process is an important general principle for producing any successful new product, such that 85 to 90 percent of its overall costs will have already been committed by decisions made before detailed design or production engineering begins.[2] This figure may be even greater for medical devices because modifying a design at a late stage means retracing expensive and time-consuming regulatory processes.

9.2.2   Reimbursement issues

Most medical technology products are quite different from consumer products because very few are available over the counter, and patients do not often pay the full cost of a device-related medical procedure or diagnostic test themselves. Instead, a medical device manufacturer will seek reimbursement from the patient's healthcare provider, who may be a public institution like a national health service or a private body such as a private clinic or health insurer. Reimbursement mechanisms vary greatly between different countries (and even between regions within some countries) and are quite complex in terms of the different mixes of public and private systems and product types covered. Access to specific healthcare products, especially in private care regimes, may also depend on patient factors such as age, income, and employment status. From the technology manufacturer's point of view, this presents an unhelpfully fragmented and nonharmonized environment to market products in.

As highlighted in a comprehensive clinical report on reimbursement by Bromley et al., devices are being heavily implicated as major cost drivers for healthcare, and clinical evidence is increasingly being called for by the payers to justify these costs, whether it is for large capital items used for diagnostics or personal devices.[3] Papatheofanis outlines five economic variables that payers typically use for health economic analysis: cost of therapy, cost of side effects, costs avoided as a result of treatment, costs utilized based on information from diagnostics or referral of a patient, and costs utilized or saved during extended years of life as a result of therapy.[4] For the design engineer, this necessitates some understanding of both treatment and funding pathways that a device will be involved in and of the decision-making processes involved in having a device accepted for reimbursement. Detailed coverage of reimbursement methods is beyond the scope of this chapter, but it is worth mentioning the growth in popularity of a formal method known as Diagnostic Related Groups (DRG) that was introduced by the major U.S. health insurer Medicare and now has variants in many countries of the European Union (EU) and in Japan and Australia. DRGs classify hospital case types into groups expected to have similar hospital resource use, which code for a tariff that is calculated from a fixed base rate along with a weighting to adjust for hospital size and other factors. An important characteristic of the DRG method of reimbursement is that it is not based on the cost of a device alone, but rather on the entire cost of hospital resources made up by the clinical interventions and services required for diagnosis and treatment of a disease, plus its possible complications or comorbidities. Tariffs are intended to be updated on a regular basis (at least every few years) so that they represent the true cost of the procedure. DRGs introduce a high degree of transparency into the reimbursement process, but their existence makes it all the more important to predict how a device will map to all of its intended uses with their various reimbursement codes and to use this knowledge to inform early decision making in its development.

9.2.3   Health technology and the role of HTA bodies

HTA aims to provide information to support healthcare decisions and policymaking at local, national, and international levels. The University of York's Centre for Reviews and Dissemination (CRD) maintains an international database on behalf of the International Network of Agencies for Health Technology Assessment including a list of HTA bodies worldwide and records of ongoing projects.[5] One example of an HTA body, the U.K.'s National Health Service (NHS) R&D division HTA program, usefully defines health technology as follows:[6]

> Health Technology is an internationally recognised term that covers any method used to promote health, prevent and treat disease and improve rehabilitation or long-term care. "Technologies" in this context are not confined to new drugs or pieces of sophisticated equipment, but include procedures, settings of care and screening programmes.

Many HTA bodies are closely involved with reimbursement and the economics of healthcare delivery. They typically provide a speed of review that is faster when compared to the time required for rigorous surveys of clinical evidence, and in practice draw on existing surveys. The use of HTA in practice is typified by its fundamental role in the National Institute for Health and Clinical Excellence (NICE) of England and Wales that was established in 1999 to appraise appropriate use of both new and existing health technologies, which include pharmaceuticals, medical devices, diagnostic techniques, clinical procedures, and health promotion.[7] Each NICE appraisal invites evidence from stakeholders including professionals, patient groups, and manufacturers. Alongside this a technology assessment report on clinical and cost-effectiveness is commissioned from the national HTA program. NICE then uses the combined information from stakeholders and reports to inform its policymaking decisions. HTA gives government organizations like NICE considerable power over the introduction of new technologies and removal of existing ones if the technologies come to be regarded as clinically ineffective or too costly. Therefore the growing prevalence and influence of HTA is something that healthcare technology developers should recognize. The next section covers the equally important area of medical device regulation and standards.

9.3   Medical devices and regulation

9.3.1   Definitions, principles, and classes

Medical devices are highly regulated. This is essential to ensure patient and practitioner safety, good performance and quality of manufacture in the premarket stage, correct listing and advertising of the device that is being placed on the market, and fulfillment of postmarket obligations such as detecting and alerting users of any problems and monitoring of a device's clinical performance. The following sections introduce the essential premarket regulatory processes and some of the postmarket obligations with respect to the planning and realization of a device.

9.3.1.1 Definitions

The first step in the regulatory process toward marketing a medical technology product is knowing that it is a medical device. Definitions and nomenclatures for medical devices are not internationally agreed upon although they are being worked on by the Global Harmonization Task Force (GHTF), which was founded in 1992 by the EU, United States, Canada, Australia, and Japan.[8] These efforts have been assisted by the passing of EU-wide harmonization legislation, the U.S. Federal Drug Administration (FDA) Modernization Act, and substantial adoption of GHTF-recommended models by the Therapeutic Goods Administration (TGA) in Australia, the Therapeutic Products Directorate (TPD) in Canada, and by Japan's Ministry of Health, Labor, and Welfare (MHLW).

Naming of devices is becoming standardized via ISO 15225 (Nomenclature—Specification for a nomenclature system for medical devices for the purposes of regulatory data exchange), which is helping to facilitate international consensus on a global medical device nomenclature (GMDN) that is endorsed by the GHTF. The GHTF has proposed a harmonized definition for medical devices (document SG1/N029R11),[8] but it is instructive to see the current differences between the United States and the EU. Therefore, definitions for medical devices in the two regions will now be considered.

The FDA, via its Center for Devices and Radiological Health, describes a medical device as:[9]

> an instrument, apparatus, implement, machine, contrivance, implant, in vitro reagent, or other similar or related article, including a component part, or accessory which is: recognized in the official National Formulary, or the United States Pharmacopoeia, or any supplement to them, intended for use in the diagnosis of disease or other conditions, or in the cure, mitigation, treatment, or prevention of disease, in man or other animals, or intended to affect the structure or any function of the body of man or other animals, and which does not achieve any of its primary intended purposes through chemical action within or on the body of man or other animals and which is not dependent upon being metabolized for the achievement of any of its primary intended purposes.

Within the EU, medical devices are covered by three important pieces of legislation composed of the Medical Devices Directive (MDD), the Active Implantable Medical Devices Directive (AIMDD), and the In Vitro Diagnostics Directive (IVDD). The AIMDD applies to all active devices and related accessories intended to be permanently implanted in humans; the IVDD applies to all devices and kits used away from the patient to make a diagnosis of patient medical conditions; and the MDD covers all other devices that are not in the two special categories. Nondevice products on the other hand are covered by a directive known as the Community Code on Medicinal Products for Human Use.

The MDD (93/42/EEC) defines a medical device as:[10]

> any instrument, apparatus, appliance, material or other article, whether used alone or in combination, including the software necessary for its proper application, intended by the manufacturer to be used for human beings for the purposes of:
>
> - diagnosis, prevention, monitoring, treatment or alleviation of disease,
> - diagnosis, monitoring, treatment or alleviation of or compensation for an injury or handicap,
> - investigation, replacement or modification of the anatomy or of a physiological process,
> - control of conception,
>
> and which does not achieve its principal intended action in or on the human body by pharmacological, immunological or metabolic means, but which may be assisted in its function by such means.

Part of the rationale behind the form of these definitions is to distinguish devices from pharmaceutical products. From a patient's point of view, devices and drugs are complementary parts of managing their health, and there are some important differences from HTA and other perspectives as shown in Table 9.1.[7] Within MDD definitions an infusion pump that supplies a pharmaceutical is a device whereas a conventional pill capsule is not, and some drug-device combinations are also covered as devices. It is worth noting that some recent developments such as biologic coatings, tissue-engineered products, and other drug-device combinations further push at the borderline, so that clarifications to legislation have been and continue to be necessary.

9.3.1.2 Essential principles

In addition to definitions, the MDD provides Minimum Essential Requirements (Annex I) for the design and manufacture of medical devices to ensure the health and safety of patients, users, and third parties, which makes safety a central principle in the legislation.[11] The requirements include (although this list is not exhaustive):

1. A general requirement for inherently safe design and construction
2. Safety and compatibility of materials, including medicines if acting as an ancillary device
3. No adverse effect on characteristics or performance from normal use, transportation, or storage

Table 9.1 Comparison of Medical Devices and Drugs

	Devices	Drugs
Principal action	Other than pharmacological, immunological, or metabolic means Often mechanical, electrical, or materials engineering based	Pharmacological, immunological, or metabolic means Chemical based
Product life cycle	Constantly evolving, incremental modification Often short life cycle	Unchanging compound Long life cycle
Clinical evaluation	Lower requirement for clinical investigation (class dependent) Difficult to blind (to find a placebo), crossover common Difficult to stabilize for conventional RCT Multiple end users	More formalized evaluation and extensive clinical investigation Easy to blind with a placebo, crossover rare Uses conventional RCT Usually one end user
Use issues	Results vary with operator skill Wide range of over-the-counter availability Often require intensive training to use Complications decrease with use	Results unrelated to physician skill Majority prescribed Lower training requirement Complications increase with use
Diversity	A few large companies and many small companies (SMEs)	Mainly large multinational companies
Costs	Varying overheads with slow return Higher distribution cost Higher maintenance and disposal/recycling costs	High overheads, but quicker return Lower distribution cost Lower disposal/recycling costs

4. Elimination, minimization, protection against and informing of risk, ensuring residual risk is acceptable when weighed against benefit to the patient. Risks include those posed by:
   a. Substances leaking into or out of the device
   b. Infection and microbial contamination
   c. Tissues of animal origin
   d. Ionizing radiation
   e. A device being connected to, or equipped with, an energy source (electrical, mechanical, thermal risks)
   f. Combination of devices with, and connection to, other systems
   g. Environmental conditions including interference with other devices
   h. Aging of a device
5. Ergonomic design
6. Sufficient accuracy, if a measuring device
7. Adequate product marking and user instructions

The MDD essential requirements have for the most part been adopted by the GHTF Essential Principles of Safety and Performance of Medical Devices (SG1-N020R5)[8] and are therefore likely to be the basis for future international consensus.

9.3.1.3 Device classes

Once you know your product is a medical device, the next step is to know its class because this determines the route to conformity. Again, although not universally agreed upon by regulators, there is a degree of harmonization whereby medical devices are classified into three or four classes according to level of risk, use, and degree of invasiveness. The higher the risk, the higher the class number. Devices may also be classified during development. For example, the FDA classifies investigational devices in two classes: significant risk (SR) or nonsignificant risk (NSR).

Currently, however, there are alternative approaches to actually determining the device class for a new device. In the EU this is done by applying a set of eighteen rules (as specified in MDD Annex IX) with additional special rules for active implantable medical devices and in vitro diagnostic devices, whereas in the United States classification is carried out by government expert panels under Title 21 of the Code of Federal Regulations (CFR) whereby a device is generally assumed to be high risk unless it can be shown otherwise. The panel method can introduce uncertainty into the process, which can be especially worrying if the device is bordering on high risk because this greatly increases the cost of obtaining conformity. On the other hand, a rule-based method can be difficult to apply, although for the MDD there are some useful additional guideline documents.[12] Table 9.2 shows the types of devices in the various classes that result from the MDD rules. Nonactive and accessory equipment are classified along with the active devices that are of most interest to pervasive healthcare designers. Further discussion about classification in the EU and United States and global trends can be found in Davey et al.[13]

Table 9.2 Examples of Product Classifications Arising from MDD Rules (Collated from Guidance Document MEDDEV 2.4/1—rev. 8 Part 2: July 2001[12])

Device Class	Examples
I, low risk	Active diagnostic devices intended to illuminate or view the patient's body in the visible spectrum (e.g., examination lights, surgical microscopes, dental curing lights).
	Active diagnostic devices intended for thermography; devices for recording, processing, or viewing of diagnostic images.
	Devices intended in general for external patient support (e.g., hospital beds, patient hoists, walking aids, wheelchairs, stretchers, dental patient chairs).
	Incontinence pads, nonsterile dressings, plaster of paris, corrective glasses, noninvasive electrodes (for EEG or ECG), image-intensifying screens, cups and spoons, syringes without needles, dentistry mirrors, gloves, reusable scalpels, saw blades, tubing for transient use.
IIa, medium risk	Active therapeutic devices intended to administer or exchange energy, unless doing so in a potentially hazardous way: electrical, magnetic, electromagnetic (muscle stimulators and external bone growth stimulators, TENS devices and eye, electromagnets, electrical acupuncture); thermal (cryosurgery equipment, heat exchangers, except the types described below); mechanical (powered dermatomes, powered drills and dental handpieces); light (phototherapy for skin treatment and for neonatal care); sound (hearing aids; ultrasound: equipment for physiotherapy).
	Many active devices intended for diagnosis (e.g., imaging devices such as MRI, evoked response, diagnostic ultrasound, gamma cameras, PET, SPECT), ECG, EEG, electronic thermometers, stethoscopes, and blood pressure measuring equipment. Also x-ray films and photostimulable phosphor plates.
	Active devices intended to administer or remove medicines, body liquids, or other substances to or from the body, unless potentially hazardous (e.g., suction equipment, feeding pumps, jet injectors for vaccination, nondose-critical nebulizers).
	Channels for active drug delivery and blood, syringes for infusion pumps, devices for temporary storage of transplant organs or long-term storage of bodily materials, filtering machines and centrifuges (mechanical separation). Most tubing for short-term use (e.g., urinary catheters, tracheal tubes, stents). Wound dressings managing the microenvironment (e.g., polymer film). Dental prostheses.
	Many surgically invasive devices intended for transient use (e.g., needles used for suturing, needles of syringes, single-use scalpels/blades, drill bits connected to active devices, trial implants).
	Devices intended specifically to be used for disinfecting medical devices, except contact lens solution.

IIb, elevated risk	Active therapeutic devices intended to administer or exchange energy, doing so in a potentially hazardous way: kinetic energy (lung ventilators); thermal energy (incubators, warming blankets, blood warmers, heat exchangers); electrical energy (high frequency electrosurgical generators, electrocautery equipment and electrodes, external pacemakers, external defibrillators, electroconvulsive therapy equipment); coherent light (surgical lasers); ultrasound (lithotriptors, surgical ultrasound devices); ionizing radiation (radioactive sources for afterloading therapy, therapeutic cyclotrons, linear accelerators, therapeutic x-ray sources). Also active devices intended to control and monitor the performance of active therapeutical devices in Class IIb (e.g., external feedback systems). Also, diagnostic x-ray sources. Intensive care monitoring and alarm devices (for, e.g., blood pressure, temperature, oxygen saturation), biological sensors, blood gas analyzers used in open heart surgery, cardioscopes and apnea monitors, including apnea monitors in home care. Active devices intended to administer or remove medicines, body liquids, or other substances to or from the body, unless potentially hazardous: infusion pumps, ventilators, anesthesia machines, anesthetic vaporizers, dialysis equipment, blood pumps for heart-lung machines, hyperbaric chambers, pressure regulators for medical gases, medical gas mixers, moisture exchangers in breathing circuits if used on unconscious or nonspontaneously breathing patients, dose-critical nebulae. Catheters incorporating sealed radioisotopes, excluding the central circulatory system (CCS), tubing for long-term use (e.g., urethral stents, wound dressings with secondary healing properties). Many implantable devices (e.g., prosthetic joint replacements, nonimplantable devices used for contraception or the prevention of the transmission of sexually transmitted diseases such as condoms and diaphragms). Dose-critical medicine applicators (e.g., insulin pens, surgically invasive adhesives, contact lens solutions).
III, high risk	Cardiovascular catheters (e.g., angioplasty balloon catheters including guidewires, dedicated disposable cardiovascular surgical instruments). Catheters incorporating sealed radioisotopes for the CCS. Implantable devices to be used in direct contact with the heart, the CCS, or the central nervous system (CNS) (e.g., heart valves, spinal stents, CNS electrodes). Implantable contraceptives (e.g., intrauterine devices). Bioactive implantable devices (e.g., absorbable sutures and biological adhesives). Rechargeable nonactive drug delivery systems. All devices incorporating a medicinal product as an integral part with an action ancillary to that of the devices: antibiotic bone cements, condoms with spermicide, heparin-coated catheters, endodontic materials with antibiotics, some ophthalmic irrigation solutions, dressings incorporating an antimicrobial agent to provide ancillary action on wounds.

9.3.2   Regulatory matters

9.3.2.1   Conformity

Conformity is the successful outcome of the compulsory regulatory processes by which a medical device will be accepted for marketing in a particular region. In the EU, conformity is shown by obtaining *CE marking*, which means the device is certified to have met regulatory requirements. This includes adherence to the essential principles of the directives as assessed by an accredited third party (notified body) in the case of Class II/III devices or via self-assessment in the case of a Class I low-risk device (provided the company is registered as meeting a quality systems management standard). In the United Kingdom, notified bodies are audited by the Medicines and Healthcare products Regulatory Agency (MHRA) of the Department of Health. In the United States, conformity is shown by obtaining FDA approval from the government by going through one of two processes. Premarket approval (PMA) involves demonstration of reasonable safety and effectiveness, which is required for a high-risk or new kind of device. A less stringent premarket notification 510(K) process is allowed for devices that are substantially equivalent (SE) to a device already on the market, known as a predicate device. According to the FDA:[9]

> a device is deemed to be SE if, in comparison to a predicate device it: has the same intended use as the predicate device; and has the same technological characteristics as the predicate device; or has different technological characteristics, that do not raise new questions of safety and effectiveness, and the sponsor demonstrates that the device is as safe and effective as the legally marketed device. A claim of substantial equivalence does not mean the new and predicate devices must be identical. Substantial equivalence is established with respect to intended use, design, energy used or delivered, materials, performance, safety, effectiveness, labeling, biocompatibility, standards, and other applicable characteristics.

Some 1,700 devices are already classified by the FDA so it is essential for the innovator of a device to check for substantial equivalence in determining the likely route to compliance for marketing it in the United States. An illustrative example of SE devices are 3-D navigation systems for computer-aided orthopedic surgery (CAOS) that are predicated to stereotaxic equipment originally developed for neurosurgery. Because of this, several manufacturers of CAOS navigation systems have obtained approval via the 510(K) route. In contrast, companies found it difficult to obtain FDA approval for robotic orthopedic surgery systems due in part to the lack of a suitable predicate device, which therefore necessitated the PMA route.

9.3.2.2 Standards

Adherence to product standards is a vital component of the regulatory process, and although premarket standards vary in format in different regions, they universally apply to three main areas, according to the World Health Organization:[14]

1. Device attributes: Product safety and performance
2. Manufacturing: Quality system
3. Labeling: Description and instructions for use

There are a large number of international and regional standards relating to medical devices.[15] (For an overview, see ISO TR16142 [Medical devices—Guidance on the selection of standards in support of recognized essential principles of safety and performance of medical devices].) There is also a fairly recent comprehensive reference by Fries.[16] Device attributes standards are very much related to the essence of the essential requirements as previously described for the MDD. Data standards, because they are very important to pervasive systems, will be considered later on in this section. Standards for clinical studies, although linked to safety and performance, will also be mentioned in a separate section. Manufacturing standards are for the most part covered by quality assurance (QA) systems that are required for the production and testing of all device classes. In the EU this is now covered by the medical device standard ISO 13845. In the United States, a company must follow good manufacturing practice/quality systems (GMP/QS) standards. These standards require design controls and documentation for the manufacturing process. Labeling standards contribute to safe use and tracking of products.

Rather than go into specific details, it is perhaps more useful in a changing regulatory environment to examine trends in the evolution of standards in recent years, as follows (with relevant examples):

- Regulations are subsuming many voluntary standards, which is deemed useful because it is easier to change a standard than a regulation and this also aids governments in creating regulations. An exception is the area of environment protection, which is becoming more regulated.
- Standards are becoming more sector specific. For example, the general ISO 9001 quality standard is now for the most part incorporated into ISO 13485 (Medical devices—Quality management systems—Requirements for regulatory purposes). ISO 9003 (Software engineering—Guidelines for the application of ISO 9001:2000 to computer software) and IEC 61508 (Functional safety of electrical/electronic/programmable electronic safety-related systems) are likely to be supplemented by IEC 62304 (Medical device software—software life-cycle processes).
- Standards are becoming more internationally recognized. For example, updated U.S. and Japanese GMP standards are being based on

the new ISO 13485:2003. This is intended to assist in facilitating global trade and access to new technologies. The GHTF has also proposed a documentation system to harmonize collation of all the material needed for premarket approval applications, known as a summary technical document (STED), which is currently being piloted and evaluated by its members.

- Standards are becoming less prescriptive in terms of tools and methods but instead are taking on essential principles (e.g., for performance) and a risk management philosophy that requires the manufacturer to actively take the responsibility for implementing and continually reviewing its processes to analyze, identify, and control risk. As an example, the forthcoming third edition of IEC 60101 (Medical electrical equipment) will require adherence to ISO 14971 (Medical devices—application of risk management in medical devices).

- Usability engineering is becoming less of a guideline and more of a requirement. This is related to risk management and arose from the need to understand and mitigate potential misuse. In particular, usability will be fully incorporated into the third edition of IEC 60101 and the scope of use will also change from "under medical supervision" to "all use." This is alongside recognition of the need to have tight controls on home use.

- Sustainability and environmental protection are becoming more regulated, building on voluntary environmental management standards such as ISO 14001. Example include the EU's Waste from Electrical and Electronic Equipment (WEEE) and Restriction of Certain Hazardous Substances (RoSH) directives and integrated product policy (IPP), where greater responsibility is placed on manufacturers regarding the disposal of devices, necessitating "cradle-to-grave" engineering. The Eco-design Requirements for Energy-using Products (EuP) directive is intended to make electrical devices more environmentally friendly. Japan has also passed a set of environmental laws including the Home Appliances Recycling Law (HARL), Law for the Effective Utilization of Resources (LPEUR), and Green Purchasing Law (GPL).

- Recognition is rising regarding the need to ensure safety of refurbished and denoted equipment as well as the reprocessing of devices labeled for single use. This includes equipment intended for reuse in developed countries, and, for trade and donation of equipment to developing countries (some have previously had bad experiences with used equipment).[14]

9.3.2.3   *Postmarket surveillance, vigilance, and adverse incident reporting*

Once a device is on the market and is being used by practitioners and patients, it is very important that its safety and performance continue to be assessed. Devices can be used and misused in many different and unforeseen ways (perhaps as a result of inadequate labeling and user instructions) and, in spite of premarket efforts, can fail in actual use or otherwise incur complaints from customers. Changes in clinical practice may also impact actual use of a device, while advances in state-of-the-art technology can demand changes to a design. This is quite apart from the processes of HTA, which are used to assess and reassess clinical effectiveness and cost-effectiveness for policymaking purposes.

The two main methods by which postmarket assessment is implemented are postmarket surveillance (PMS) studies by the manufacturer and by gathering and disseminating information through adverse incident reporting networks. In Europe a related term, vigilance, is used to denote the responsibility of the manufacturer to track and report on problems arising with a device.

Postmarket activities are for the most part covered by the requirements of standards and regulations. For example, under the U.S. Code of Federal Regulations (CFR), the FDA may impose postmarket evaluation and periodic reporting on a manufacturer as part of the PMA approval of a device (21 CFR Part 814) and can also order subsequent postmarket surveillance for any Class II and III device where there is a serious risk to health from device failure, as well as for long-term implants or life-sustaining devices (21 CFR Part 822). The EU MDD requires the manufacturer to have a systematic procedure to deal with devices' postmarket experiences. Additionally the relevant body (CA) must be informed of any deficiencies that resulted (or could have resulted) in death or serious deterioration of health or of the technical or medical reason for the recall of a device. There are also methods recommended for conducting PMS such as active supervision, customer surveys, inquiries of users and patients, literature reviews, and postmarket clinical follow-up (PMCF). PMCF is recommended for capturing infrequent complications and long-term performance issues that would not be detected in the premarket phase, as outlined in guidance MEDDEV 2.12/2.[12]

9.3.2.4   Data issues

A key feature of pervasive computing is the interaction of devices with ICT systems. This convergence of technologies is becoming manifest by widespread use of commercial off-the-shelf (COTS) hardware and software, such as the personal computer, and the use of standards-based communications such as the wired and wireless ethernet as outlined by Cohen.[17] As Cohen points out, COTS and data communication technologies can together facilitate automatic data collection, analysis, reporting, dynamic reconfiguration, and remote software upgrading for medical devices. The move to electronic patient records is likely to go hand in hand with increased demand for devices with a computer interface.

There are a number of standards and initiatives in the area of data communications that are specific to medical technologies. One such industry standard is digital imaging and communication in medicine (DICOM), which originates from the need to transfer medical images from computerized tomography (CT) and magnetic resonance imaging (MRI) scanners. The other well-established industry standard is High Level 7 (HL7), designed for hospital administrative data exchange. IEEE 1073/ISO 11073 (Health informatics—Point-of-care medical device communication) is an international standard that is currently at the draft stage. A recent initiative emanating from the United States is integrating the hospital environment (IHE), which aims to bring together usage of DICOM, HL7, and Internet protocols in healthcare systems within a set of technical frameworks for various healthcare domains (e.g., IT infrastructure, cardiology, laboratory, radiology).[18] Furthermore, the U.S. National Academies recently published an in-depth report highlighting the need for partnership between engineering and medicine in order to apply the best systems engineering approach to healthcare delivery that is "safe, effective, timely, patient-centered, efficient and equitable."[19]

Data security is an issue that is receiving particular attention with the introduction of EPRs. In the United States, the Health Insurance Portability and Accountability Act (HIPAA) requires, as of 21 April 2005, that healthcare providers adopt a security procedure to ensure the integrity, availability, and confidentiality of information maintained and transmitted by medical devices.[20] From a service provision perspective, Grimes outlines the need to shift from management of discrete devices toward an overall policy of safeguarding information and lists a range of device and system requirements that might be used to mitigate security threats.[21] These threats include hardware failures or errors, user misuse or abuse of a device, erroneous data entry, malicious assault, unauthorized data access or modification, and environmental effects such as electromagnetic interference (EMI) or interruption of utilities such as power. These and other issues related to data handling and protection must therefore feed into the design process for pervasive healthcare devices and systems.

9.3.3 Clinical studies

A major aspect within the premarket processes for approving medical devices is the gathering and use of data concerning a device's clinical efficacy and safety. For certain classes of devices, clinical approval resulting from the outcomes of testing on human subjects is mandatory for its market certification. This might involve actual clinical studies or the compilation and critical analysis of existing evidence from the literature and other sources concerning similar devices.[22] Even for low-risk devices, clinical data may be important for proving the need and benefits of a product to users and payers.

Premarket clinical studies typically take place after detailed design of the device is completed. Data that are intended to be used for predicting clinical effectiveness on living humans, especially for invasive devices, are also commonly obtained in the development stage from animal experiments, human cadavers, and tissue. Preclinical information may also be gained from systematic consultation of individuals or groups of healthcare professionals and patients, as well as from modeling exercises.

If a clinical trial is necessary due to insufficient existing evidence, the manufacturer is required to follow procedures in applying for permission to do so and for the running of the trial. The processes for doing this are regulated by the same bodies as in Section 9.3.2.1, and various standards are applicable. Devices at this stage are legislated for their clinical study in the United States through the investigative device exemption (IDE) regulation (21 CFR 812) and in the EU through clinical investigation requirements in the MDD, with guidance issued via MEDDEV 2.7 (Clinical investigation, clinical evaluation: Guide for manufacturers and notified bodies).[12] International standard ISO 14155 (Clinical investigation of medical devices for human subjects) covers recognized practices for conducting trials. In the United States and elsewhere, clinical and laboratory practices are governed by requirements known as good clinical practices (GCP) and good laboratory practices (GLP).

Although the details of clinical trial methods and approval processes are outside the scope of this chapter, the remainder of this section will concentrate on ethical approval aspects because this is also relevant to preclinical studies that device designers may also be involved in. Some developments in postmarket clinical trial design that are intended to be more suitable for medical device evaluation from an HTA perspective will also be introduced.

9.3.3.1   Ethics

Before conducting human clinical studies, and indeed even before systematic preclinical consultations with professional and patients (or healthy volunteers), it is a normal requirement to have obtained ethical approval from the institutions involved. This typically involves interaction with healthcare service providers. By way of an example, within the United Kingdom's National Health Service there are local research ethics committees (LRECs) that make decisions about single-center studies and multisite research ethics committees (MRECs) for streamlining studies that span multiple sites.[23] A developer must apply to obtain the necessary approval from these committees by supplying the detailed plan (protocol) of the experiment to be carried out and, preferably, advance permission from all of the participating centers. LRECs are guided to consider the following general criteria and any other criteria specific to the experiment:[24]

1. Has the scientific merit of the proposal been properly assessed?
2. How will the health of the research subjects be affected?
3. Are there possible hazards and, if so, adequate facilities to deal with them?
4. What degree of discomfort or distress is foreseen?
5. Is the investigation adequately supervised and is the supervisor responsible for the project adequately qualified and experienced?
6. What monetary or other inducements are being offered to the NHS body, doctors, researchers, subjects, or anyone else involved?
7. Are there proper procedures for obtaining consent from the subjects or, where necessary, their parents or guardians?
8. Has an appropriate information sheet for the subjects been prepared?

An illustration of ethics committee workings for FDA-supported research can be found via the U.S. Office of Human Research Protections.[25] This explains the role of institutional review boards (IRBs), the equivalent of RECs in the United States.

Research ethics have a considerable impact on the methods by which evidence on the efficacy of a medical device can be obtained. Protection of human subjects and patients is paramount, and stringent legislation covering all human subjects is accompanied by special rules to protect specific groups, for example, children, prisoners, and cognitively disabled persons. Furthermore, in recent years special research ethics rules and guidelines have been devised for experiments involving genetics, human and animal tissues, and organs. These guidelines may become more relevant to device designers as the number of drug-device combination products proliferates. Animal experimentation is receiving renewed ethical consideration in many countries as a result of greater attention to animal welfare issues, the question of efficacy of animal models and cost, and especially due to a more concerted adoption of the principles of the 3Rs—replacement, reduction, and refinement of animal use. In Australia, this has become most sophisticated by means of the animal ethics committees (AECs) that have been introduced into animal welfare legislation through a revised Australian code of practice for the care and use of animals for scientific purposes.[26] The AECs act as decision-making bodies in a similar manner to RECs. In the U.K., a government-funded National Centre for the 3Rs has recently been established whose stated aim is the ultimate replacement of animal use.[27] Further information about research ethics for both humans and animals can be obtained via the Online Ethics Center Web site.[28]

In addition to RECs there are a number of other types of ethics committees. In the U.K., for example, clinical ethics committees (CECs) provide advice and support on ethical issues arising from clinical practice and patient care within healthcare organizations.[29] Although these committees are not decision-making bodies like RECs, they may influence aspects of healthcare such as hospital procedures and equity of access, which may be quite relevant to a technology developer. Finally, ethics in business are experiencing a growth in concern within the medical device sector, which is relevant to the collaboration of companies with healthcare professionals (HCPs). For example, the U.S. medical devices trade association AdvaMed has recently adopted a voluntary code of ethics to facilitate ethical interactions with HCPs who may "purchase, lease, recommend, use, arrange for the purchase or lease of, or prescribe Members' medical technology products in the U.S."[30]

9.3.3.2 *Advances in clinical trials for devices*

The gold standard for clinical trials is the randomized controlled trial (RCT) that was developed for evaluating drug treatments but is now used for the full range of clinical procedures. However, there are some problems associated with RCTs for medical devices because of the differences between devices and drugs as shown previously in Table 9.1. In a useful illustration of this, Lilford et al. have outlined some of the issues involved in surgical trials (which often involve devices) to show how application of the RCT can be problematic.[31] These issues include ethics of intervention versus nonintervention, the problem of blinding of participants (surgeons, patients, and hospital staff), surgeon- and technique-related variables, timing of trials, patient preferences, entry criteria, appropriate outcome measures, and a number of statistical considerations. As a result, less powerful clinical studies, such as cohort studies and case studies, are quite prevalent in the literature for medical devices. However, the value of studies lower down in the hierarchy of evidence can be difficult to assess from the HTA perspective. Lilford et al. argue strongly for randomization and in a separate paper propose tracker trials for comparing fast-changing technologies like devices.[32] Tracker trials include the contemporary examples of treatments employed by clinicians but allow for the addition of new devices and treatments into the study as the trial progresses. This is to help ensure early adoption if a new procedure is found to be superior, as well as for removal of a procedure if it is shown to perform poorly or if it has been superseded by technological developments. These are still randomized trials, but they do not have preset and rigid protocols and are ongoing studies rather than the one-off event that normally characterizes an RCT. New developments of this kind may become influential in the future of medical device assessments.

9.4 Routes for medical devices innovation

Having covered some of the nuts and bolts of medical device development and their accompanying regulations and standards in the previous two sections, we are now in a position to look at the overall innovation process.

9.4.1 New product development processes

New product development (NPD) methods typically break the innovation process down into a series of stages with associated decision making called stage-gate systems.[33] Such processes help to reduce development time and improve quality through standardization of practice and subsequent capture of best practice. Rochford and Rudelius have studied medical devices in this context, comparing "new-to-the-world" products and product modifications using a twelve-stage process.[34]

Members of the author's Multidisciplinary Assessment of Technology Centre for Healthcare (MATCH) have recently completed a consultation with our industrial partners that resulted in a simplified four-stage generic process for medical devices, giving a snapshot of the current approach to product development in the industry.[35,36] This process is summarized in Figure 9.1. Much of the progression of stages will be familiar to project managers and product developers. However, there are some points of special note. First, there will be more stages in practice. The four-stage model is intended to be a high-level representation of the decision-making steps

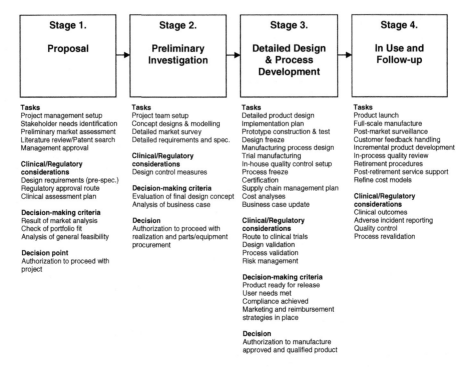

Stage 1.	Stage 2.	Stage 3.	Stage 4.
Proposal	**Preliminary Investigation**	**Detailed Design & Process Development**	**In Use and Follow-up**

Tasks
Project management setup
Stakeholder needs identification
Preliminary market assessment
Literature review/Patent search
Management approval

Clinical/Regulatory considerations
Design requirements (pre-spec.)
Regulatory approval route
Clinical assessment plan

Decision-making criteria
Result of market analysis
Check of portfolio fit
Analysis of general feasibility

Decision point
Authorization to proceed with project

Tasks
Project team setup
Concept designs & modelling
Detailed market survey
Detailed requirements and spec.

Clinical/Regulatory considerations
Design control measures

Decision-making criteria
Evaluation of final design concept
Analysis of business case

Decision
Authorization to proceed with realization and parts/equipment procurement

Tasks
Detailed product design
Implementation plan
Prototype construction & test
Design freeze
Manufacturing process design
Trial manufacturing
In-house quality control setup
Process freeze
Certification
Supply chain management plan
Cost analyses
Business case update

Clinical/Regulatory considerations
Route to clinical trials
Design validation
Process validation
Risk management

Decision-making criteria
Product ready for release
User needs met
Compliance achieved
Marketing and reimbursement strategies in place

Decision
Authorization to manufacture approved and qualified product

Tasks
Product launch
Full-scale manufacture
Post-market surveillance
Customer feedback handling
Incremental product development
In-process quality review
Retirement procedures
Post-retirement service support
Refine cost models

Clinical/Regulatory considerations
Clinical outcomes
Adverse incident reporting
Quality control
Process revalidation

Figure 9.1 A four-stage innovation process for medical device development. (Adapted from Dixon et al.[35] and Eatock et al.[36] with permission.)

involved. Second, as a result of the regulatory processes that must be undertaken, the first three stages involve a great deal of planning before detailed product design can be attempted, after which strict design controls must be in place to track and document all modifications. This is quite typical of safety-critical systems. These controls and the validation processes that close the iterative loops during Stage 3 have received attention by others in terms of guidance to the medical device industry (e.g., the Cambridge Engineering Design Centre's design for validation approach).[37]

A notable omission in documented current practice is specific planning for HTA and reimbursement as mentioned earlier in the chapter. This is being addressed in MATCH by the promotion of health economics modeling,[38] which would benefit from its earliest adoption in Stage 1 with subsequent refinement at other stages, such as during clinical trials.[39] The other main omissions are the environmental and sustainability strategies that will have to be put in place as newer regulations are introduced. These will require the industry to go beyond dealing with retirement of obsolete devices and toward advance planning for disposal at the point when detailed requirements are drawn up and design decisions involving choices of materials and energy use are made. Sourcing, procurement, reporting, servicing, and repair are all areas that will be affected by such legislation.[40]

A further development area is the approach to stakeholder requirements, especially in the area of user needs capture to ensure devices are fit-for-purpose. This is another area that is moving from guidance toward regulation and will benefit from review throughout the life cycle, suggesting the introduction of more formal human-factors methods into the concept design phase of Stage 2 and specific planning for usability testing in Stage 3.

9.4.2   Special considerations for pervasive medical device development

For devices involving data, which are inherent to pervasive computing in healthcare, the communication and security issues mentioned earlier will require specific planning because they will impact on design choices, user testing, and postmarket surveillance. Because regulations and standards are in continual development it will be most beneficial for pervasive healthcare designers to engage with the bodies concerned with these. Also, interaction with patient groups should help provide insights into public concerns about protection of personal records and other societal issues.

Because many pervasive healthcare devices are likely to be used in the home, this raises obvious issues in the area of maintenance and disposal or recycling as mentioned above. Comparison with the experience of deploying equipment for disabled users by healthcare providers would be useful in this respect. As well as those supplied by professionals, many pervasive healthcare devices are likely to be purchased over the counter, and this will raise further issues such as ensuring adequate training.

Finally, what will be the best combined methods to use for user needs capture of pervasive healthcare devices in conjunction with the ICT systems they interact with is still an open question that would benefit from further research.[41] Ethnomethodologies, scenario-based design, and usability engineering are some methods that are already being used in computer science fields such as human–computer action and computer-supported cooperative work and most recently in ubiquitous and pervasive computing.

9.5   Conclusions

The environment of the pervasive healthcare innovator is a multidisciplinary one, bringing together the roles of the clinical designer/engineer and ICT specialist. It is hoped that this chapter has provided insight into the processes of healthcare technology development. On the device side especially there is a need for understanding the regulatory processes within the product life cycle, plus other aspects that might normally be outside a designer's remit such as reimbursement and HTA processes. Bringing this knowledge into the design process should help to improve decision making in medical device innovation.

In a large medical devices organization, there will most likely be expert individuals and departments in some of the specific areas outlined in this chapter, such as clinical trials and regulations, so it will not be necessary for designers to have detailed knowledge of these. Of course, this is the benefit of a multidisciplinary team. For a small company or university research group, however, it is more likely that individuals will need to become multidisciplinary to some extent.

Medical device regulations and standards are presently going through great changes and the Web has become an important medium for accessing the latest information and advice, highlighted by the large number of online references in this chapter. Official sources of information on regulations and extensive guidance include the Medical Devices Directives[10] and MEDDEV[12] in the European Union, Device Advice[9] in the United States, and via the Web sites of the other members of the Global Harmonization Task Force.[8] Industry-supporting Web sites and magazines such as *Medical Devicelink*,[42] *Medical Device Technology*,[43] and publications of professional bodies such as the IEEE Engineering in Medicine and Biology Society[44] are further useful sources of advice and information.

Pervasive computing is an exciting field that should address and solve many of the issues involved in the development of new healthcare technologies and bringing them to patients.

Acknowledgments

The author acknowledges support of this work through the MATCH Programme (EPSRC Grant GR/S29874/01), although the views expressed are entirely his. Many thanks are due to MATCH researchers Shirley Davey, Julie Eatock, and Sukhvinder Johal for their valuable comments on the initial draft of the manuscript and to the editors and reviewers for their subsequent remarks.

References

1. Booth, A. *The ScHARR Guide to Evidence-based Practice*. ScHARR (School of Health and Related Research) Occasional Paper No. 97/2, University of Sheffield, 1997. See also http://www.nettingtheevidence.org.uk/ (accessed 21 March 2005).
2. Baxter, M. *Product Design: A Practical Guide to Systematic Methods of New Product Development*. London: Chapman and Hall, 1995.
3. Bromley, A., Bensenane, J., Riley, K., Sullivan, K., Mazzei, L., Guglieri, A., Arrizabalaga-Moreno, F., Alfonsel, M., and Rozynski, E. *Gaining Reimbursement for Medical Devices and Diagnostics*. Clinical Report CBS924, Strategic Management Reports Series, PJB Publications Ltd., March 2004.
4. Papatheofanis, F.J. Estimating cost in health economic analyses. *IEEE Engineering in Medicine and Biology Magazine*, 22(5), 14–15, 2003.
5. Centre for Reviews and Dissemination (CRD) Web site. University of York, U.K., http://www.york.ac.uk/inst/crd/ (accessed 15 March 2005).
6. NHS HTA Programme Web site. National Coordinating Centre for Health Technology Assessment (NCCHTA), U.K., http://www.ncchta.org/ (accessed 10 March 2005).

7. Taylor, R. National Institute for Clinical Excellence (NICE). HTA rhyme and reason? *International Journal of Technology Assessment in Health Care*, 18(2), 166–70, 2002.
8. Global Harmonization Task Force (GHTF) Web site. http://www.ghtf.org/ (accessed 14 March 2005).
9. Device Advice Web site. Center for Devices and Radiological Health, Food and Drug Administration (FDA), U.S., http://www.fda.gov/cdrh/devadvice/ (accessed 15 March 2005).
10. Europa Enterprise—Medical Devices Web site. http://europa.eu.int/comm/enterprise/medical_devices/ (accessed 12 March 2005).
11. Higson, G.R. *Medical Device Safety: The Regulation of Medical Devices for Public Health and Safety*. Bristol: IOP Publishing, 2002.
12. MEDDEV. *Guidelines Relating to Medical Devices Directives*. Europa Enterprise—Medical Devices, http://europa.eu.int/comm/enterprise/medical_devices/meddev/ (accessed 12 March 2005).
13. Davey, S., Anderson, J., and Meenan, B. *An Overview of Current Classification Systems for Healthcare Devices and Their Limitations*. Multidisciplinary Assessment of Technology Centre for Healthcare (MATCH), May 2005. Available from http://www.match.ac.uk/.
14. World Health Organization (WHO). *Medical Device Regulations: Global Overview and Guiding Principles*. Geneva: World Health Organization, 2003.
15. International Standards Organisation (ISO) Web site. http://www.iso.org/ (accessed 15 March 2005).
16. Fries, R.C. *Reliable Design of Medical Devices*. New York: Marcel Dekker, 1997.
17. Cohen, T. Medical and information technologies converge: the impact on clinical engineering. *IEEE Engineering in Medicine and Biology Magazine*, 23(3), 59–65, 2004.
18. Integrating the Healthcare Enterprise (IHE) Web site. http://www.ihe.net/ (accessed 23 March 2005).
19. Reid, P.P., Compton, W.D., Grosman, J.H., and Fanjiang G. (Eds.). *Building a Better Delivery System: A New Engineering/Health Care Partnership*, National Academic of Engineering and Institute of Medicine, The National Academies Press, http://www.nap.edu/books/030909643X/html (accessed 10 December 2005).
20. Office for Civil Rights, U.S. Department of Health and Human Services Web site. http://www.hhs.gov/ocr/hipaa/ (accessed 23 March 2005).
21. Grimes, S.L. Security: a new clinical engineering paradigm. *IEEE Engineering in Medicine and Biology Magazine*, 23(4), 80–82, 2004.
22. Boutrand, J.P. When to conduct a clinical trial: medical device technology (MDT). Octomedia Ltd., October 2004: 28–31. Available from http://www.medicaldevicesonline.com/.
23. Central Office for Research Ethics Committees (COREC) Web site. National Patient Safety Agency, National Health Service, U.K., http://www.corec.org.uk/ (accessed 29 March 2005).
24. *Local Research Ethics Committees*, HSG(91)5, 1991. Department of Health, Heywood, Lancs., U.K.. http://www.dh.gov.uk/PublicationsAndStatistics/Publications/PublicationsLibrary/fs/en/ (accessed 21 March 2005).
25. *Code of Federal Regulations, Protection of Human Subjects*, 45 CFR Part 46, 2001. Office of Human Research Protections, U.S. Department of Health and Human Services Web site, http://www.hhs.gov/ohrp/humansubjects/guidance/45cfr46.htm (accessed 14 March 2005).

26. Animal Ethics Web page. National Health and Medical Research Council, Ethical Issues, Australia, http://www.nhmrc.gov.au/issues/animal ethics.htm (accessed 14 March 2005).
27. National Centre for the Replacement, Refinement, and Reduction of Animals in Research (NC3R) Web site, U.K. http://www.nc3rs.org.uk/ (accessed 14 March 2005).
28. Online Ethics Center (OEC), Reference Materials on Research Ethics, http:// onlineethics.org/reseth/reference.html (accessed 14 March 2005).
29. Clinical Ethics Network (CEN) Web site. http://www.ethics-network.org.uk/ (accessed 21 March 2005).
30. *Code of Ethics for Interactions with Health Care Professionals*, Advanced Medical Technology Association (AdvaMed). http://www.advamed.org/ publicdocs/coe.html (accessed 14 March 2005).
31. Lilford, R., Braunholtz, D., Harris, J., and Gill, T. Trials in surgery. *British Journal of Surgery*, 91, 6–16, 2004.
32. Lilford, R.J., Braunholtz, D.A., Greenhalgh, R., and Edwards, S.J.L. Trials and fast changing technologies: the case for tracker studies. *British Medical Journal*, 320, 43–46, 2000.
33. Cooper, R.G. Stage-gate systems: a new tool for managing new products. *Business Horizons*, 33(3), 44–54, 1990.
34. Rochford, L. and Rudelius, W. Stages and success in the medical products industry. *Industrial Marketing Management*, 26, 67–84, 1997.
35. Dixon, D., Eatock, J., Meenan, B., Young, T., and Anderson, J. *MATCH Guide on Product Development Processes in the Medical Device Industry*. Multidisciplinary Assessment of Technology Centre for Healthcare (MATCH), May 2005. Available from http://www.match.ac.uk/.
36. Eatock, J., Dixon, D., Meenan, B., Young, T., and Anderson, J. *Report on Medical Device Product Development Processes: Some Illustrative Examples*. Multidisciplinary Assessment of Technology Centre for Healthcare (MATCH), April 2005. Available from http://www.match.ac.uk/.
37. Healthcare Web page. Cambridge Engineering Design Centre, University of Cambridge, http://www-edc.eng.cam.ac.uk/healthcare/ (accessed 22 March 2005).
38. Health Economics Research Group (HERG) Web site. Brunel University, U.K., http://www.brunel.ac.uk/about/acad/herg/ (accessed 22 March 2005).
39. Johnston, K., Buxton, M.J., Jones, D.R., and Fitzpatrick, R. Assessing the costs of healthcare technologies in clinical trials. *Health Technology Assessment*, 3(6), 1–76, 1999.
40. Wong, C. On the horizon: new rules for medical devices: preparing for new environmental regulations and product compliance. *Agile SME Solutions, News*, August 23, 2004, http://sme.agile.com/news/2004/biosci-082304.pdf (accessed 22 March 2005).
41. Craven, M.P. and Martin, J.L. How does the healthcare industry involve users in medical device development?—Pointers for UbiHealth. Paper presented at UbiHealth 2004: The 3rd International Workshop on Ubiquitous Computing for Pervasive Healthcare Applications, at UbiComp 2004, the Sixth International Conference on Ubiquitous Computing, Nottingham, England, 7–10 September 2004. http://www.pervasivehealthcare.com/ubicomp2004/ papers/ (accessed 22 March 2005).

42. Medical Devicelink. http://www.devicelink.com/ (accessed 31 March 2005).
43. Medical Device Technology (MDT). http://www.medicaldevicesonline.com/ (accessed 15 March 2005).
44. Institute of Electrical and Electronic Engineers (IEEE), Engineering in Medicine and Biology Society, *Engineering in Medicine and Biology Magazine.* http://www.ieee.org/organizations/pubs/magazines/emb.htm (accessed 31 March 2005).

chapter ten

User evaluation in pervasive healthcare

Tim Adlam, Roger Orpwood, and Teresa Dunn
Bath Institute of Medical Engineering, United Kingdom

Contents

10.1 Introduction 244
10.2 The roles of user evaluation in pervasive healthcare 245
10.3 Evaluation in user-centered engineering design............................. 246
    10.3.1 Design methods 246
    10.3.2 The conventional engineering design method 247
    10.3.3 User-centered design.. 248
    10.3.4 Implementing user evaluation in design.............................. 250
10.4 Forming a team... 250
10.5 Recruiting and selecting evaluators 251
    10.5.1 Ethics, informed consent, and people
             with cognitive disabilities 253
10.6 An initial user survey 254
10.7 Validated outcome measures 256
    10.7.1 The Quebec User Evaluation of Satisfaction
             with Assistive Technology (QUEST 2.0)............................... 257
    10.7.2 Psychological Impact of Assistive Devices Scale (PIADS) .. 257
10.8 Intellectual property protection 258
10.9 The first evaluation 258
    10.9.1 Beginning the first evaluation 259
    10.9.2 Concluding the first evaluation... 260
10.10 Installing systems in users' homes... 260
    10.10.1 Preinstallation survey... 260
    10.10.2 Installation guidelines 260

10.10.3 Installing systems in the homes of people with
         cognitive disabilities 261
10.11 Supporting users and devices 262
10.12 Interim evaluations 263
10.13 The final evaluation 263
10.14 Preproduction evaluation.. 264
10.15 Conclusions... 264
10.16 Case studies.. 265
      10.16.1 Case study: A cooker monitor for people with
               dementia 265
               10.16.1.1 Observations from ENABLE............................... 266
      10.16.2 Case study: An evaluation of end user sensor
               installation in a domestic environment......................... 267
               10.16.2.1 Observations from the Home Energy Tutor...... 268
      10.16.3 Case study: Evaluating Turvy—the Wizard
               of Oz method 268
               10.16.3.1 Conclusions from evaluating Turvy.................. 269
               10.16.3.2 Applying Wizard of Oz to the
                          healthcare context 270
      10.16.4 Case study: Evaluating the CareNet display
               in situ.. 270
               10.16.4.1 Analysis... 272
10.17 Regulatory device evaluation and adverse
      incident evaluation... 272
      10.17.1 Product assessment... 272
      10.17.2 Adverse incident investigation 272
10.18 Conclusions... 273
References 273
Bibliography 274

10.1   Introduction

Pervasive technologies are moving into health and social care. This move is supporting the transfer of care from structured institutions to the community. The increased ability to provide patients with effective and complex homecare is delaying and reducing institutional admissions and allowing earlier discharge of patients.

This dispersal of care from the institution to the community places new demands on care staff. Patients are monitored in their own homes using simple monitoring devices and the resulting data must be collected and analyzed. Risk reduction technologies[1] are being installed in homes, needing assessment and installation. Information systems make medical records available when and where they are needed.[2] Other devices and systems enhance communication between vulnerable people, their friends, and their families.[3]

Within healthcare institutions, pervasive technology is making care providers more efficient through the fast distribution of medical information. Pervasive technologies are also being used for tracking surgical instruments and other artifacts within operating theaters and around hospitals.[4]

Designing for healthcare presents a greater challenge than designing for industry because many of the intended users of a healthcare device are people with physical, neurological, or cognitive disabilities. Because people are complex and unique, it is not possible for a designer to simply imagine what it is like to be a user to be able to design appropriately for that user. This can be especially true when designing for impaired populations. The designer must obtain a thorough understanding of the users, their disabilities, their environments, and their problems. The greatest challenge to the designer is not solving the problem but understanding the problem. User evaluation is an essential tool for obtaining understanding.

Technology developed for use by people has an interface that is composed of the parts of the device or software that the user interacts with, and enables the user to communicate with the device. This user interface should be designed to ensure that it is accessible, usable, and useful to its intended users.

Accessibility: n. The ease with which something is approached or entered.

It is essential that the user interface be accessed by its users. For example, a text-only interface is inaccessible to blind users.

Usability: n. The effectiveness, efficiency, and satisfaction with which users can achieve tasks using the product.

High usability means a system is (1) easy to learn and remember; (2) efficient, visually pleasing, and fun to use; and (3) quick to recover from errors.[5]

Usefulness: n. The degree to which a device or system enables a user to perform a previously difficult or impossible task that the user desires to perform.

Useful devices are used by users. We all have tools that are useful and are well designed for their purpose.

10.2   The roles of user evaluation in pervasive healthcare

User evaluation is used in several distinct contexts. These are:

1. As a tool for determining users' preferences and capabilities within the design process
2. As a means of assessing usability within a formal regulatory evaluation
3. As part of an investigation into an adverse incident due to human error

Different techniques and methods are employed in each. In particular, the design context requires several different methods of user evaluation depending on the depth and type of information required.

In order to discover how an interface should be designed, it is necessary to consult potential users and conduct evaluations to determine and validate the design. Evaluations within a design context are diverse in their objectives and in the methods used, ranging from an informal discussion to the use of formal outcome measures, questionnaires, and interviews.

Products are often tested and evaluated by regulatory bodies such as the Food and Drug Administration (FDA) in the United States and the NHS Purchasing and Supply Agency (NHS PASA) in the United Kingdom. The NHS PASA runs the Medical Device Evaluation Service. Devices are evaluated and tested to ensure that they are safe, comply with standards, and work as advertised by the manufacturer. Additionally, the Medicines and Healthcare Regulatory Agency (MHRA) investigates adverse incidents, and the National Institute for Clinical Excellence (NICE) evaluates new clinical treatments for efficacy. These evaluations where technologies and devices are assessed formally for their clinical effectiveness and safety are not the primary subject of this chapter. Therefore they are considered briefly at the end of the chapter.

Poor interface design can lead to accidents when controls and feedback are misunderstood by the user. When such incidents occur, an investigation is required to determine the cause of the incident. User evaluation may form a part of such an investigation. In these cases (in the U.K.) evaluation is carried out by the MHRA or is subcontracted by the MHRA to other organizations.

This chapter is primarily concerned with user evaluation as part of an iterative design process, rather than its use for clinical evaluation, technology assessment, or postmarket monitoring.

10.3   Evaluation in user-centered engineering design

A thorough evaluation is not the sole objective of user evaluation in design. User evaluation is a very powerful tool for discovering required elements of a design by incorporating input from intended users. In user-centered design, the design process is continually modified by the refinement of a specification through successive user evaluations. As such, conventional engineering design techniques are not applicable in this context.

10.3.1   Design methods

Engineering design presents special challenges and requirements when compared to other evaluation contexts. Evaluation enables the discovery of the nature of the problem being solved, followed by the evolution of a solution in response to increased understanding of the problem. The purpose of evaluation in this context is *not* for the gathering of knowledge for its own sake, but rather for finding out the essential nature of the problem the user experiences so that it can be mitigated effectively by an appropriately designed device.

Engineering design is a complex, creative process that explores an identified problem and then develops a technological solution. Effective design depends on the definition of the problem to be solved: if the problem is not understood, the designer will solve the wrong problem or solve the correct problem inadequately. Discovery of the true nature of the problem is essential.

10.3.2 The conventional engineering design method

Conventional engineering design has three stages:

Stage 1—Development of a specification. The specification is used to clearly define and quantify the problem that needs to be solved in a way that the designers can understand. It also describes and quantifies the functions that the device must be able to perform. It is obtained through investigating the problem prior to design and must contain all the information the designer needs to make design decisions once the design process has begun. Before beginning design, the specification is "frozen" so that the designer has a fixed target to achieve.

Stage 2—Design of the device. This is a creative process of problem solving and is guided by the specification. Several concepts will be created and evaluated against the specification for their ability to solve the problem. A single concept will be selected for further development. The engineer will develop the selected concept into a detailed design, incorporating features that fulfill the original specification. Frequent reference will be made to the specification as design progresses, ensuring that the developing solution is compliant with the frozen specification.

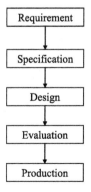

Figure 10.1 The conventional engineering design process.

Stage 3—Design completion. After the first prototype has been built, an evaluation of the prototype is required to confirm that the device solves the problem and meets the functional requirements outlined in the specification. The performance of the device will be compared to the specification and changes made to the design where necessary. These changes are likely to be implemented in a production version or, in the case of a custom-built one-off machine, in the prototype.

In this context, design and evaluation are carried out against a pre-defined specification that does not change as it completely describes the requirement. This design process is not appropriate for the design of health-care devices. The design of a medical device presents special challenges because of the complexity of the human mind and body.

10.3.3   User-centered design

A human being is a complex adaptive system that is not easily defined or measured. Our materials are nonlinear composites and our systems are complex and interdependent. Each person is an individual with his or her own unique set of behaviors, reactions, and needs. For these reasons it is not possible to use the conventional engineering design method and define a complete specification for a device that interacts with a person, prior to device design and evaluation.

Designing a healthcare device is a large task that requires an iterative process involving engineers, users, and clinical professionals. User-centered design aims to design a device that identifies and meets users' needs as fully as possible.

Design begins by discovering the users' requirement and the compilation of a functional specification, which may be revised later. Users sometimes ask for a particular type of device that they think will solve a problem, when in fact careful investigation will show that the real problem may not be solved by such a device at all. The "problem" is the users' perception of the

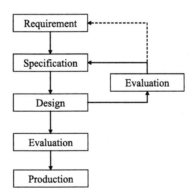

Figure 10.2 The user-centered design process.

need that is to be addressed by a device. The "requirement" is the true nature of the users' need, and it is not always immediately obvious. The process of requirement discovery reveals the true nature of the problem to the designer and sometimes to the user as well. The user's expression of the problem may not be detailed and may be expressed in inexact terms by a nontechnical user. The expressed problem is not usually sufficient to begin design. Further investigation is required to identify which needs the user is seeking to have met by the proposed device. A preliminary evaluation by users may be used at this stage and is often conducted with a simple prototype representing an aspect of the function of the proposed device. This method of requirement discovery may provide information that could not easily be obtained through questioning.[6]

Having obtained a specification that describes the users' need, work on the design can begin. At this stage the designer is aware that she or he does not fully understand the users' needs. Therefore the first prototype will be designed quickly and with no unnecessary complexity as it is likely to be heavily modified during subsequent iterations. Evaluation of the first prototype with potential users will yield large amounts of information. This can be a very exciting stage of the design process as this may be the first time users will be able to independently perform a difficult task or believe there is a technical solution to their problem.

Information from users will take many forms: some structured and anticipated, some unstructured and unexpected. Examples of structured information include results from questionnaires and interviews, formal measurements of the ability to perform a task, and physiological measurements. Unstructured information includes comments made by users after an evaluation and observations of a user attempting a task without the prototype or using the prototype in a way unintended by the designer. Structured information provides the details of the specification such as dimensions for components and operating procedures. It allows detailed design to proceed. Unstructured information provides insight into the problem. It suggests alternative solutions where no existing solution seems appropriate. It can help develop a stronger relationship with a user, leading to more candid user feedback in the future.

The design of the second prototype incorporates what has been learned from the first evaluation. It is likely that the first prototype will be discarded as many shortcomings will now have become obvious. It is for this reason that the designer should not put more effort into the first prototype than necessary to create a functional and safe design. The second evaluation will yield further details about the problem and the effectiveness of the second prototype. The users will have become more familiar with the evaluation process and, having a stronger relationship with the evaluating team, are more likely to be frank and honest about their opinions on the device when interviewed or given questionnaires.

Third and subsequent iterations will yield greater detail and more precise assessments of the usability and accessibility of the device. At some point the designer must decide when the benefits of new information and insights from user evaluations no longer outweigh the time and material costs of conducting the evaluations. At this point the designer will freeze the design and proceed to production development. However, before committing the design to production, it is advisable to carry out a final preproduction evaluation.

10.3.4   *Implementing user evaluation in design*

Conducting a user evaluation within healthcare involves many diversely skilled people working together as a team. To be successful, this team requires strong technical and clinical skills, interpersonal skills, knowledge of statistical methods, knowledge of organizational structures, and the ability to adapt and react quickly to changing situations.

The following description and analysis of the process of pervasive healthcare device and system design has been broken down into these eleven sections:

1. Forming a team
2. Recruiting and selecting evaluators
3. An initial user survey
4. Validated outcome measures
5. Intellectual property protection
6. The first evaluation
7. Installing systems in users' homes
8. Supporting users and devices
9. Interim evaluations
10. The final evaluation
11. Preproduction evaluation

These sections generally follow in the order in which they would occur during the design and evaluation process, although as will be seen this order should not be taken as a completely rigid structure.

10.4   *Forming a team*

Evaluation is nearly always a team or collaborative activity because of the diverse skill and knowledge set required by the participants. Typical members of a team for a pervasive healthcare application might include software, electronics, and mechanical engineers; an occupational therapist, doctor, nurse, or other clinician depending on the application; a researcher; and, of course, the user evaluators themselves. It is important that the team members are able to work together and that communications between the team members are unimpeded.

Although there are many variations to the composition of a team, a common one is as follows:

- **The engineers** are responsible for the design of the device or system. They must be able to integrate incoming data from the evaluations into the design process. It is also likely that they will provide technical support to the users during the evaluation, which requires good interpersonal skills.
- **The clinicians** provide expertise relating to the disability or health problem being addressed by the proposed device. They are often the originators of a project. Initially they provide the interface between the users and the technical members of the team. Good interpersonal and observational skills are essential, as they will work closely with both the users and the engineers. Depending on the type of device being designed, they may be physicians, surgeons, nurses, or therapists.
- **The researcher** often coordinates the evaluations and acts as a personal point of contact for the users. This is particularly helpful in a larger project with many users; however, this role may be fulfilled by the clinicians and engineers in smaller projects or where access to a patient is limited to frontline care or clinical staff because of the needs of the patient. The researcher may also be responsible for statistical analysis of evaluation results. The researcher will do much of the face-to-face work with the users by visiting and telephoning them. A social care or psychological background can be valuable because it provides the researcher with the excellent interpersonal skills needed to make the most of contact with users.

Having formed the team for the design and evaluation work, the project can begin. One of the first tasks is the recruitment of evaluators.

10.5   Recruiting and selecting evaluators

Evaluators are the members of the team who will try out iterations of prototypes and provide feedback to the designer on specific design aspects as well as on the overall usefulness and user satisfaction of the device. Although these are the members of the team that are the most difficult to recruit and retain, they are also the most important and will provide information about the usability, accessibility, and usefulness of a prototype. Recruitment is not always a straightforward process and the selection of evaluators should not always be at random from the population of potential users. Although random selection will recruit a more representative population of evaluators, not all evaluators will be suitable to work in the iterative healthcare design process. For example, where initial evaluations are carried out using crude first-stage prototypes, some technical knowledge or background can enable an evaluator to ignore the minimal attention given to cosmesis and concentrate on analysis of the functionality of the prototype.

One method to reduce recruitment difficulties is to form a relationship with a preexisting group that represents the population of potential device users. For example, if developing systems for people with stroke or dementia, it would be advantageous to work with the local group of the Stroke Association or the Alzheimer's Society. If they feel that the work the designer is doing is to the advantage of their members, such associations can forward questionnaires to their members and print letters and articles in their newsletters. The circulation of such newsletters may not be large, but they are very well targeted at the potential user population. Because of the great benefits to be had, a collaborative relationship with a user association should be formed as early as possible to facilitate initial survey work and begin the process of informing potential user evaluators about the proposed work. As development progresses, the association can provide updates on progress to its members, provide a forum for comment, and be used to recruit additional evaluators if necessary. In the authors' experience, once such a working relationship has been established, recruitment of evaluators is not difficult.

Alternatively, evaluators can be recruited through existing medical or therapeutic channels. Medical and therapeutic practitioners with whom the team already has contact can be asked to approach their patients or clients to ask if they would be willing to take part in an evaluation. Obviously where practitioners have longstanding relationships with their clients, recruitment is easier as the practitioners are trusted to a greater extent. An advantage to this recruitment method is that once the group is established, communication to the group is not difficult. However, expanding the size of the group can be difficult as there may not be many people within a clinician's patient population that are appropriate for any given evaluation.

A final means of recruitment is to use local media such as newspapers and radio. This type of appeal for user evaluators is much broader and reaches many more people than the previous two examples. A disadvantage is that the recruitment request comes from a stranger with whom the potential user evaluator has not built any trust, resulting in a much lower response rate.

In the design context there are several evaluatory roles that users carry out. It is important to select appropriate people for those roles. For example, it is not possible to evaluate the design of every prototype iteration with a full-scale randomized controlled trial. When the engineers are designing a system or device, it can be useful to have a small group of well-informed users work closely with the engineers and evaluate the ability of a device to perform its given function as the prototype design progresses. This group may have a technical background and are likely to be people that can see past crude prototypical casings and housings to the device inside, evaluating specific features when needed. Such people are not easy to find, but they can assist the designer immensely by providing rapid answers to questions of functionality and usability.

When evaluating the interface of a device, it may be more appropriate to evaluate it with naïve users who have little experience of information technology interface conventions. Although technically naïve users are easier to find, they are harder to recruit to the project and need greater reassurance that their contributions will be useful and valued.

The authors have found that potential users like to feel they are involved in the design work, not just occasionally testing something they don't really understand for a person they don't really know. For this reason, evaluators are invited to join the evaluation team and be a part of a research project rather than be experimental subjects. Evaluators are included and involved in the process of research rather than just being instruments in the hands of the researcher. This involvement is not illusory. Evaluators make an essential and valuable contribution to the design of the device or system.

10.5.1   Ethics, informed consent, and people with cognitive disabilities

When recruiting people for an evaluation program, remember at all times that they are not employees and have *no obligations* to the evaluation team. It must be made clear to the evaluators that they are free to leave the evaluation at any time without question or justification; that they are not obliged to use the devices given to them; and that their opinions and personal information will remain confidential within the project.

It is important to obtain *informed consent* for an evaluation from users. (Indeed in some contexts it is mandatory. In the U.K., for example, if an evaluation is carried out using NHS employees or patients, the evaluation must achieve formal ethics approval from an NHS ethics committee.) The consent of users who are not fully informed about the program they are agreeing to participate in is not meaningful or sufficient for involvement to proceed. Obtaining informed consent requires a clear explanation of the evaluation program using terms users can understand, careful listening to users' questions about the program, truthful and complete answers to those questions, and the distribution of clear and comprehensive materials describing the evaluation program that potential evaluators can read and understand at their leisure. The evaluation team must be sure that the users understand what they are consenting to. Asking a potential user to describe her perception of the evaluation program and her role in it can reveal misconceptions that need to be gently corrected. It is helpful, and often necessary, to ask the potential user to sign a written consent form. This suggests to the user that the decision about consent is a serious matter worthy of proper consideration. The form can also set out the responsibilities and expectations of the evaluation team as well as the rights of the user evaluator within the context of the evaluation.

Special consideration should be given to recruiting and working with users with cognitive disabilities and memory impairments such as Alzheimer's disease, stroke impairments, or traumatic brain injury. Unlike any other disabled user group, users with a cognitive disability may have a compromised ability to grant or withhold informed consent. Where users have more severe cognitive disabilities, it is not possible for them to give informed consent. When working with people with such disabilities, informed consent must be obtained from a person who is suitably qualified and empowered to make that decision. For example, a person with dementia may have granted enduring power of attorney to a relative or close friend when he was still able to make such decisions. This person will be somebody that the person with dementia trusts and who is considered to be acting in the best interests of the person with dementia.

Obviously, degrees of cognitive disability are not measurable in finite terms but span a continuum from none to a level that completely prevents a person from understanding the evaluation program. Whatever the degree of disability, the evaluation team should endeavor to enable users to understand as much as they are able to about the evaluation program and to allow users to give or withhold their own consent based on that understanding. Depending on the degree of the disability, a greater or lesser weight should be placed on the opinion of close carers when judging whether informed consent has been granted. Often when making these decisions it is important to consult professional and unpaid caregivers that are close to the users. People with severe brain injury or dementia are still people and deserve the dignity of being able to determine their own futures as much as they are able. It is in situations like these that the interpersonal skills often referred to in this chapter become absolutely essential for working with and understanding disabled people and their carers.

The user evaluators are an integral part of the project team and should be treated as such. When technical support is requested by an evaluator it should be provided as quickly as possible. Cultural sensitivities should be respected and taken into account. Things such as small cultural differences can have a big impact on the effectiveness of an evaluation. The ENABLE[6,7] evaluation described as a case study later in this chapter contained such an incident, where a slight difference in the meaning of the word "evening" resulted in a new version of software for a device being hastily developed.

The key to a successful evaluation is to make evaluators feel like the valuable members of the team that they are. It is vital to respect them and listen to them, giving them time and attention when needed.

10.6 An initial user survey

The initial user survey is a unique part of the user evaluation process. This is often the first contact the evaluators have with the design team and it is the step where the first bits of information about a new design are gathered. It is here at the beginning that first impressions are made and the building of relationships begins. If the design team intends to use formal outcome measures as part of their evaluation, then the decision as to which measures are to be used should be made here at the beginning, allowing appropriate baseline measures to be made as the study progresses.

The initial survey is designed to discover as much as possible about the users' perceptions of the identified problem and to narrow the focus of the design effort to specific clearly identified problems. The initial survey will not necessarily identify the true nature of the problem, just the users' perceptions of it. Commonly used tools for initial surveys are questionnaires and interviews. The design of questionnaires is covered in detail in many other books and papers and therefore will not be covered in detail here.

It is likely that the research and design team will want to use a customized initial user questionnaire to question a large group of users about their perceptions of the problem to be addressed and their experiences of living with this problem. The researchers will try to find out not only the details about what the problem is and how it manifests itself, but also its physical and relational context. The following list identifies information that can be obtained from an initial survey that will be useful to the design team at the beginning and throughout the project:

1. A description of the problem by the users
2. Tools and methods that the users have tried and whether these worked or not
3. People who help the users solve the problem
4. The type of solution the users expect the designers to produce
5. What the users expect the solution to enable them to do that they could not do before

This is not an exhaustive list. The project team should think carefully about what information they need to make a well-informed start on the project and about what baseline information they need when analyzing results later on.

An initial survey questionnaire should use short, neutral, and unambiguous questions that are easy to understand and do not presuppose an expected answer. It should also use formal outcome measures, where the questionnaire is structured, validated, and unchanged.

Numbered rating scales, known as Likert scales, can be useful in questionnaires, enabling the researcher to quantify users' opinions and feelings about various aspects of a problem. Respondents are asked to indicate a number that represents their feelings about the question asked. For example, a question about the users' existing solution to a problem might be presented as in Figure 10.3. In this example, the 2 circled by the user indicates that this user finds his existing solution is only marginally effective at solving his problem.

How effective do you find your current solution to the problem?

Ineffective 1 (2) 3   4   5   Effective

Figure 10.3 An example of a numbered rating scale.

Interviews take much longer to carry out than questionnaires and are usually used with a much smaller group of people. Interviewers must be patient and able to listen. The authors have found that the most effective interviews are semi-structured, with a list of topics to be covered in a conversational style. As with questionnaires, interviews that will form part of a formal outcome measure should be carried out as prescribed in the outcome measure method, to try and ensure that comparable data is collected from each queried user and that the data collected are valid for the outcome measure. Keeping the interview informal and mostly unstructured allows interviewees to relax and feel at ease with the interviewer. They are more likely to be honest about the problem in these circumstances, rather than giving the answer they think the interviewer wants. The interviewer must make it clear to the interviewee that the interview is not an examination or a test. The interviewees' opinions and feelings are sought and, as such, there are no correct answers.

Initial surveys are an important part of an evaluation. This starting point yields details for the design and establishes a way of measuring progress toward the final solution objective.

10.7 *Validated outcome measures*

This is an appropriate place to consider validated outcome measures. These are carefully designed and validated instruments that measure, for example, changes in quality of life, activity limitation, psychosocial impact, and so on. In a user-centered design context, especially in a healthcare context, they are one of the few tools available that provide a validated, quantified measure of whether a system or device is effective at solving the problem it was designed and built for. Validated outcome measures are not usually appropriate for use *during* the design process. They can require many hours of work to carry out and do not provide the detailed guidance required by a design team making decisions about specific aspects of a design. A validated outcome measure is more of an overall measure of the device. As such, it may assess whether a system has improved the quality of life of a user or whether users are able to perform a greater range of daily tasks more effectively with the device. However, validated outcome measures are too unwieldy and take too long to apply for them to be employed when making specific design choices such as deciding whether the symbology on a user interface is appropriate or whether a particular control button is in the right place.

There are many different measures available. Not all measures are appropriate for every project, so they should be carefully selected to fit the context in which they are to be used. Some can provide an absolute measure of a parameter, allowing comparison of outcomes between different projects or aspects of a project. Others measure only changes and do not give an absolute measure of a parameter. The designers of outcome measures are usually willing to advise on the suitability of a particular measure for a particular purpose.

Two examples of validated outcome measures that may be appropriate for use in a pervasive healthcare technology design context are given below.

10.7.1   The Quebec User Evaluation of Satisfaction with Assistive Technology (QUEST 2.0)

QUEST 2.0 is a measure of users' satisfaction with a device, *not* of the performance or effectiveness of a device.[8] It provides a validated absolute measure of user satisfaction with an assistive technology device. QUEST 2.0 measures user satisfaction with the following features of a device: dimensions, weight, durability, comfort, adjustment, safety, simplicity of use, and effectiveness. It also measures user satisfaction with the following service features: service delivery, repairs and servicing, professional services, and follow-up services. The instrument produces a single score for a device and its accompanying service that is between one (dissatisfied) and five (very satisfied). QUEST 2.0 was designed for use in a face-to-face interview but it can also be used in a postal survey.

10.7.2   Psychological Impact of Assistive Devices Scale (PIADS)

PIADS, an outcome measurement instrument, measures competence, adaptability, and self-esteem.[9] It is self-reporting and is designed for adults and children above the age of ten. It is quick and can be completed by telephone. PIADS is essentially a quality-of-life measure that provides a standardized assessment of the psychosocial impact of a device. It can be used across a broad range of different devices and contexts and has some of the following advantages:

- It predicts device retention and abandonment.
- It can be used to assess stigma of use.
- It allows psychosocial impact to be distinguished from health and condition issues.
- It gives good agreement between self-reporters and caregivers.
- There is no charge for its use.

PIADS has some disadvantages:

- It can only be used on one device at a time.
- It measures abstract concepts, not the practical aspects of a device.
- It requires proxy reporting for children and people with learning and cognitive disabilities.
- It is influenced by user expectations.

There is no single outcome measure that is appropriate to measure all aspects of the impact of a device on a person's life. Measurement instruments appropriate to the required measurements must be chosen in each case.

Factors influencing this decision are the cost of the instrument materials, ease of application, and the appropriateness of the instrument to the population to be measured.

Validated outcome measures are powerful tools for assessing the effectiveness and impact of an assistive device. They enable comparison of effectiveness between different devices in an integrated system and quantify the impact of devices on key issues such as quality of life and the ability to perform tasks efficiently. If they are going to be used in a project, the decision about which measures to use should be made in the early stages of detailed project planning. Some outcome measures require an initial baseline measurement to be made, which must be carried out before the evaluator has sight of a proposed solution that may influence the initial perception of the problem.

10.8   Intellectual property protection

Once design begins, intellectual property will be generated and will need to be protected if the completed design is to be licensed to a manufacturer. If prototypes are evaluated in the community, they are disclosed, compromising any subsequent patent application. There is a conflict here. A design cannot be fully protected until it is finalized, but it cannot be finalized without an evaluation. Thus some evaluation will always have to preclude full protection. A decision must be made about when to protect and what to protect. It may be possible to protect the concept of the device at the beginning of the evaluation process and then protect the detailed design of the concept progressively as features are tested and confirmed. Internal device technology and concealed user interface features may be protected with a nondisclosure agreement between the design team and the evaluators until the design is finalized.

The degree of protection desired will vary from project to project and from team to team. Applying intellectual property protection to a device strengthens its value to a manufacturer, but it can make the process of doing the evaluation and building trust more difficult. The balance between these two conflicting desires must be found for each project and each team.

10.9   The first evaluation

After carrying out the initial survey, accurately identifying the scope and objectives of the project, and designing and building the first prototype, it is time to set up the first user evaluation. This evaluation will probably be carried out by a small group of users who have been identified through the initial survey questionnaire. The first evaluation is the most important. It will yield more data more quickly and will have a greater influence on the design of the device or system than any of the other evaluations. It is also critical relationally: if the users' trust and confidence in the project team is compromised at this stage, further evaluations will yield compromised data or will not happen at all.

10.9.1 Beginning the first evaluation

Much is learned during the first evaluation by the users and by the researcher, clinician, or engineer conducting it. When the first prototype is delivered the user may feel many different ways: disappointment at its crude appearance, anticipation and excitement at the prospect of the problem being at least partially solved, and nervousness about testing a unique and perhaps expensive piece of equipment. The beginning of the first evaluation is a good time for the designer to accompany the researcher or clinician and meet the users, see their reactions, and hear their comments firsthand. Having the designer present also enables him to explain the use of the device accurately and answer any questions that might arise. The user is able to meet the person who is designing the device or system, strengthening the feeling of being part of the team. Some instant feedback will be given to the designer about the appearance and feel of the device. On some occasions the author has been to a user's house to deliver a device and then returned to his office with it on the same day: as soon as the user tried the device for the first time a serious deficiency in the design became apparent that had not been anticipated before the evaluation. The first evaluation was over within five minutes.

Often at first evaluations, there are family members and carers present. They sometimes try to speak and act on behalf of a disabled user. The team should try to interact directly with the user as much as possible. Although questions and explanations should be directed first to the user, it is necessary and appropriate to make sure carers and family members also understand the evaluation and have their own questions answered. After all, they are often supporting the user who is trying out a new and imperfect device.

By the end of the first meeting, the user has been further integrated into the team. She has a new device to try out and a new role to learn. The designer has met one of the people he is designing for and has firsthand information about the problem being addressed. At the end of the meeting, the project team should have made sure that the user has:

- The prototype device
- Full written and illustrated instructions on the safe and correct use of the device
- A sheet of contact numbers and addresses for the designer, technical support provider (if different), and the researcher

When handing over a prototype to a user, one of the authors, who is an engineer and provides technical support, also leaves his work, home, and mobile phone numbers with user evaluators. He makes sure that the user knows he is able to contact the author at any time. If the device is installed in a fixed place, a contact sheet is placed near or on the device.

10.9.2   Concluding the first evaluation

When the researcher and possibly the engineer return to the evaluator to collect the results of the evaluation, the evaluator will want to tell the researcher or engineer about his experiences with the prototype. In these circumstances evaluators often want to say what they think the researcher and engineer want to hear and are reluctant to criticize the device. The researcher and engineer should encourage the evaluators to be critical and candid about their experiences and opinions, explaining that they need the evaluator to be frank so that they can ensure that the design of the device is appropriate to the evaluator's needs.

First evaluations are an exciting and memorable part of the project. It is on occasions like these, where a user gains hope that their problem will be solved, that working in this field becomes so rewarding.

10.10   Installing systems in users' homes

Many pervasive healthcare devices will be handheld or at least portable, needing no infrastructure to be installed in the user's environment. Some systems, however, need more time-consuming installation with the consequential disruption of the user's household and life.

10.10.1   Preinstallation survey

The installation of complex systems needs careful planning and sometimes an additional visit to the user's house to do a preinstallation survey of the installation site. The survey is used to:

- Check the availability of electricity and other required utilities.
- Identify and check the dimensions of the installation site.
- Check the location and compatibility of any other appliances that the system must work with such as the telephone or television.

After consent is given, the preinstallation survey can be done by the researcher during the evaluation setup visit. The most appropriate person to do the survey is the designer, who fully understands what is needed for a successful installation.

10.10.2   Installation guidelines

Some of the points made in this section may seem obvious; however, it is easy to allow standards of thorough preparation and conscientious working to slip when performing the same installation task many times. Maintaining high standards can save time and frustration, while instilling the user with a sense of confidence in a professional and competent research team.

Installation should take place at a time that is convenient for the evaluator, which may not be during normal office hours. The installation time should be arranged well in advance, making sure that any additional team members needed during the installation are available. The evaluator may also wish to have a primary carer present.

When preparing for the installation, the installer should prepare a checklist of items needed for the installation. This list can also be used as an inventory when leaving to make sure that nothing is unintentionally left behind. Before leaving for the installation, the installer should go through the checklist and make sure that all the necessary maps, contact details, tools, materials, device components, passes, and identification and user information are present. If any of these items is missing, the installation may have to be aborted.

The authors have found the following guidelines to be helpful when installing in users' homes:

- On arrival at the installation site, the installer should check that consent is still granted before beginning work.
- Work should be carried out in a way that minimizes the amount of damage caused to the evaluator's house or apartment. It is the responsibility of the project team to rectify any damage caused during the installation.
- Questions about the installation should be answered truthfully and conscientiously.
- If, at any time during installation, user consent for the installation is withdrawn, the installer should remove what has been installed and leave the site in its original condition.
- Social engagement and conversation with the evaluator *during the evaluation* engenders trust and strengthens the relationship between the evaluator and the project team.

These are guidelines, not rigid rules. The most important thing is for the team to act with integrity and diligence and to treat all evaluators with dignity and respect at all times. It is easy at the end of a long and difficult day to become irritated by a person with dementia who is asking for the tenth time who one is, why one is there, and what this thing is for. Although the irritation cannot be prevented, speaking and acting irritably has no place in this context.

10.10.3 *Installing systems in the homes of people with cognitive disabilities*

When working with people with cognitive disabilities, special consideration should be given to the needs of the user evaluator. Some special measures may need to be taken to ensure that the installation is completed with a minimum of disruption to the user and the installers.

In addition to the points made above, the following guidelines[1] are presented for evaluation teams working with people with cognitive disabilities:

1. Always send two people to an installation. Sending two people allows one of the installers to leave the building while the other remains to prevent a loss of access to the site and work in progress. This can occur if the evaluator forgets who the installers are and decides to deny access to her home when the absent installer returns.
2. Be patient. People with dementia may ask the same question many times. This may be because they do not remember asking it previously or they cannot remember the answer given. Installers should answer each time as though it is the first time they have been asked. During evaluations in the U.K. and Europe, the authors found that evaluators with moderate dementia did sometimes remember the purpose of an installation after having it explained clearly to them six or seven times during the installation process.
3. Listen to the evaluator. As well as asking the same questions many times over, evaluators may tell the same stories many times over. Such life stories can provide insight later in the evaluation into why the evaluator is reacting to a situation in a particular way.
4. Be sensitive to the evaluator's state of mind. Try to observe cues that may indicate the evaluator's state of mind such as repeated actions, facial expressions, tone of voice, persistent questioning about a particular aspect of the installation, and so on. If the installers are unsure if the evaluator is distressed, they should ask the evaluator first, then a carer, if present also. If the evaluator is becoming distressed, stop the installation in time to allow the site to be returned to a fully functional state before leaving.

The effective evaluation of complex systems for people with cognitive disabilities by people with cognitive disabilities is possible and valuable; however, these evaluations are not as straightforward as evaluations by people with no such disabilities. Care and consideration of their needs must be taken into account at all stages of the project over and above what would normally be applied during an evaluation.

10.11 Supporting users and devices

Effective and rapid support of evaluators is important because, inevitably, prototypical devices are unreliable. This "feature" of prototypes must be clearly explained to evaluators before the evaluation. The evaluators may become dependent on a device, so rectifying its loss due to unreliability carries a high priority when allocating resources. Engineers providing technical support should be able to work with evaluators sensitively and efficiently. For a particularly troublesome prototype, they may be the evaluator's most frequent point of contact with the project team and as such should be prepared to collect and distribute any findings that are relevant to the evaluation.

Where possible, systems should self-report faults. If a system has even a low level of intelligence, remote reporting of faults and failures is easy to achieve with a mobile phone text-messaging module. Such modules are readily available from major mobile phone manufacturers, are inexpensive, and are easy to use and control with a simple USB or RS232 interface. They do not use the user's own telephone landline, providing a protected independent link to rapid and informed technical support. It should be noted that while text-messaging services are usually instantaneous, there can be delays in message delivery during peak times, therefore text messaging should not be used for safety critical messaging where a rapid response is essential.

If a technical fault cannot be rectified quickly onsite, then a swap-in/swap-out method provides the best service option for the user. The engineer responding to the identified fault brings a spare device to the evaluation site and substitutes it for the faulty device if it cannot be repaired immediately.

Users may also need support themselves during an evaluation. Occasionally the presence of a device addressing a fundamental human need can trigger strong emotional reactions. The project team should make sure that the evaluator's usual sources of emotional support are aware of the evaluation and are willing and able to provide support should it become necessary.

10.12   Interim evaluations

Subsequent evaluations usually have a less dramatic impact on the user. The evaluators should understand the operation and concept of the device and know the team well enough to feel uninhibited about criticizing the design. There may be communications between the evaluators and the design team during the interim evaluations for the purpose of clarifying the use or implementation of a new feature, for the evaluator to ask for urgent changes to be made, or for the designer to ask the evaluators' opinion on a potential new feature. In a design context, interim evaluations are less structured temporally than they would be for a formal product evaluation, with the duration of the evaluation depending on the needs of the user.

10.13   The final evaluation

Making the decision to conclude the design and evaluation process is not always straightforward. It is likely that not all the evaluators will be satisfied with the design. Some compromises will have been made for the sake of manufacturability or to accommodate the needs of the greater proportion of users at the expense of an unusual minority. Nevertheless at some point the decision to end the process must be made. This is likely to be when further design and evaluation is not producing significant changes to the design, when *most* evaluators are satisfied that the design meets their needs, and when the device is deemed to be manufacturable at reasonable cost with some production development.

Before the design effort can be concluded, a decision must be made to freeze the design and proceed to production development. The final evaluation will be used more to validate a stabilized design than to investigate new features. It is often a short evaluation, as by this time the users are familiar with the device and can quickly assess whether it works well or not.

It is advisable to recruit some naïve evaluators at this stage who do not have previous experience with using the device. When the device becomes a product, new users will not have used it before, so it is important to check that the device is appropriate for new users as well as with those who are familiar with its operation and may have adopted ways of dealing with some of its shortcomings.

A formal outcome measure applied at this point in development provides validated proof that the device improves quality of life and has a positive impact on the users' lives. Such proof is persuasive when approaching companies for manufacture, sales, and distribution agreements.

10.14 Preproduction evaluation

The preproduction evaluation is not part of the design process, but it is essential for the validation of the design following modifications for ease of production. This is also an evaluation that should in part be done with naïve users to ensure that people who are unfamiliar with the device will benefit from its use.

10.15 Conclusions

The process of evaluation as a part of user-centered design is complex and multidisciplinary. Above all, however, it is a process of listening and understanding people and their needs and wishes. When users' needs and abilities are given their proper place at the center of the design effort, it creates devices and systems that are useful, usable, and accessible. This approach to the design of devices that interact with people in complex and sometimes intimate contexts is also commercially sound: it results in products that can be confidently marketed as meeting the correctly identified needs of the intended users.

It is also important to emphasize that user evaluation should not be abandoned even where the intended users are technically naïve or physically or cognitively disabled. A lack of knowledge about the internal workings of a device and modern digital interface conventions, or a disability that reduces the scope for human interaction with a device, should not impede user evaluation. If a person has a need that can be met with technology, then that person is capable at some level of interacting with prototypes of a device and influencing the design of the final product.

10.16   Case studies

This section contains four case studies of user evaluations that illustrate evaluation techniques that can be usefully applied to the design of pervasive healthcare devices and systems. Two of the studies (with the Home Energy Tutor and Turvy) are taken from applications outside of the field of pervasive healthcare; however, the techniques illustrated are very appropriate to the healthcare environment.

10.16.1   Case study: A cooker monitor for people with dementia

The ENABLE[7] project sought to measure how technology for people with dementia impacted quality of life.[1] It was active in five European countries: the U.K., Ireland, Norway, Finland, and Lithuania. Five devices were evaluated by people with dementia in their own homes and the Dementia Quality of Life Instrument (DQoL)[10] was used as the outcome measure to assess the impact of those devices on the evaluators' quality of life.

Some of the devices evaluated were finished products, while others were prototypes still under development. Although it was primarily an evaluation of the impact of the devices on quality of life, ENABLE unintentionally became an iterative design evaluation as well, because it identified and rectified faults and problems with the prototypes being evaluated.

One of the devices evaluated was a cooker monitoring system for people with dementia. It used smoke, heat, and natural gas sensors to detect potentially dangerous problems within a cooker. If a problem was detected, the system actively turned off the cooker knobs. If the problem did not go away after this intervention, the cooker monitor would isolate the cooker from its gas supply and then call for help from an external carer. The system had a feature that later proved valuable. It was able to self-check and report identified technical faults to the engineers directly using mobile phone text messaging.

In the U.K., the evaluations were set up by a researcher who was trained to work with people with dementia and had previous experience doing research with people with dementia. Evaluators were selected by the mental health service of the local social services department.

Evaluators and their primary carers were introduced to the project initially by the researcher and the evaluators' community mental health nurses or occupational therapists. A subsequent visit was attended by the researcher, an engineer, and possibly the mental health nurse or occupational therapist. At this second visit, the device and its functions were described in detail by the engineer, consent for the evaluation was requested, and, if consent was given, an initial site survey was carried out. Occasionally the site survey was carried out a few days later.

Installation of the system typically took six hours and was carried out by two installers who were engineers working on the design team. As was referred to earlier in this chapter, two engineers were sent after an early experience when a lone installer left the installation site and was, on his return, refused access to the site by the evaluator. After a short negotiation, access was granted and work continued. Subsequently two installers attended installations and one always remained at the site.

Evaluations were structured to last for a year. The researcher visited the evaluators every month to discuss the progress of the evaluation, the evaluator's thoughts about the system, and to carry out a DQoL measure. The DQoL was designed to be used as a questionnaire, but it was found during the evaluation it was more successful when used conversationally. The DQoL questions were woven into a conversation rather than asked sequentially with no conversational structure.

The cooker monitor proved to be unreliable. Some of the installations needed frequent visits to rectify problems that occurred with the sensors and mechanisms that turned off the cooker knobs. Many of the evaluators of the cooker monitor dropped out of the evaluation because of the unreliability of the device or because they were transferred to hospital or residential care for unrelated reasons. The ENABLE cooker monitor evaluation led to the development of a simpler, lower-cost, easier-to-install version called the cooker minder. All the cooker monitors were removed at the end of the ENABLE project, except for one in Lithuania.

One of the U.K. evaluators continued to work with the designers and is now evaluating a cooker minder. This evaluation is a less formal, long-term evaluation that has been running for a year. The cooker minder uses voice messaging to prompt the user when a problem is detected. Although there is no active control over the knobs, the device can still isolate the cooker from the gas supply if the problem does not abate. The cooker minder is fitted with a data logger developed for the project that logs sensor, isolation, and reset events, time stamps them, and stores them in flash memory for future analysis. The data logger can be collected by a researcher or engineer when he visits the evaluator or it can be easily removed from the cooker minder and mailed to the designer by the evaluator.

10.16.1.1 Observations from ENABLE

The ENABLE evaluation project highlighted the need to make sure that prototypes intended for people with dementia are further developed than usual because of the difficulty this population has in adapting to change such as technical faults with equipment. When working with people with dementia, reliability is of great importance. This slows down development by increasing the time to first prototype and by increasing the design as well as the amount of internal technical evaluation time needed between evaluation iterations.

ENABLE also demonstrated the feasibility of working directly with evaluators with dementia, so long as those in contact with the evaluators relate to them sensitively and appropriately. Personal relationships have been the key to successful evaluations in the ENABLE project. Where relationships were nurtured, evaluators had better recall of the project team, which reduced the need for repetition and increased the amount and quality of feedback from the evaluators.

Some of the quality-of-life results from ENABLE were inconclusive, particularly device evaluations where the dropout rate was high. While some evaluations did show a positive impact on quality of life, one of the major outcomes of the project was the experience gained working with people with dementia and their carers in a device evaluation environment.

10.16.2 Case study: An evaluation of end user sensor installation in a domestic environment

This evaluation illustrates the use of very low-cost, low-technology proxy devices to evaluate the ability of users to install a distributed sensor network in their own homes.[6] Although this example is for an energy monitoring application, the technique described could usefully be employed in a health-care context. Distributed sensor networks might be usefully employed to assess the well-being and disease progression of a person with Alzheimer's disease or a physically frail older person.

Evaluators were recruited by a market research firm and were screened to be representative of the target user group of the Home Energy Tutor. Evaluators with a technical background were not recruited. Each evaluator received $75 for participating. A package was delivered to users containing the Home Energy Tutor: this was a mock-up of an energy use monitoring system that used a wireless network of various sensors positioned in key places around the home. One-to-one scale models of different sensors were supplied along with instructions for their installation.

Two observers delivered the package to each evaluator's home. The observers remained to observe and assess the evaluators' reactions to the package and its contents and the accuracy of the evaluators' positioning of the sensors based solely upon the written instructions supplied. No help or coaching was given by the observers. Data were collected by the observers in the form of notes, photographs, and a questionnaire. After the installation was completed, the evaluators were interviewed to determine their level of understanding of the requirements of each sensor.

The evaluation yielded the key results quoted here:

- Make appropriate use of user conceptual models for familiar technologies: use appearances, interfaces, and controls that the user understands and is familiar with.
- Balance installation usability with domestic concerns.
- Avoid the use of cameras, microphones, and highly directional sensors if possible. Users are not good at positioning such sensors accurately.
- Detect incorrect installation of sensors and provide value for partial installations. Partial installations can still yield useful information about the capabilities of the evaluators and the design of the device. Make sure that the assessment tools used can capture this information.
- Educate the user about data collection, storage, and transmission.

10.16.2.1 Observations from the Home Energy Tutor

This evaluation shows the value of a quick, low-technology evaluation performed early in the design process. It yielded useful results with little investment in time and materials. The information gained will guide the future development of domestic distributed sensor installations and showed that home installation by users is a realistic proposition provided that the users understand the purpose and nature of the sensors being employed, not just where they should be positioned. The initial naïvety of the evaluators was important here as naïve users were those most likely to experience problems with the installation. Recruitment actively selected naïve users in preference to those who were technically aware. Where the user will be a person with a cognitive disability, it may be more appropriate for a carer or healthcare worker to install the sensor network. It is still likely that this person will be technically naïve.

10.16.3 Case study: Evaluating Turvy—the Wizard of Oz method

This case study illustrates the "Wizard of Oz" method of evaluation admirably and for this reason it is included in this chapter.[11] The Wizard of Oz[12,13] method employs a human to simulate an intelligent artificial system in an evaluation of an interface between the user and the intelligent system. The substitution of direct human control of an interface for control by an artificial system means that usefulness and interaction methods of intelligent systems can be evaluated and modified before the development of the complex software needed to run them. Fundamental changes to the system's behavior can be implemented without having to rewrite software, and hardware development is minimized and confined to the interface itself.

Turvy was a simulation of an instructible software agent. It was designed to learn how to perform repetitive text-based tasks through user-guided examples and instructions that are input by the user though a speech recognition interface and mouse cursor pointing.

The objective of building and evaluating the simulation was to learn how users respond to a software agent and what interface paradigms would be appropriate in this context. Turvy, the simulation of the agent, was very carefully designed and defined before beginning the evaluation. Tasks were standardized and the amount of information introduced into the systems by the user was minimized by making the task purely manipulative of existing data. Turvy worked by observing an example, making a plausible generalization, and then revising it as further examples and instructions were given.

Users sat at a computer and worked on the text-based task, in this case reformatting bibliographic entries. Next to the user sat a facilitator, and nearby sat Turvy, played by the system designer, who was also equipped with a keyboard, screen, and mouse attached to the same computer as the user. The system designer was to one side and slightly behind the user and was not visible unless the user turned away from her own computer screen.

Having learned how to do the reformatting, the user was asked to teach Turvy how to perform the same tasks. At this point, the Wizard (the person playing Turvy) becomes active and interacts with the user through speech and shared control of the word processing software. Turvy's speech and grammar and responses were strongly structured and controlled, simulating the limited vocabulary, grammar, and decision-making ability that a software agent might employ. It was found that users quickly adapted to communicating with Turvy. They spoke differently to the agent, mirroring its clipped sentences, and referred to the Wizard and Turvy as separate entities.

10.16.3.1 Conclusions from evaluating Turvy
The evaluation of Turvy yielded a rich set of measured and observed data on user interaction with software agents. Users adopted Turvy's vocabulary and preferred to use speech to describe a new focus of attention rather than point with the mouse cursor.

The Wizard of Oz evaluation answered the designers' questions about user interaction with a software agent and also raised some additional questions about users' style of interaction with Turvy. Turvy's designers compiled guidelines for working with Wizard of Oz evaluations:

1. *Prior implementation experience is invaluable.* The experience of Turvy's designers suggests restrictions should be placed on the capabilities of the simulated intelligent system.
2. *The agent's behavior should be based on an algorithm.* This keeps the simulation honest and ensures consistent behavior and experimental repeatability.
3. *The agent's dialogue capabilities should be based on a constrained interaction model.* Interaction with the simulated system agent must be constrained by an explicit list of instructions and feedback that the agent can work with.
4. *It is possible to build real systems derived from studies of verbal Wizard-human discourse.*
5. *The designer benefits from becoming the Wizard.* The designer was trained by playing the role of agent. Revisions were strongly motivated by the designer's responsibility for the users' confusion and discomfort. Simulating an incomplete design reveals its deficiencies.
6. *Qualitative results are the most valuable.* By acting as Wizard, interviewer, and facilitator, the project team becomes immersed in the work and many important results become obvious.
7. *Interviews are essential and video recordings are useful.* Questioning and listening to users is the most efficient way of finding out what works, what doesn't work, and also the users' own internal models of how a system works. Video recordings of interviews allow analyses of users' speech and gestures.

10.16.3.2 *Applying Wizard of Oz to the healthcare context*

Turvy was implemented in the laboratory. Most pervasive healthcare devices will be evaluated in the field. This can be made possible with remote communications technology between a small call center and the evaluation site or by carrying out short-term snapshot evaluations by bringing the necessary equipment to the evaluation site and taking it away again after the evaluation. Examples of healthcare contexts in which the Wizard of Oz technique could be applied include the prompting of people with cognitive disability using a navigational aid or a responsive voice control interface for a profoundly physically disabled person.

10.16.4   *Case study: Evaluating the CareNet display in situ*

The CareNet Display is an ambient display showing a photograph of an elder and providing information about the elder's daily life to his local carers (the local members of his care network).[3,14] Devices were placed in the homes of key local care network members, providing them with updated and relevant information about the elder's daily life and needs. They could also have been placed in the elder's home, though this was not necessary for the carers' devices to be functional and was not done in the evaluation.

The CareNet Display consists of a display placed in a wooden picture frame, connected via a GPRS wireless modem to a remote Web server. The display shows a picture of the elder, surrounded by several icons representing different aspects of the elder's life. The display is touch sensitive and interactive. Additional information about the elder's life becomes visible when the icons are touched. The seven information categories displayed are meals, medications, outings, activities, mood, falls, and calendar. The icons change to indicate that something significant has happened in the category they represent.

The information represented on the CareNet Display would be collected by a distributed sensor network in the elder's house; however, because this evaluation was in the early stages of development, the information was collected by people. The designers spoke to experts on distributed sensor networks to ensure that the information displayed could realistically be collected by a sensor network now or in the near future. This evaluation was used to inform the future selection of sensors in the distributed sensor network.

Prior to the finalization of the design for the evaluation, the information to be displayed was determined through a roundtable discussion with seventeen care network members. Twenty types of information were ranked using a card-sorting exercise. The top seven types of information that could be gathered by phone call (the selected means of gathering information) were selected and incorporated into the CareNet Display devices. The elder who is the source of the information to be displayed chose which of the care network members could see which information as some of the information was potentially sensitive. The information shown on the display was collected by researchers who telephoned the elders several times every day and updated the displays remotely.

The design team conducted several months of group discussions and interviews with their target users before finalizing the design and installing the displays for evaluation.

In the in situ evaluations, prototypes of the CareNet Display were used for three weeks. There were thirteen evaluators: four elders, and two or three family members per elder who provided regular care. Each family member had a prototype of the CareNet Display. All the evaluators were interviewed before and after the evaluation, and the nine family members also completed a questionnaire halfway through the evaluation.

Evaluators were recruited by several different methods: talks at geriatric care conferences, posters in day centers, and through working with local eldercare experts. In addition to the recruited elders and care network members, other family members (children and spouses) also contributed to the evaluation.

The displays were installed for three weeks at a time (this seemed to be the maximum commitment of time the already busy caregivers were prepared to make). There were no special verbal instructions given as to how to use them, though a help booklet was left for the evaluator to read. The displays were mounted in a custom beechwood frame. Elders were telephoned between three and six times a day and were always asked at the end of each call if it was okay to share the information gathered with the display users. The caregivers were paid $150 to participate and the elders and other data providers were each given between $75 and $300.

Evaluators were interviewed for between sixty and ninety minutes before and after the evaluation, which was also documented with researcher notes, questionnaires, audio recordings, and photographs.

The evaluation validated the effort put into discovering the nature of the problem being addressed by this device (isolation) before it was installed. The results were positive and encouraging for further development with real improvements in care and reductions in stress levels being achieved. For example, it was discovered that one of the elders was eating the same food day after day. She had diabetes and mild dementia, so a poor diet was particularly unhelpful for her. Another example of a finding that could only have been made through user evaluation was that in a darkened room the display ceased to be an ambient device fitting in unobtrusively into a room; it became obtrusive because of its brightness. The subtle differences in icon color were found not to be an effective means of communicating information as the changes were not often noticed. It was also found that most of the elders liked being telephoned several times every day by a researcher. This was an unexpected finding. Many caregivers wanted "human touch" information to be included on the display, which would be difficult to gather with a sensor network. A challenge for future work is to incorporate information such as for *whom* Mrs. Jones is knitting, not just that she is knitting.

10.16.4.1 Analysis

This evaluation was very successful and met its objectives of informing the design team about the expectations and abilities of the elders and their care networks. It also successfully informed the team about sensor network design, although some additional challenges were raised that had not been considered. The value of the effort put into the initial interviews and discussions with elders and caregivers became evident when the displays were installed and reviewed. There were no major design problems to overcome during the evaluation and no dropouts. All the evaluators said that they would use the displays if they were available and most said that they would buy one at reasonable cost.

This case study has not described all the findings in the paper referenced for the sake of brevity; however, this was a particularly effective evaluation of an early prototypical pervasive device that is worthy of further study.

10.17 Regulatory device evaluation and adverse incident evaluation

User evaluation is also used by test and evaluation centers funded by regulatory bodies such as the FDA, the NHS PASA, and MHRA. User evaluation is employed to check the usability of devices and systems. Reports on devices are written and published to guide healthcare purchasing decisions.

10.17.1 Product assessment

User evaluation and usability assessment against consensual "standards" of good usability design by experts are used as tools for the formal assessment of a product by healthcare agencies. Reports generated from such evaluations are used by healthcare providers to facilitate purchasing decisions. The evaluations must be thorough, unbiased, and provide the information that potential users and purchasers require to make decisions.

The validation process can also lead to design changes after the launch of a product. Manufacturers of medical devices are required to keep a vigilance file for each product. The identification of adverse incidents that have been the result of inappropriate design may lead to the device being iteratively redesigned and evaluated to ensure that the design fault has been removed. This is particularly so if the cause of adverse incidents was a poorly designed interface that led to human error rather than a technical fault with the design.

10.17.2 Adverse incident investigation

An important role for government-funded test and evaluation centers is to investigate adverse incidents where harm has been caused to a user or patient or where harm has been averted by exceptional user vigilance. The cause of the adverse incident must be determined, whether it is, for example, operator error, device malfunction, or use of the device in an inappropriate environment. It is necessary, if possible, to determine what the root cause of the incident is and how future similar incidents can be avoided. If human error may be involved in the incident, user evaluations will be conducted to determine how people tend to use the device and how the error that led to the incident occurred.

Particular care is now being taken to observe how users interact with their environment, colleagues, the interface of the device, and the device as a component of a larger system, possibly including other devices. System-wide changes may be necessary to avoid recurrence of similar incidents. Examples of remedies include changes in means of communication between team members, changes in the design of user interfaces, changes to the environment in which the device is used, user training, and so on.

10.18 Conclusions

User evaluation is an essential part of the successful design of pervasive healthcare devices and systems. Evaluation is a wide-ranging activity requiring a team with many different skills, both technical and interpersonal. It can only be effective when the design team engages with the evaluators at a relational level while ensuring at the same time that evaluations are thorough and rigorous. The guidelines and case studies in this chapter illustrate that it is possible to evaluate effectively, even with people with cognitive disabilities.

Participating in user evaluation brings the designer and user closer together and changes both people for the better. Designers begin to understand the people they are designing for and also the users' needs, fears, and desires. The users become people with faces rather than a faceless specification. This is a strong motivating factor toward good design. Users become more acquainted with the challenges of solving technical problems and the huge amount of work that is put into a well-designed product. In learning to work together, designers, evaluators, and, ultimately, all product users benefit. User evaluation is not just a useful tool for a designer in pervasive healthcare but it is also key to the design and production of devices that work.

References

1. Adlam, T. et al. The installation and support of internationally distributed equipment for people with dementia. *IEEE Transactions on Information Technology in Biomedicine*, 8, 253, 2004.
2. Bardram, J. *Hospitals of the Future—Ubiquitous Computing Support for Medical Work in Hospitals*. UbiHealth 2003, Seattle, http://www.pervasivehealthcare.com/ubicomp2003/papers/ (accessed December 2005).
3. Consolvo, S., Roessler, P., and Shelton, B. The CareNet display: lessons learned from an in-home evaluation of an ambient display. In *Proceedings of the 6th International Conference of UbiComp*, N. Davies, E. Mynatt, and I. Siio, eds., Berlin, Germany: Springer, 2004.

4. Fishkin, K. et al. Ubiquitous computing support for skills assessment in medical school. *UbiHealth 2004*, Nottingham, http://www.pervasivehealthcare.com/ubicomp2004/papers/ (accessed December 2005).
5. Howe, D. The free on-line dictionary of computing. http://www.foldoc.org/ (accessed November 2004).
6. Beckmann, C., Consolvo, S., and LaMarca, A. Some assembly required: supporting end-user sensor installation in domestic ubiquitous computing environments. In *Proceedings of the 6th International Conference of UbiComp*, N. Davies, E. Mynatt, and I. Siio, eds., Berlin, Germany: Springer, 2004.
7. Gilliard, J. and Hagen, I. Enabling technologies for people with dementia—cross-national analysis report. European Commission document number QLK6-CT-2000-00653, August 2004, http://www.enableproject.org (accessed December 2005).
8. Demers, L., Weiss-Lambrou, R., and Ska, B. The Quebec user evaluation of satisfaction with assistive technology (QUEST 2.0): An overview and recent progress. In *Technology and Disability: The Assessment of Assistive Technology Outcomes, Effects and Costs*, G.J. Gelderblom and L. de Witte, eds., Amsterdam: IOS Press, 2002.
9. Jutai, J. and Day, H. Psychosocial impact of assistive devices scale (PIADS). In *Technology and Disability: The Assessment of Assistive Technology Outcomes, Effects and Costs*, G.J. Gelderblom and L. de Witte, eds., Amsterdam: IOS Press, 2002.
10. Brod, M., Stewart, A.L., Sands, L., and Walton, P. Conceptualization and measurement of quality of life in dementia: the dementia quality of life instrument. *The Gerontologist*, 39, 1999, p. 25.
11. Maulsby, D., Greenberg, S., and Mander, R. Prototyping an intelligent agent through Wizard of Oz. *Proc. InterChi93*, 1993, 277.
12. Gould, J.D, Conti, T., and Hovanyecz, T. Composing letters with a simulated listening typewriter. *Proc. ACM CHI '82*, 1982, 367.
13. Wilson, J. and Rosenberg, D. Rapid prototyping for human interface design. In *Handbook of Human–Computer Interaction*, M. Helander, ed., New York: North Holland, 1988.
14. Consolvo, S. and Towle, J. Evaluating an ambient display for the home. *Proc. CHI2005*, ACM, Portland, Oregon, 2005.

Bibliography

Orpwood, R. Design methodology for aids for the disabled. *J. Med. Eng. Technology*, 14, 2, 1990.
U.K. NHS Purchasing and Supply Agency Evaluation Service (NHS PASA). http://www.pasa.nhs.uk/evaluation/ (accessed April 2006).
U.K. Medicines and Healthcare Products Regulatory Agency (MHRA). http://www.mhra.gov.uk/ (accessed April 2006).
U.S. Food and Drug Administration (FDA) Device Evaluation. http://www.fda.gov/cdrh/ode/ (accessed April 2006).

chapter eleven

The business of pervasive healthcare

Dadong Wan
Accenture Technology Labs, Chicago, Illinois
Luis E. Taveras
Accenture, Florham Park, New Jersey

Contents

11.1   Introduction 276
11.2   Enabling technologies 277
    11.2.1   Health devices 277
    11.2.2   Networks 279
    11.2.3   Analytics 279
    11.2.4   Interactions 280
11.3   The business case 280
    11.3.1   Key business drivers 281
    11.3.2   Calls for a continuous care model 284
    11.3.3   Case study: Chronic care initiatives at Medicare 285
11.4   Examples of commercial applications 287
    11.4.1   Remote cardiac monitoring 287
    11.4.2   Health Buddy system 289
    11.4.3   Philips Motiva 291
    11.4.4   QuietCare system 292
11.5   Opportunities and challenges 294
11.6   Summary and conclusions 296
References 297

11.1 Introduction

The global healthcare industry is facing an unprecedented crisis. With the fast growth of an aging population in the United States and around the world, the demand for affordable but high-quality healthcare is accelerating. According to the U.S. Department of Health and Human Services, by 2050, the retiring baby boomers will require the population of caregivers to more than triple to 6.5 million workers, even though healthcare already accounts for about 15 percent of U.S. gross national product (GNP).[1] This rising demand is placing a tremendous economic burden on governments, private employers, and individual consumers alike. It also becomes a strain on the available capacity of skilled care professionals and nursing homes. Meanwhile, technologies such as personal computers, cell phones, and the Internet continue to transform every aspect of human existence, including healthcare. The advent of miniaturized sensors, wireless networks, and mobile devices is making traditional medical and consumer health devices smarter, cheaper, easier to use, and more ubiquitous. Telehealth and remote monitoring of patients is rapidly becoming an essential part of the new healthcare reality. Together with the increasing adoption of electronic medical records among provider organizations, these technologies hold great potential to transform the healthcare landscape, helping contain cost, improve the quality of care, and enable new classes of services.

The Internet has been a driving force behind the continuing trend of consumer-centric healthcare by opening up the floodgates of consumer health information. About two-thirds of Internet users search for health information.[2] However, an informed consumer is just the beginning. To be effective, health information needs to flow not just from institutions to patients but also from patients to healthcare providers, especially while patients are not in a care facility. In today's healthcare system, whether or not the patient receives the right level of care depends heavily on where the person is. Patients usually get the attention they need as long as they are physically in a hospital, a physician's office, or another professional care facility. As soon as they move out of these facilities, care providers lose touch with these patients, knowing little about what is going on with them and what care they might need. This problem becomes especially evident for people with chronic conditions (e.g., diabetics) and the elderly, who typically live normal lives at their own homes a majority of the time. In this case, the lack of timely information about the patient status and required interventions could lead to costly emergency room visits, hospitalization, and even death.

The recent rise of disease management is a direct response to some of these deficiencies in the current healthcare system by introducing structured, proactive interventions for the targeted population (i.e., chronic, heavy users of healthcare services). Most disease management companies today rely on a "high-touch" approach by having specially trained nurses call individual patients on a periodic basis, providing necessary counseling, coaching, and education. This approach has led to significant improvements in compliance and reduction in hospitalization.[3] However, its effectiveness is somewhat limited by the lack of real-time visibility to the current status of the patient population.

Until recently, it has been quite costly to support the continuous flow of patient data from patient homes to care providers, largely because doing so requires a dedicated communication and device infrastructure. With the increasing availability of home broadband, wireless networks, and a wide range of consumer health electronics, a de facto end-to-end infrastructure has begun to emerge. This, together with the widespread use of mobile devices, makes it feasible to carry out two-way, continuous interactions between patients and their care providers, regardless of where they are physically located. We use the term "pervasive healthcare" to refer to this new reality. Enabled by a smart environment, pervasive healthcare is always on, always active, and always aware. It encompasses a broad range of telehealth applications including remote patient monitoring, virtual visits, and telemanagement. We believe that the emergence of pervasive healthcare presents both opportunities and challenges for key players in the healthcare ecosystem, including payers, hospitals, clinicians, employers, device manufacturers and, of course, patients.

This chapter explores the potential business implications of pervasive healthcare technologies. First, we provide an overview of the key enabling technologies for pervasive healthcare including consumer health devices, wireless networks, analytic engines, and interaction technologies. Second, we describe the business case: that is, how pervasive computing enables a new type of care model that meets the critical needs of the chronically ill and the rapidly growing elderly population by reducing emergency room visits and shortening hospital stays. Third, we highlight four examples of commercial applications and how they are used to deliver both business values and improved patient outcomes. Finally, we discuss the opportunities pervasive healthcare brings to various stakeholders, including providers, payers, employers, device manufacturers, and patients. We also describe major challenges ahead, including reimbursement and standardization.

11.2 Enabling technologies

Pervasive healthcare is an emerging reality that is made possible by a number of recent technological developments. Figure 11.1 summarizes the four main enabling technologies for pervasive healthcare: health devices, networks, analytics, and interactions.

11.2.1 Health devices

The availability of miniaturized and inexpensive health devices allows the seamless capture of what is currently going on with individual patients and their environments while they are at home and away from a professional

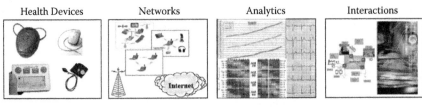

Health Devices	Networks	Analytics	Interactions

- Wearable health devices
- Implanted devices
- Home health electronics
- Home security
- Cameras

- Low-power networks
  - Bluetooth
  - ZigBee
  - UWB
- Home Wi-Fi
- GPRS/3G cellular
- Home broadband

- Trending
- Aggregation
- Synthesis
- Interpretation
- Prediction

- Visualization
- Decision-support
- EMR integration
- Collaboration

Figure 11.1 Four enabling technologies for pervasive healthcare.

care facility. Examples of the types of information that can be collected include the current location of the patient, activity level, pulse rate, blood pressure level, and sleep and stress patterns. While home health devices like blood pressure cuffs and bathroom scales have been around for years, new computing and communications capabilities not only make these devices smarter but also introduce an entirely new class of devices to the market.

Four areas of device advancement are of particular interest to successful pervasive healthcare. The first area is the increasing digital capacity of health devices. Thanks to Moore's Law (the power of microprocessor technology doubles every eighteen months) and the continuing microelectronics revolution, traditional home health devices have become increasingly digital, equipped with their own microprocessors and capable of actively communicating wired or wirelessly. Some of these devices (e.g., DynaPulse 200M blood pressure cuff) come with software that allows the user to perform sophisticated trend analysis. One can also easily connect these devices to the Internet and securely share the information with doctors and care providers anywhere in the world. For example, A&D Medical manufactures a number of Bluetooth-enabled home devices, including weight scales and blood pressure cuffs, allowing these devices to wirelessly communicate data anywhere in the house and to care providers.

The second area is the miniaturization and consumerization of medical devices. Professional medical services such as ECG and cholesterol testing are moving out of laboratories and into consumer homes. The ecg@Home from HealthFrontier, for example, measures about 4 inches by 3 inches and weighs only 100g. It records and stores electrical heart signals that can be obtained noninvasively by using two built-in electrodes on which the thumbs are placed. With these smaller, more affordable, and easier-to-use devices, consumers can perform the same types of measurements or monitoring that used to be carried out only at clinics or hospitals within the comfort of their own homes.

The third area is the emergence of a new class of health devices that have their roots in the fitness community but have recently begun to appear in the broader healthcare market. The HealthWear armband from BodyMedia is a good example. Worn on the back of the upper right arm, the armband continuously gathers detailed physiological data like movements, heat flux, skin temperature, and galvanic skin response. Such low-level data provide the basis for inferring important lifestyle patterns, such as energy expenditure, sleep, and exercise, which are useful in a wide range of treatment programs, including weight, sleep, chronic disease, and wellness management. Recognizing this huge growth potential, traditional consumer electronics giants ranging from Samsung to Best Buy have placed a large bet on the consumer health device market by opening new business units and stores dedicated to these products.

Finally, the availability of better and cheaper environmental sensors and cameras has begun to transform homes into intelligent environments where the health status of its residents, not just safety, can be closely monitored. In addition to turnkey solutions from traditional home security service providers, consumers can also purchase such off-the-shelf products as Motorola's HomeSight, which can be seamlessly integrated with a home wireless and broadband infrastructure, allowing family members to view what is going on at their homes through webcams and to receive automatic notifications via email or cell phone in critical situations such as when a smoke detector goes off.

11.2.2   Networks

Closely related to health monitors and electronics are recent developments in wireless networks, including home Wi-Fi, Bluetooth, ZigBee, and Ultra-Wide-Band (UWB). While Internet broadband provides the backbone between the home and remote care providers, the missing link for seamless health monitoring and care delivery lies in the link between different devices around the home, which is aptly called the "last foot" problem. Traditional health devices are stand-alone and do not need to communicate with other devices or applications. However, to enable effective pervasive healthcare, different monitoring devices need to coordinate and transmit data. New devices such as Bodymedia and Fitsense use proprietary, lower-power wireless networks. Early signs indicate that the market at large is beginning to embrace Bluetooth as a de facto standard for in-home health device communication.

11.2.3   Analytics

Once the physiological and behavioral data about an individual's health and their immediate environment are gathered and transmitted to the right place, the next challenge becomes how to make sense of it so that appropriate actions can be taken. For example, with cameras at home, it is possible to monitor the movement and activities of an elderly person in a house. To discern a significant health event (e.g., a fall) requires analytics that interpret the raw visual input from the cameras and other sensory sources. Ultimately, individualized models are needed for each patient based on the longitudinal data so that predictions can be made about how likely it is that the individual might fall. Thus, preventive measures could be taken before an acute situation develops. Some predictive modeling techniques based on sensory data have already been developed and commercially applied in other industries. Smart Signal, for instance, has successfully used a similarity-based modeling (SBM) application for detecting anomalies and predicting potential failures in complex systems such as aircraft and city bus engines and nuclear power reactors.[4] In doing so, they are able to demonstrate significant savings in maintenance costs while simultaneously reducing downtime for airlines and power companies. As more detailed physiological data about the patient become readily available as a result of the proliferation of health and other sensory devices, a similar analytic approach may be applied to detect anomalies and predict failures in human health, especially for people with chronic conditions. Therefore, these predictive capabilities could help reduce the number of potential emergency room visits and hospitalizations.

11.2.4 Interactions

Healthcare delivery involves a complex web of players, including primary care physicians, specialists, nurses, pharmacists, hospitals, insurers, dieticians, family caregivers, and patients. Each of these players has their unique roles and corresponding informational needs. Even if all the data about a patient are readily available, the hurdle still remains of how to enable the right people to make the best use of the information and insight to make timely and good decisions. The final piece of the technology puzzle for pervasive healthcare is the development of advanced interaction technologies, which include user modeling, visualization, decision support, and collaboration tools. These technologies help people manage their attention and online environments so that they stay focused on what is truly important. They also provide the necessary intelligence that determines the real-time routing of information so that the right people get the right information at the right level of detail.

11.3 The business case

In recent years, tremendous progress has been made on the technology front in terms of improved economics, power, functionality, and usability. Despite that, these technologies are unlikely to gain widespread acceptance by the healthcare industry unless it is shown that they help improve clinical outcomes and provide positive return on investment (ROI). In this section, we focus on the latter. Specifically, we argue that pervasive healthcare offers a promising means for reducing healthcare cost and for addressing the shortage of provider resources. More important, it will be a key enabler for a new model of care that is necessary to address the root cause of the current global healthcare crisis (i.e., the prevalence of chronic diseases and global aging demographics).

11.3.1   Key business drivers

Perhaps the most important force behind the current and potential widespread use of pervasive healthcare is the need to drive down or at least contain skyrocketing healthcare costs, which affect not just private health insurers but also governments at all levels, employers, and individual consumers. For example, in April 2005 General Motors announced that its earnings for the fiscal year fell far short of expectations, largely due to its increasing healthcare obligations: GM paid in excess of $5 billion in 2005 to insure its workers and retirees. This translates into about $1,500 for every GM car and truck sold.[5] Even as private businesses are willing to take on their fair share of healthcare costs, employees must still shoulder significant increases in both health premiums and out-of-pocket payments in the form of deductibles and co-pays. As shown in Figure 11.2, despite a moderate slowdown in 2004, healthcare premiums in the United States continue to grow at double-digit rates, far exceeding that of overall inflation and workers' compensation.[6]

Early evidence shows that pervasive healthcare can help significantly reduce the cost of care by reducing potential emergency room visits and hospitalizations. An analysis by the New England Healthcare Institute reveals that compared to standard outpatient care, using remote monitoring for heart failure patients reduces rehospitalization rate by 32 percent following a heart failure hospitalization. Based on the average rate of $9,700 for a 5.5-day stay in the hospital for each hospitalization, this amounts to a net savings of $1,861 per patient for the six-month study period.[7] Similar cost savings were also reported by Partners Healthcare and the Veterans Health

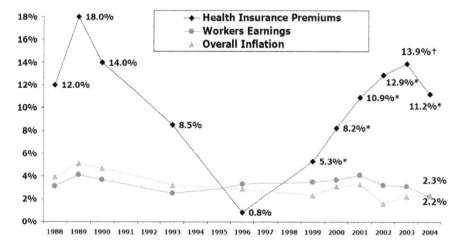

Figure 11.2 Increases in health insurance premiums compared to other indicators.

Administration (VA).[8,9] The VA trial on patients with diabetes, lung disease, and heart failure reported a 35 percent reduction in readmissions and a 60 percent drop in emergency room visits. This translated into a net savings of $23 million across the study population.[10,11]

The second driver for pervasive healthcare is to help improve productivity among provider organizations by addressing increasing shortages of healthcare professionals, ranging from medical technologists to nurses. The worldwide nursing shortage is perhaps most noticeable and critical, because nurses are the primary source of care and support for patients at the most vulnerable points in their lives. Figure 11.3 shows the continuing rise in demand for registered nurses in the United States compared to the dwindling supply. By 2015, demand will exceed supply by 20 percent, quadrupling that of 2000.[12] The problem is especially acute in senior and home care. According to the Center for Medicare and Medicaid Services, 90 percent of long-term care organizations lack sufficient nurse staffing to provide even the most basic care.[13] At these levels, home care agencies are being forced to refuse new admissions due to lack of staffing.

Although most attention about pervasive healthcare currently focuses on cost reduction, pervasive technology can play a vital role in improving caregiver productivity. For example, when a homebound patient can take his or her own vital signs, each visit by a care professional can be shortened by fifteen to twenty minutes by doing the test before the nurse arrives. Additionally, the technology also could eliminate needless weekly visits, with visits prompted only when a monitoring device detects a problem. A recent study by Partners Healthcare shows telemonitored congestive heart failure (CHF) patients required 40 percent fewer nursing visits.[8] In another study, by using its remote cardiac monitoring network, Medtronic reports a 50 percent reduction in the length of follow-up visits.[14] As a result, home care agencies and cardiac clinics are able to handle significantly more patients with the same or even a fewer number of staff.

The third and final driver for pervasive healthcare has to do with the quality of patient care. In its landmark report on issues surrounding the

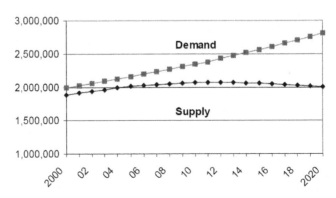

Figure 11.3 Projected RNs supply and demand in the United States, 2000–2010.

quality of care delivery in the United States, the Institute of Medicine (IOM) noted that hospital discharge often signals an abrupt drop in care quality because, when left on their own, patients frequently lack the means, information, discipline, and oversight necessary to care for themselves.[15] One example is patient compliance with prescribed medication plans. A UCLA study of patients immediately after their hospital discharge reveals that 73 percent failed to use at least one medication according to physician instructions and, of all the drugs ordered at discharge, only 32 percent were taken at all.[16] The net result of such noncompliance is poor outcomes in terms of delayed recoveries, complications, and even rehospitalization.

One of the major premises of pervasive healthcare is that it allows patients to continue to receive professional attention and care while they are at home and away from a medical facility. In this continuous care environment, real-time remote monitoring devices can diminish noncompliance behaviors and abnormal physiological developments can be detected promptly. As a result, appropriate intervention action can be taken quickly so that an acute or emergency situation might be averted. This assertion was supported by the results of a six-month, 600-patient study by the VA that showed not just significant cost savings but also a more than 90 percent patient satisfaction rating. Furthermore, patients also reported they were more educated, secure, and better able to manage their own healthcare needs.[11] This increased sense of patients' security was also supported by the result from Medtronic CareLink, which showed that patients and their families enjoyed additional peace of mind knowing they were always remotely connected to the clinic and physicians (see also Section 11.4.1).[14]

Figure 11.4 summarizes the market forces behind the adoption of pervasive healthcare among the various stakeholders. While rising healthcare costs, provider resources shortage, and care quality are the most visible drivers, a closer examination reveals two deeper forces underlying the current crisis: the prevalence of chronic diseases and an aging global population. Because most chronic patients are also elderly, the latter trend will only exasperate the former. In the following two sections, we make a strong case that addressing these two fundamental forces calls for a continuous care model and that pervasive healthcare is an ideal technology enabler for this

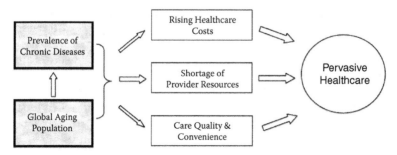

Figure 11.4 Market forces behind the adoption of pervasive healthcare.

new model. The current Medicare chronic care initiatives underway show early signs of movement in this direction, as discussed in more detail in Section 11.3.3.

11.3.2 Calls for a continuous care model

The increase in chronic diseases, including heart disease, diabetes, and asthma, is widely recognized as the number one healthcare challenge. Over 75 percent of U.S. healthcare spending goes to the care of people with chronic conditions. In 2004, nearly half of Americans, or 133 million people, were diagnosed with one or more chronic condition. With the rapidly aging baby-boomer generation and their concomitant health problems, the number of people with chronic conditions is projected to reach 157 million by 2010.[17] Yet the current healthcare system, which is based on an episodic, acute care model, is poorly equipped to meet the needs of this rapidly growing population. Instead, the episodic model is optimized toward "fixing" patients when they have acute problems. This works well for patients who have broken legs or are in need of open heart surgery, but does not work well for patients with Alzheimer's disease or diabetes.

Because chronic illnesses are ongoing conditions that are often not curable, chronically ill patients require daily management and self-care as well as coordinated and timely interventions from healthcare providers. Without appropriate guidance and behaviors such as taking the right medication at the right time, measuring vital signs regularly, and eating properly, a patient's status can deteriorate from manageable symptoms into a more serious condition such as heart failure, asthma attack, or another crisis that requires an emergency room visit or hospitalization. A poorly managed chronic condition can also lead to a number of complications or other illnesses. For example, diabetic patients are at risk of peripheral vascular disease, which can in turn lead to amputation and disability.

Although chronic diseases affect people of all ages, they strike the elderly population particularly hard. For example, congestive heart failure is a leading chronic disease for older adults and accounts for over a quarter of a million deaths each year.[17] The other prevalent aging-related chronic illness is the decline or impairment of cognitive abilities. An estimated 4.5 million Americans suffer from Alzheimer's disease. It is expected that by 2050 this number will rise to between 11.3 and 16 million people.[18] Compared with other chronic diseases, managing these two conditions requires more support and participation from caregivers because patients themselves are often frail and unable to care for themselves.

The prevalence of chronic diseases calls for a continuous care model where patients are empowered to engage in self-care and receive ongoing education and proactive interventions from healthcare providers. The key in this model is having the ability to identify early warning signs of potentially acute problems and to intervene promptly to avert unnecessary complications or trips to emergency rooms. This new model is precisely what pervasive healthcare attempts to support. With remote monitoring using home and wearable devices and always-on connectivity, real-time alerts and a plethora of other telehealth capabilities become possible. Patients monitored continually by healthcare devices can be automatically provided with reminders about what they need to do. When necessary, care providers on the remote end can also step in to make sure the patient is on the right course or to answer any specific questions. This model provides an effective means to manage patient conditions while they are at home or any other location away from their caregivers. Consequently, it can result in less patient suffering, lower costs, and better use of scarce professional care resources.

The VA is a pioneer in successfully demonstrating the effectiveness and savings in applying pervasive healthcare to chronic disease management. In 2000, the VA funded eight two-year clinical demonstration projects in Florida to test the concept of "aging-in-place" by providing disease management, care coordination, and remote monitoring of veterans in their homes. The published results from these projects showed a 40 percent drop in emergency room visits, a 60 percent decrease in hospitalizations, a 64 percent decline in nursing home admissions, an 88 percent reduction in nursing home bed days of care, and over 90 percent patient satisfaction ratings.[11] Similar positive results were reported from a later expanded program in ten other states and territories.[19] Based on the success of these early pilot programs, in 2003 the VA established the Office of Care Coordination (OCC) to oversee a national rollout of such programs, with the ultimate goal of making the service available to 1.2 million veterans by 2008.

11.3.3 Case study: Chronic care initiatives at Medicare

Medicare is the U.S. federal health insurance program and covers nearly 42 million Americans: 35.4 million seniors and 6.3 million people under the age of sixty-five who have disabilities. The program was first introduced in 1966 as part of amendments to the Social Security Act and has grown steadily over the years. In 2005, Medicare benefits amounted to about $325 billion or 13 percent of the federal budget. This number is expected to grow to $444 billion in 2010.[20] A major reason for this rapid growth in cost is the prevalence of chronic conditions among the Medicare population. Nearly 80 percent of Medicare beneficiaries suffer from at least one of the following chronic conditions: stroke, diabetes, emphysema, heart failure, hypertension, or arthritis. Almost 63 percent have two or more conditions. About 20 percent of beneficiaries have five or more chronic conditions and account for over two-thirds of Medicare spending.[20] This growing burden of chronic diseases constitutes a severe threat to the long-term solvency of the program, prompting Congress to enact the Medicare Modernization Act of 2003 (MMA). In Section 721, the Act specifically calls for voluntary chronic care improvement programs for fee-for-service beneficiaries with one or more chronic conditions.

As part of the MMA implementation, the Centers for Medicare and Medicaid Services established two major chronic improvement initiatives: Medicare Health Support (MHS) and Care Management for High-Cost Beneficiaries (CMHCB). The former is a two-phase program. Phase I, which began in August 2005 and will last for three years, is a piloting phase involving over 180,000 Medicare beneficiaries who suffer from congestive heart failure and complex diabetes. Based on a competitive bidding process, ten healthcare organizations from different geographical areas throughout the United States were selected to operate these pilots. The program is the first-ever nationwide effort to test advanced disease management techniques including the use of remote monitoring and other pervasive healthcare technologies. These techniques are intended to help chronic patients manage their own health, aid in adherence to a physician's treatment plan, offer self-care guidance and support, and enhance communication between patients and care providers. The operators are required to monitor and report on health outcomes including reductions in emergency room visits or hospital readmittance rates, beneficiary satisfaction, and cost savings. Phase II calls for the expansion of these programs or program components to more Medicare beneficiaries. However, this second phase will only move forward if the pilot has successfully achieved its preset goals including the delivery of a 5 percent net savings to the government.

The CMHCB is another three-year demonstration program to help Medicare beneficiaries with high-cost/high-risk chronic conditions to improve their quality of life, prevent complications, and reduce medical expenses. Six organizations were selected to test a direct provider model in coordinating care for such patients by offering them with care support beyond traditional settings. This approach includes the use of home monitoring devices, care coordination, self-care and caregiver support, education and outreach, health reminders, and twenty-four-hour nurse helplines. Among the six pilot organizations is the Advancing Chronic Care through E-Health Networks and Technologies (ACCENT)—a consortium of physician clinics in Oregon and Washington, a home monitoring company, and the American Medical Group Association. The ACCENT-participating medical groups are implementing technology-supported care management that encompasses home-based appliances for electronic health coaching and patient monitoring as well as decision-support tools for providers. In the fall of 2005, the program enrolled about 2,000 Medicare patients with congestive heart failure, chronic obstructive pulmonary disease, and/or diabetes. ACCENT is notable for its first implementation of remote monitoring and coaching across multiple sites and its heavy reliance on pervasive healthcare technologies. The program will document how the physicians' use of these technologies helps high-cost chronic patients to stay healthy, avoid complications, and lower medical costs.

11.4 Examples of commercial applications

In this section, we describe four commercial applications of pervasive healthcare that are already available in the marketplace. While most of them are still at an early stage, these applications offer a glimpse of how established and start-up companies alike are commercializing these technologies and using them to help improve healthcare delivery, namely providing higher quality of patient care at a lower cost.

11.4.1 Remote cardiac monitoring

Heart disease is the number one cause of deaths in the Western world. A quarter of all deaths in the United States—about one thousand deaths a day—are sudden cardiac deaths. Cardiac deaths are often linked to a condition called cardiac arrhythmia or abnormal heart rhythm. As many as 2.2 million Americans are living with one type of arrhythmia called atrial fibrillation. One of the most effective ways to extend and improve the lives of these patients is by surgically implanting small electronic medical devices including pacemakers and implantable cardioverter defibrillators (ICDs). These devices continuously monitor heart rhythms and automatically deliver appropriate electrical therapy if the heart rate becomes irregular. Based on overwhelming clinical evidence about the effectiveness of ICD therapy, the U.S. government has decided to expand ICD coverage significantly to thousands of Medicare beneficiaries who are at an elevated risk for sudden cardiac death.[21]

Once a pacemaker or ICD is implanted, follow-up visits become essential to ensure its proper functioning. Traditionally, these visits must take place in a cardiologist's office every three to six months, depending on the type of device. During these follow-up visits, the nurse or technician noninvasively uses a programmer terminal to gather data from the device. Information about the battery life, lead status, device settings, and other data is evaluated. If necessary, changes in the device settings can also be made during these visits. A physician is called upon when clinical attention is necessary. As the number of patients eligible for ICD therapy has almost doubled in recent years, follow-up visits have increasingly become a bottleneck in patient care.[14]

One of the most significant breakthroughs in arrhythmia management is the emergence of a new generation of pacemakers and ICDs that supports wireless remote monitoring and management. With this capability, physicians are able to evaluate a patient with an implantable cardiac device over the Internet and beyond clinic walls, without requiring the patient to leave home. The patient typically uses a small wireless device at home or on the road to interrogate and collect data about heart and ICD activity. The data are wirelessly transmitted via a standard telephone, cellular, or Internet connection to a remote data center, where the data are then analyzed. Clinicians access patient data by logging onto a secured Web site. Patients and family members can also view information about devices and conditions on their own secured personalized Web sites.

Remote cardiac monitoring through implanted devices provides clear benefits to device manufacturers, clinicians, and patients. Faced with potential recalls for defective ICDs and pacemakers, the medical device industry has turned to wireless monitoring to keep tabs on their products and minimize potential damages in the case of malfunctioning. For clinicians, the new remote monitoring capability helps improve operational efficiency and enables them to handle a much larger volume of patient flows with the same level of staffing. For example, the average length of time for a follow-up visit for patients using Medtronic CareLink has been reduced by half.[14] Remote cardiac monitoring also leads to better and more efficient patient care. The ready access to patient and device data allows timely problem detection and attention from clinicians. For patients, the most notable improvement is enhanced convenience. Instead of having to schedule and travel to the clinic, over half of the routine follow-ups can be performed in the form of "virtual visits" directly from the comfort of the patient's home. In addition, patients and their families enjoy the peace of mind knowing they are always remotely connected to the clinic and physicians.

The four major implanted cardiac device manufacturers (Biotronik, Guidant, Metronic, and St. Jude Medical) have marketed their own versions of FDA-approved ICDs that support remote monitoring (see Table 11.1). The Biotronik Home Monitoring system received FDA approval in 2001. Aside from being the first system of its kind on the market, two other features set it apart. First, it relies on a Global Standard for Mobile (GSM) cell-phone link to connect the patient to their service center, which means that the patient can be monitored anywhere in the world where GSM coverage is available. Second, the device can be programmed so that when certain important events take place (e.g., the onset of silent arrhythmias), the data are transmitted in real time and the physician can be notified immediately.

Unlike Biotronik Home Monitoring, the St. Jude Medical HouseCall Plus system uses standard telephone lines to transmit ICD data equivalent to a full, in-office device interrogation including electrograms, surface ECGs, delivered therapies, and stored electrograms. It also features live medical professionals to analyze the transmissions immediately and communicate with the patient. Being the latest entrant to the market, the Guidant Latitude Patient Management System features an in-home monitoring unit that can connect to third-party devices via Bluetooth, including weight scales and blood pressure cuffs. In doing so, Guidant opens its patient monitoring platform to accommodate data from sources other than its own devices.

By far, the most widely used remote ICD monitoring is Medtronic CareLink Network, which was originally approved by the FDA in 2002. It is currently available to nearly 530,000 pacemaker and ICD patients and in more than 150 electrophysiology clinics. The system is running in twenty-two VA clinics, and nearly five hundred veterans are being monitored remotely. In total, there are over 37,000 CareLink users in the United

Table 11.1 Remote Cardiac Monitoring from Four Major Medical Devices

Company	Remote Cardiac Monitoring Service	Date of FDA Approval	Number of Users	Special Features
Biotronik	Biotronik Home Monitoring	Oct., 2001	N/A	Real-time data transmission if necessary; GSM cellular network
Guidant	Latitude Patient Management	Sept., 2005	8,000	Home communicator uses Bluetooth and can connect to optional third-party devices such as scale, blood pressure cuff
Medtronic	Medtronic CareLink Network	Jan., 2002	37,000	Reimbursement agreements with major payers across the United States
St. Jude Medical	HouseCall Plus	Nov., 2003	N/A	Live medical professional available to analyze data immediately and communicate with the patient

States.[22] Medtronic attributes its huge success to its ability to work with major health plans and other payer organizations so that clinics and physicians are compensated for providing remote monitoring and "virtual office visits" under either existing or miscellaneous reimbursement codes.

11.4.2 Health Buddy system

Health Hero Network is a leading provider of technology solutions for remote health monitoring and care management. Its Health Buddy system includes monitoring devices, clinical information databases, Internet-enabled decision-support tools, content development tools, and health management programs. The centerpiece of the solution is the Health Buddy appliance, which can transmit patient data to remote healthcare professionals via a telephone line. By using one of the four buttons on the device, the patient can answer questions about vital signs, symptoms, and behavior. The Buddy Link allows third-party health devices (e.g., glucometers, weight scales, and blood pressure cuffs) to connect to the appliance, enabling the patient to gather and upload vital sign data. At the back end, the iCare Desktop is an Internet-based care management and research tool allowing care professionals to spot trends, assess risk levels, and monitor a large patient population. The Health Hero Network uses these capabilities to deliver daily monitoring and education that promote positive behavioral changes and better compliance from the patient. This is achieved through more than a dozen standard health management programs for common chronic illnesses including diabetes, congestive heart failure, and asthma. Figure 11.5 depicts how these components fit together to form an integrated system.[23]

Since 2001, a number of field pilots have demonstrated that the use of the Health Hero system could lead to cost savings and improvements in patient quality of life and provider satisfaction. The Veterans Integrated Service Network of Florida (VISN 8), for example, used the system in a six-month program involving six hundred patients with chronic conditions including heart disease, lung disease, diabetes, and mental health. The study showed a 74 percent reduction in inpatient, outpatient, and medication costs, which translated to a net savings of $23 million across the study population. Patients also reported they were more educated, more secure, and better able to manage their own healthcare needs.[9] A separate program at the Henry Ford Health System involves over two hundred heart failure patients since 2004. Early results indicate a high patient satisfaction (92 percent), improved compliance (88 percent), and a decrease in ER visits and hospitalization.[24]

The Health Hero system has been used by over 7,000 patients in a number of healthcare organizations, including the VA, Kaiser Permanente, the Henry Ford Health System, and Mercy Health Center in Texas. The VA, by far the largest user, currently uses the system as part of their OCC initiative to improve clinical outcomes and to increase the quality of healthcare for

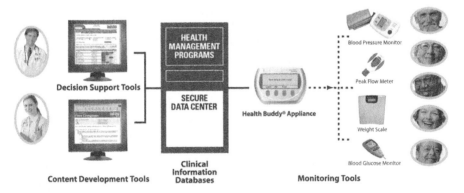

Figure 11.5 Health Buddy system. (With permission from Health Hero Network.)

veterans. More recently, Health Hero Network, together with three Washington State/Oregon-area medical groups, was awarded one of the six demonstration projects by CMS under the Care Management for High-Cost Beneficiaries Demonstration (see Section 11.3.3). Over the course of three years, the program will enroll about 2,000 Medicare patients with severe chronic illnesses including congestive heart failure, chronic obstructive pulmonary disease, and diabetes. Its goal is to employ the Health Buddy system to help improve the quality of life for these patients while significantly reducing costs by preventing hospitalizations.

McKesson—one of the largest healthcare information technology companies—has successfully integrated the Health Buddy system into two of its own solutions: Telehealth Advisor and Horizon Homecare. While Telehealth Advisor provides support for interactive disease management, Horizon Homecare is targeted at the growing number of chronically ill patients who are also in need of home care services. With the new capability, over 17,000 nurses and other clinicians who are currently using Horizon Homecare will have access to an integrated solution, including data from nurse visits and telehealth monitors.

11.4.3 Philips Motiva

Royal Philips is one of the leading providers of innovative telehealth solutions around the world. Through its Medical Systems division, Philips has been offering telemonitoring devices and services that enable disease management companies and other healthcare providers to remotely monitor chronic disease patients at home over a standard telephone line. The latest offering from Philips is called Motiva. Motiva takes remote patient management to a new level by taking advantage of technologies that are available to an increasing number of households, such as broadband connectivity and interactive television. By combining its core competencies in both consumer electronics and medical devices, Philips aims to turn the TV set into a personalized health appliance that connects patients at home with their care providers, providing them a customized healthcare channel or "virtual health coach." By supporting daily personalized interactions and rich multimedia content delivered on demand through a broadband connection to the home television, Motiva enables providers to work with patients with chronic conditions to monitor their conditions, ensure compliance, and drive desired behavioral changes.

At the center of Motiva are a set-top box and a set of wireless health monitoring devices, which are installed in the patient's home at the time of enrollment. Through a secured broadband connection, the patients receive timely reminders and personalized reinforcement messages, educational videos, and feedback on their vital signs, based on a customized care plan defined by their individual care providers. The patients can use the wireless monitoring devices to track their own vital signs including body weight, heart rate, heart rhythm, and blood pressure. Beyond patient compliance,

Motiva's clinical decision-support module helps care providers to better coordinate care delivery and offers more cost-effective patient management.

Motiva is still in a pilot testing and early rollout stage. In June 2005, Philips announced the completion of a four-month pilot study of Motiva in the United States, which involved thirty chronic heart failure patients selected by a New Jersey–based physician group.[25] The key reason for selecting heart failure patients is that most of these patients are over sixty-five, live at home, and want to remain independent for as long as possible. So far, the study is showing overall positive results among both patients and care providers, including:

- Patients show a broad acceptance of using TV to get personalized health information.
- Patients find the TV interface easy to use and find it helps them establish an effective daily routine.
- Patients like getting daily feedback about their vital signs and education videos about how to better manage their diseases.
- Doctors and nurses feel the system helps improve their connection with their patients and makes them more aware of the patients' health status.

To further quantify the clinical and business benefits of Motiva, Philips and Achmea, one of the largest health insurers in the Netherlands, recently started a large-scale clinical evaluation of the system on a randomized population of 630 chronic heart failure patients from eight hospitals in the Rotterdam area of the Netherlands. The twelve-month pilot study will test the usability of the Motiva platform and its impact on hospitalizations, quality of life, mortality, and utilization rates. In May 2006, Philips announced the commercial availability of Motiva in the United States. It expects to roll out the system later in the year in selected European countries.

11.4.4   QuietCare system

Today, over thirty million seniors older than sixty-five years live alone in the United States, and this number is rising daily. According to the Centers for Disease Control and Prevention, one in every three seniors falls in the home every year. The annual medical expense related to falls alone amounts to more than $20 billion in the United States and is projected to climb up to $32 billion by 2010.[24] Furthermore, falls also cause tremendous emotional stress on the elderly and their families as a result of unnecessary hospitalization, premature nursing house placement, and increased dependency. To help ease the financial and emotional burden of elderly care, a new kind of elderly monitoring service is making rapid inroads in the marketplace, and QuietCare is a notable example.

QuietCare is an unobtrusive home health security system first launched in 2002 by Living Independently following more than twelve years of research and development funded partially by the National Institutes of Health and

Aging.[26] At the center of QuietCare are five or six small, wireless activity sensors that are strategically placed in an elderly person's home (i.e., in the kitchen, in the bathroom, near the medication counter, and in the bedroom). Each sensor continuously gathers data about the individual's activities of daily living (ADLs), including waking up, bathroom usage, potential bathroom falls, meal preparation, medication handling, and the room temperature. For example, an alert is generated if a senior fails to visit the medicine cabinet during the day or goes to the bathroom and doesn't come out within an hour. The information is transmitted to a local hub device that periodically relays it to a remote server over the standard telephone line. One unique feature of the system is its ability to learn about each individual's normal ADL patterns over time. Information is analyzed in real time at the data center so that someone at the remote response center can react in a timely fashion to emergencies or emerging problems before they become critical. When certain significant changes or events occur, caregivers can be notified via e-mail, text message, pager, or phone call. Historical information about the user is accessible by authorized personnel via a secured Web site.

Initial results from installations at assisted living facilities have shown that QuietCare is a valuable early detection and warning mechanism. The alerts generated by the system quite accurately reflect what actually takes place in the residents' apartments and can lead to prompt interventions. For example, the alerts because of increased bathroom usage for two residents led to the timely diagnoses and treatment of a urinary tract infection and a bladder problem. Besides helping identify problems before they become emergencies, QuietCare also offers a number of other benefits to care facilities, patients, and patients' families. It enables facilities to utilize resources more optimally by offering 24/7 coverage for certain residents without the cost or intrusion of additional staff. It also provides necessary documentation to detail client needs and staff responses. As a result, the system helps to improve staff accountability and regulatory compliance. Because the system is unobtrusive, it is able to offer enhanced peace of mind to seniors, their family members, and professional caregivers, all without requiring seniors to wear anything or change their normal routines.

QuietCare is one of the first commercially available and affordable systems of its kind. The benefits it has demonstrated in early installations have led ADT—one of the largest home security service providers in North America—to license the technology and market it to U.S. and Canadian consumers as a home health security and medical alert service. ADT's entrance is especially significant because its well-known brand lends instant credibility to this emerging market. Additionally, ADT's extensive experience in customer service, network of contact centers, and very large existing customer base can not only help the initial uptake but also the ultimate scalability of this service to thousands of homes.

ADT offers two levels of monitoring service: QuietCare and QuietCare Plus. Both are designed to help seniors stay independent, avoid medical emergencies, and live in their own home longer, while giving their family

Table 11.2 Cost Comparison between QuietCare and other Modes of Elderly Care[27]

	Home Health Aide	Assisted Living Facility	Nursing Home	Hospitalization
QuietCare				
$2.99/day	$15/hr	$50/day	$345/day	$1,200/day
	30 hrs/week	$250/week	$2,415/week	8 days/stay
				$9,600/total
				stay
$89.95/	$2,100/	$3,500/	$2,524/	
month	month	month	month	
$1,079/year	$23,400/year	$42,000/year	$30,288/year	

caregivers peace of mind. At the basic level, ADT monitoring specialists respond to system alerts about a possible bathroom fall, no morning bedroom exit, or unsafe room temperature by both calling the customer's residence and notifying family caregivers via a secured Web page, e-mail, or pager. To provide additional security, seniors and family members can subscribe to QuietCare Plus, which includes Companion Services—ADT's personal emergency response system (PERS). Companion Services include a waterproof button on a personal pendant or wristband and a two-way voice communication system that doubles as a speakerphone with push-button answering. In the case of an emergency, seniors simply push the button. Upon receiving the alarm signal, ADT will send help if needed. In comparison with other means of care, QuietCare offers a cost-effective alternative in helping elderly persons avoid medical emergencies while retaining their independence and dignity, as shown in Table 11.2.

11.5 Opportunities and challenges

Pervasive healthcare applications such as the examples described above offer key stakeholders great opportunities to address their respective priorities. By increasing patient visibility and enabling care providers to take early actions, they provide a lower cost and better quality of care, resulting in a winning proposition for everyone involved in healthcare: patients, clinicians, health plans, employers, government, and family members. Patients will benefit from improved health and quality of life and reduced out-of-pocket costs as a result of less frequent emergency room visits and hospital stays. Better education and support enable the patient to have a significantly increased ability to understand and manage their overall health and well-being. Healthcare providers benefit from improved decision making and enhanced work productivity. Equipped with both a comprehensive patient history and a current picture of a patient's condition, doctors could have more productive and comprehensive office visits with patients. Other potential benefits include more manageable patient loads and better utilization of scarce provider resources, including nurses and hospital beds. Health plans could reduce insurance premiums and payouts by ensuring that patients are treated proactively. Through unobtrusive monitoring and objective documentation, pervasive healthcare could also help increase compliance and prevent fraudulent claims by providing health insurers with an accurate account of what is actually happening with the patient. Employers win by controlling rising healthcare costs for current and retired employees and their families. Furthermore, better health services ultimately translate into higher worker productivity through lower absenteeism. Governments could ease the burden on taxpayers, who ultimately pay for medical entitlement programs. In 2002, for example, the average per-person healthcare cost in the United States was $5,000, with about 45 percent financed from government funds such as Medicare and Medicaid.[28] Finally, family members can enjoy the healthy company of their loved ones longer and have the peace of mind of knowing that they are receiving any necessary medical attention on a continuous basis.

Pervasive healthcare also provides a tremendous business opportunity for technology companies. The demand for new innovative wearable and home health devices and services will likely continue to grow at a rapid rate. To fully leverage their technology prowess and brand power, well-established firms like Intel, Samsung, Panasonic, and Philips have aggressively expanded into the health device market. Retailers such as Best Buy have opened separate stores to sell such devices. Meanwhile, many start-ups are coming out of the gate to offer a wide range of niche solutions (see Section 11.2.1). Traditional medical device manufacturers, including Medtronic and Guidant, have also added telemetry capabilities to their new generation of devices. As a result, they are able to better differentiate their products as well as create new revenue streams through services (see Section 11.4).

Despite the vast potential of pervasive healthcare, a number of major challenges still lie ahead. Perhaps the foremost barrier is reimbursement. In other words, who will pay for the technology and the services it enables (e.g., remote monitoring and virtual visits)? As long as doctors and hospitals remain unsure about whether they will get paid for these new services, pervasive healthcare is unlikely to be widely adopted. In the current system, the logical answer is funding through health plans and insurers. However, these parties are generally slow in providing coverage for new technologies. Consumers have increasing control over more healthcare dollars through flexible spending and medical saving accounts. However, recent studies show that consumers are reluctant to pay for these services out of their own pockets.[29] In addition to the financing hurdle, this new way of enhancing the interaction between caregiver and patients faces a cultural challenge as well because many doctors and other caregivers have shown resistance to change. This resistance becomes particularly strong when new services require changes in workflow. In addition, some patients, especially those who are elderly and frail, may feel uncomfortable with technology and prefer having doctors and nurses physically present at their bedsides instead of virtually from another location.

Beyond the abovementioned organizational and cultural barriers, there are also a number of technical obstacles including standards, data security, and usability. Due to the large number of parties involved in healthcare delivery, standards are critical to achieving interoperability and seamless services. Currently, there is still a lack of standards in many areas including wireless communications among home and wearable devices, data privacy, and data formats governing information sharing among patients, clinicians, and health plans. As a result, healthcare organizations must choose between incompatible solutions available on the market and must guess which service they feel might eventually win. This uncertainty discourages early adopters and slows down the uptake of these technologies. The other major technical barrier is related to usability, which relates to simpler software user interfaces, form factors, battery lifetime, and so on. Usability is especially important for pervasive healthcare applications because the typical user is an elderly patient with a chronic illness whose goal is to get well, not to learn new technologies.

11.6 Summary and conclusions

This chapter covered business issues surrounding pervasive healthcare. We began by reviewing four key enabling technologies including health monitoring devices and consumer health electronics, wireless networks, analytic engines, and advanced interaction technologies. Next, we presented a detailed business case for pervasive healthcare, discussing how we may apply such technologies to address not just the industry-pressing problems of cost, nursing shortage, and quality but also the root causes of the current healthcare crisis, including the prevalence of chronic conditions and global aging demographics. We then selected four commercial applications to highlight how pervasive healthcare technologies are being used today and what business values and patient outcomes are being delivered. We finished with a discussion of major economic, cultural, and technical challenges that must be overcome for a broader adoption of the technology.

Given the evidence we have provided in this chapter about the value of pervasive healthcare, we believe that this technology will eventually take hold in the marketplace. The continuing commoditization of computing and communications hardware and software and the maturation of sensory devices will make these technologies ever more powerful and affordable. The tremendous economic potential and the resulting improvement in health outcomes are not being overlooked by high-tech companies such as Intel, who are making huge investments and also bringing their aggressive cultures into this market. At the same time, pressured by rising healthcare costs, governments at all levels and private employers as well have finally begun to take actions. The Connecting for Health initiative by the National Health Service (NHS) in the U.K., the chronic care improvement programs at Medicare and the VA, and the telehealth pilot at the Henry Ford Health

System are just a few examples. However, the real catalyst for the broad adoption of pervasive healthcare will be the aging baby boomers. Unlike previous generations, baby boomers are well educated, health-conscious, self-reliant, and truly demanding. They are and will continue to be the key driving force behind consumer-centered healthcare. Because they have been surrounded by computers, the Internet, and cell phones during most of their working years, this generation is also much more technology savvy. Perhaps most important, the baby boomers' control of purse strings is expected to expand significantly as employers and governments continue to shift a larger share of healthcare costs to consumers. This combination of self-determination, technology savviness, aging-related healthcare needs, and spending power among baby boomers will ultimately help make pervasive healthcare a way of life.

References

1. Administration on Aging, Department of Health and Human Services. *Statistics on Aging in the 21st Century.* See aoa.gov/prof/Statistics/future_growth/aging21/health.asp, 2004.
2. Fox, Sosannah and Rainie, Lee. *Vital decisions: How Internet users decide what information to trust when they or their loved ones are sick.* Pew Internet and American Life Project, 2002.
3. Weingarten, Scott R. et al. Interventions used in disease management programs for patients with chronic illness—which ones work? Meta-analysis of published reports. *British Medical Journal,* October 26, 2002, 325: 925.
4. Wegerich, Stephan. Similar-based modeling of vibration features for fault detection and identification. *Sensor Review,* June 2005, 25: 2, 114–122.
5. Speer, Jack. Health care costs pose challenges for GM. *All Things Considered,* National Public Radio, April 8, 2005.
6. Kaiser Family Foundation and Health Research and Education Trust. *Employer health benefits: 2004 annual surveys.* Henry J. Kaiser Foundation, 2004.
7. New England Healthcare Institute. *Remote physiological monitoring: Innovation in the management of health failure.* July 2004.
8. Partners Telemedicine. See www.connected-health.org.
9. Fischman, Josh. House calls. *U.S. News and World Report,* August 1, 2005.
10. Health Hero Networks, Veterans Health Administration announces better patient care with significant cost savings in telemedicine project. See www.healthhero.com/press/press_releases/pr_06_04_01.html.
11. Myer, Marlis, Kobb, Rita, and Ryan, Patricia. Virtually healthy: chronic disease management in the home. *Disease Management,* 2002, 5(2): 87–94.
12. National Center for Health Workforce Analysis, U.S. Department of Health and Human Services. *Projected supply, demand, and shortage of registered nurses: 2000–2012,* July 2002.
13. Centers for Medicare and Medicaid Services. *Minimum nurse staffing ratios in nursing homes,* April 2002.
14. Groves, Reggie. An aid to cost-effectively delivering the right care at the right time. See www.academyhealth.org/nhpc/2004/groves.ppt.

15. Kohn, K.T., Corrigan, J.M., and Donaldson, M.S., eds. *To err is human: Building a safer healthcare system.* Institute of Medicine, National Academies Press, 2000.
16. Beers, M.H., Sliwkowski, J., and Brooks, J. Compliance with medication orders among the elderly after hospital discharge. *Hospital Formulary,* July 1992, 27(7): 720–724.
17. Partnership for Solutions: Johns Hopkins University, Baltimore, MD, for the Robert Wood Johnson Foundation (September 2004 update). *Chronic conditions: Making the case for ongoing care.* 2004.
18. Alzheimer's Association. See www.alz.org.
19. Cherry, Julie et al. Opening a window of opportunity through technology and coordination: a multi-site case study. *Telemedicine Journal and e-health* 2003, 9(3): 265–271.
20. Centers for Medicare and Medicaid Services. See www.cms.gov.
21. Centers for Medicare and Medicaid Services. Medicare announces its intention to expand coverage of implantable cardioverter defibrillators. 2003. See www.medicare.gov.
22. Arndt, Michael. Picking up the pace. *Business Week,* March 6, 2006.
23. Health Hero Network. Health Buddy system. See www.healthhero.com.
24. Cheitlin-Cherry, Julie, and Ehrman, Jonathan K. Henry Ford experience using the Health Buddy system. *Tenth Annual Meeting & Exposition of the American Telemedicine Association,* Denver, CO, April 17–20, 2005.
25. Royal Philips. U.S. study shows chronic disease patients embrace Philips personalized TV-based interactive healthcare platform to manage disease from home. See www.newscenter.philips.com/About/article-15024.html.
26. Living Independently, Inc. Quietcare: The advanced early warning system that helps people live with greater safety and independence in their own homes. See www.quietcaresystems.com.
27. Lee, Kenneth S. Policy recommendation. *White House Conference on Aging.* March 12, 2004, Philadelphia, PA.
28. Krugman, Paul. The medical money pit. *New York Times,* April 15, 2005.
29. Boehm, Elizabeth W. *Who Pays for Healthcare Unbound.* Forrester Research, July 2004.

Index

A

AbleLink system, 84
Accuracy experiments, 97
Activities of daily living (ADL), 80–81.
  See also Assisted cognition (AC)
    home activities monitoring, 91–100
    independent living technologies and,
      127–129
    modeling, 92–93
    QuietCare system and, 292–294
    recognition
        coarse-grained, 94
        fine-grained, 94–95
    tracking and support, 82–84
Acute conditions, 6–7
    management and early discharge,
      124–127
ADT security, 293–294
Advancing Chronic Care through
  E-Health Networks and Technologies
  (ACCENT), 286
Adverse incident reporting, 230–231,
  272–273
Aging of the population, 4–6, 205–206.
  See also Assisted cognition (AC)
    activities of daily living (ADL) and,
      80–81
AirWatch, 123–124
Ambulatory blood pressure monitoring
  (ABPM), 146–148
Ambulatory cardiac monitoring, 148–149
Animal ethics committees (AECs), 234
Anoto, 65–66
Artificial intelligence (AI)
    assisted cognition (AC) and, 80–81
    representing knowledge, 29–30
Assisted-care communities and institutions
    and technology, 15–16

Assisted cognition (AC)
    accuracy experiments, 97
    activities of daily living (ADL) tracking
      and support in, 82–84
    ADL recognition in, 94–95
    Autominder system, 83
    COACH system, 83
    definition, 80–81
    home activities, 91–100
    improving robustness in, 97–100
    IMP system, 82
    location, navigation, and wayfinding
      in, 81
    modeling activities and, 92–93
    modeling choices and, 95–96
    Opportunity Knocks and, 84–91
    outdoor navigation and, 84–91
    PDA-based reminding systems, 84
    research on, 81–84
    smart environments and, 175–179
    technologies, 172–175, 209–210
    wandering alert systems, 82
    wearable activity recognition
      systems, 84
Assistive technologies
    cognitive, 172–175
    communication, 162–165
    design considerations for, 179–184
    mobility, 165–168
    pervasive computing and, 29, 100–101
    smart environments, 175–179
    for specific impairments, 180–181
    user evaluations of, 257–258
    wayfinding, 81, 169–172
    wheelchair and walker, 166–168
Auditory technologies, 179, 182
Augmentative and alternative
  communication (AAC), 162–165
Authentication, user, 41, 69

Authorization in pervasive computing applications, 41–42
Automated planning and pervasive computing, 37–39
Automation, home, 178–179, 182–183
Autominder, 83
AWARE architecture, 67–68
AWARE home, 82–83

B

Baby boomers, aging of, 4–5, 276
Back-end systems, 108
Barista, 81, 84, 94–95
Bayesian networks (BN), 31–32
    dynamic (DBN), 34, 36, 96, 99–100
Blood pressure, 116–117, 120
    monitoring, ambulatory (ABPM), 146–148
Bluetooth, 87, 279
Bodybugg calorie management system, 117–118
BodyMedia, 279
Business, healthcare
    case studies, 285–286
    continuous care model and, 284–285
    key drivers in, 281–284
    and Medicare, 285–286
    return on investment (ROI) and, 280–281
    value of, 280–281, 296–297

C

Campus Aware, 81
Cardiac emergencies, 125–127, 287–289
CardioPocket, 126
Caregivers and pervasive computing, 28
Care Management for High-Cost Beneficiaries (CMHCB), 286
CareNet Display, 270–272
Case studies
    CareNet Display, 270–272
    chronic care initiatives at Medicare, 285–286
    cooker monitor, 265–267
    sensor network, 267–268
    Turvy software, 268–270
Caveat user, 42
Cell phones, 24, 25, 87
Centers for Disease Control, 6, 7, 292
Centre for Reviews and Dissemination (CRD), 221

Chronic conditions, 6–7, 115–116, 137, 279
    continuous care model and, 284–285
    management, 119–124
    wearable biomedical systems for, 145–151
    Web-based applications use by patients with, 203–204
Classification of sensors, 145–146
Clinical decision support systems (CDSS), 55
Clinical studies, 232–233, 234–235
COACH (Cognitive Orthosis for Assisting aCtivities in the Home), 83, 177
Coarse-grained ADL recognition, 94
Code of Federal Regulation (CFR), 225, 231
Cognition, assisted. *See* Assisted cognition
Cognitive behavioral treatment (CBT), 117
Cognitive disabilities and evaluation programs, 253–254, 261–262
    case studies, 265–267
COGORTH, 175
Collaboration and coordination among hospital workers, 56–57, 73–74
    groupware and, 66–68
Commercial applications of pervasive healthcare
    Health Buddy system, 289–291
    Motiva system, 291–292
    QuietCare, 292–294
    remote cardiac monitoring, 287–289
Commercial off-the-shelf (COTS) hardware and software, 231
Communication assistance, 162–165, 182
Community-based healthcare, 9–11
Computerized patient record (CPR) systems, 193–195
Computerized physician order entry (CPOE), 55, 156
Conformity, 228
Congestive heart failure (CHF), 122
Connectivity and pervasive computing, 24
Constraint-satisfaction processing, 38
Context-sensitive reminding, 176–177
Context switching, rapid, 58, 201
Continuing care retirement communities (CCRCs), 16
Continuous care model and chronic conditions, 284–285
Continuous glucose monitoring, 149–151
Continuous monitoring
    healthcare professional view, 139
    patient view, 138–139
    pharmaceutical industry view, 139
    as a way of life, 139–140
Conventional engineering design, 247–248

Cooker monitors, 265–267
Coordination and collaboration among
	hospital workers, 56–57, 73–74
	groupware and, 66–68
Costs
	of bad lifestyle choices, 116
	healthcare, 136–137, 276–277, 295
	as a key driver in healthcare
		management, 281–284
	reduced by pervasive computing, 28
Crossbow Mica Mote radio, 92
Cryptography, 40
Cyber Crumbs, 171

D

Data
	collection, 199
	issues regulation, 231–232
Decentralization and pervasive computing,
	24
Decision making
	Markov process, 36, 83
	mobile and personal health/wellness
		management systems and, 107–108
	monitoring health status and, 208
Dementia, 253–254, 261–262
	cooker monitors and, 265–267
Dementia Voice, 83
Demographics and healthcare, 4–8
Design
	considerations for assistive technology,
		179–184
	pervasive healthcare technology,
		244–245
	user-centered, 248–250
	user-centered engineering,
		246–247
Desktop computers
	heterogeneity between mobile devices
		and, 58, 60
	isolated, 60
	seamless context-aware systems, 63
Devices, medical, 22–23, 25–26, 218
	advancements in, 277–279
	classes of, 225, 226–227
	clinical studies, 232–233, 234–235
	conformity and, 228
	data issues in, 231–232
	definitions, 222–223
	essential principles of, 223–225
	ethics and, 233–234
	fitness and wellness, 278–279
	innovation routes for, 235–237

installation in users' homes, 260–262
	Medical Devices Directive (MDD),
		222–223
	miniaturization and consumerization
		of, 278
	postmarket surveillance, vigilance, and
		adverse incident reporting, 230–231
	standards, 229–230
	user evaluations of, 257–258, 272–273
Diabetes mellitus, 116, 138, 282
	continuous glucose monitoring and,
		149–151
Diagnostic Related Groups (DRG), 220
Digital Angel, 82
Digital imaging and communication in
	medicine (DICOM), 231–232
Digital pens, 65–66
Disabilities. *See* Assistive technologies
Discharge, early, 124–127
Disease
	acute, 6–7, 124–127
	among aging baby boomers, 4–7
	assistive technologies for specific,
		180–181t
	chronic, 6–7, 115–116, 119–124, 145–151,
		203–204, 284–285
	gender and, 7–8
	socioeconomic status and, 7–8
Distribution, probability, 31, 32f
Diversification and pervasive computing, 24
Dynamic Bayesian networks (DBNs), 34, 36,
	96, 99–100
Dynamic models in pervasive computing,
	33–34

E

Early discharge, 124–127
Electronic health record (EHR), 12, 53–54
Elite Care, 82
Emergency
	override capabilities, 201
	response systems, personal, 292–294
ENABLE project, 265–267
Enabling technologies for pervasive
	healthcare, 140–142, 277–280
Encryption, 40–41
Engine, inference, 88–90
Engineering design, user-centered,
	246–250
Environments, smart, 175–179
Error
	prevention and handling, 200
	reporting, 230–231

Ethics
  evaluation program, 253–254
  medical device development, 233–234
Evaluation. *See* User evaluations
Evaluation heuristic, 38
Evidence-based medicine (EBM), 110, 120,
  125, 130, 219
Executive functions, 173
Experiments, accuracy, 97

F

Failure backups, 201
Filtering, particle, 34
Final evaluations, 263–264
Fine-grained ADL recognition, 81, 84, 94–95
First evaluations, 258–260
Fitness and wellness technologies, 113–115,
  278–279
Fitsense, 279
Flexible spending accounts (FSA), 116
Food and Drug Administration (FDA), 222,
  228, 246
FRWD Outdoor Sports Computer, 114–115
Functional independence, 210

G

Gender and disease, 7–8
General Motors, 139–140, 281
General Packet Radio System (GPRS), 85–87
Georgia Tech Wearable Motherboard,
  152–155
Global Harmonization Task Force (GHTF),
  222
Global positioning system (GPS), 24, 26, 80,
  170–171
  outdoor navigation and, 84–85
GlucoWatch Analyzer, 150
GlucoWatch Biographer, 150
Goals of pervasive computing, 23–24,
  192–193
Groupware, hospital, 66–68
Guardian RT System, 150–151
GUIDE, 81

H

Hardware, mobile, 60
Hash function, 42
Head-mounted displays (HMDs), 211
Health Buddy system, 289–291

Healthcare. *See also* Business, healthcare;
  Pervasive healthcare
  community-based, 9–11
  costs, 136–137, 276–277, 295
  demographics, disease, and disability
    and, 4–8
  home-based, 9–11
  industry challenges, 137–140
  outpatient, 11–13, 281–282
  patient-centric pervasive, 140–145
  pervasive computing applications to,
    28–29, 137–140
  spending, 136–137
  stakeholders in, 108–110, 183–184, 280,
    283–284
  technology and, 8, 108–110, 277–280
  usage models in pervasive, 111–129
Health Hero Network, 289–291
Health Insurance Portability and
  Accountability Act (HIPAA), 232
Health technology assessment (HTA)
  bodies, 221
  evidence-based medicine (EBM) and,
    219
  reimbursement issues and, 220
Hearing impairments, 179, 182
Hidden Markov models (HMMs), 33–34, 36,
  95–96
Holter monitors, 148–149
Home activities
  ADL monitoring and, 91
  sensing using RFID, 92
Home automation systems, 178–179, 182–183
Home-based healthcare, 9–11
  ADL tracking and support, 82–84
  assisted cognition (AC) in, 81–84
  installation of systems for, 260–262
  location, navigation, and wayfinding,
    81–82
  wandering alert systems, 82
Home Energy Tutor, 267–268
Home monitoring, 10–11
Hospitals
  challenges for computer technology in,
    55–59, 70–75
  collaboration and coordination in,
    56–57, 66–68
  computer technology use in, 13–15,
    52–55
  early discharge, 124–127
  electronic health record (EHR), 12, 53–54
  groupware, 66–68
  human-computer interaction
    technology in, 65–66
  information systems, 53–54

mobile and pervasive computing in, 60–63

mobility among heterogeneous devices in, 57–58

nomadic work in, 55–56, 73

patient monitoring in, 52–53

picture archiving and communication systems (PACS), 54–55

rapid context switching in, 58

user authentication and security in, 69

wireless patient monitoring in, 52–53, 64–65

HTML (Hypertext Markup Language), 27

HTTP (Hypertext Transfer Protocol), 27

Human–computer interaction (HCI)

analysis of pervasive applications in healthcare, 203–210, 212–213

chronic conditions and, 203–204

the elderly and, 205–206

in healthcare and pervasive computing, 193–195, 210–212

monitoring health status and, 207–208

PDA use by physicians and, 56, 61, 203–204

safety promotion and, 208–209

smartcard technology and, 204–205

and the system development life cycle, 195–196

technology, 65–66, 80

usability engineering and, 197–202

workflow modeling, 202

I

iMetrikus, 122–123

IMP, 82

Implantable cardioverter defibrillators (ICDs), 287–289

Independence, functional, 210

Inference

engines, 88–90

probabilistic, 32

Information overload, 60

Information processing, pervasive, 143–145

Information systems, hospital, 53–54

Information technology (IT)

in assisted-care communities and institutions, 15–16

challenges in hospitals, 55–59, 70–75

growth in spending on, 15

and the healthcare model, 8–17

in home- and community-based health, 9–11

in hospitals, 13–15

integration with the physical world, 59

mobility among heterogeneous devices in, 57–58, 60

in outpatient care, 11–13

studies and testing of implementation of, 17

use among young versus older people, 7

Informed consent, 253–254

Initial user surveys, 254–256

Inspection, usability, 200–202

Installation of systems in users' homes, 260–262

Institutional review boards (IRBs), 234

Integrated multiparameter biomedical monitoring system, 151–156

Integration of the digital and physical worlds, 59

Integrity in pervasive computing applications, 42

Intellectual property protection, 258

Intelligent biomedical clothing (IBC), 131

Intelligent Hospital project, 67

Intel Research, 92

Interim evaluations, 263

Internet

and chronically ill healthcare consumers, 203–204

health monitoring and, 279

information resources on, 9–10, 276

weight management, 117–118

INTERNIST system, 30

Interviews, user, 256

iPod, 26

Isolated devices, 60

IST Vivago WristCare, 128–129

J

Java bytecode, 27

K

Kalman filtering, 89–90

Key strategies in cryptography, 40–41

Knowledge-intensive applications and artificial intelligence (AI), 29–30

L

Learning

machine, 34–35

reinforcement, 36–37

supervised, 35
unsupervised, 35–36
Life cycles, system development, 195–196
LifeShirt System, 122–123
Likert scales, 255
Location-sensing technologies, 81
Long-distance communication, 108

M

Machine learning, 34–37
Market forces behind the adoption of
  pervasive healthcare, 281–284
Markov decision process, 36, 83, 95–96
Medical savings accounts (MSA), 116
Medicare, 285–286
Medication errors, 11
Medtronic CareLink Network, 288–289
Medtronic reports, 282, 283
MemoJog, 175
Memory
  aids, electronic, 174–175
  prospective, 173
  smart environments and, 176–177
Mild cognitive impairment (MCI), 80
Miniaturization and consumerization of
  medical devices, 278
Minimally invasive sensors, 145–146
MIT House_n project, 83
Mobile and personal health/wellness
  management systems
  acute disease management, 124–127
  chronic disease management,
    119–124
  fitness and wellness, 113–115
  independent living, 127–129
  key building blocks, 107–108
  risk management in, 115–119
  sensors, 107
  stakeholders in, 108–110
  usage models, 111–129
  wheelchairs and, 165–168
Mobile devices
  hardware, 60
  heterogeneity among, 57–58, 60
  pervasive computing and, 60–63
  variety of, 24–25
Mobile digital music players, 25
Mobile phones, 24, 25, 87
Mobility
  among heterogeneous devices,
    57–58
  assistance, 165–168
  wheelchairs and, 166–168

Modeling
  activities, 92–93
  acute disease management and early
    discharge, 124–127
  choices, 95–97
  chronic disease management, 119–124
  fitness and wellness, 113–115
  independent living, 127–129
  risk management and prevention,
    115–119
  similarity-based, 280
  usage, 111–129
  workflow, 202
Monitoring
  ambulatory blood pressure, 145–148
  ambulatory cardiac, 148–149
  cardiac, 287–289
  continuous, 138–140
  continuous glucose, 149–151
  health status and HCI, 207–208
  home, 10–11
  integrated multiparameter biomedical,
    151–156
  patient, 52–53, 64–65, 138–139
  remote, 12–13, 121–122, 287–289,
    291–292
  wireless, 52–53, 64–65
Moore's Law, 278
Morbidity and mortality, 6–7. *See also* Disease
Motiva system, 291–292
Multidisciplinary Assessment of Technology
  Centre for Healthcare (MATCH), 218,
  235–237
MYCIN system, 30

N

National Center for Health Statistics, 16
National Health Service (United Kingdom),
  221
National Institute for Health and Clinical
  Excellence (NICE), 221
National Institutes of Health and Aging,
  292–293
National Sleep Foundation, 116
Navigation technologies, 81–82
  outdoor, 84–91
  wheelchair, 167–169
Near-optimal policy, 36
Network protocols, 27
New Millennium Research Council, 7
New product development (NPD) methods,
  235–237
Nokia 6600 cell phone, 87

Nomadic work, 55–56, 73
Noninvasive sensors, 145–146, 149–150
Nonprofit disease-oriented lay
  organizations, 9–10
Nursebot, 81–82

O

Obesity, 116, 117–119, 279
OnStar Vehicle Diagnostics, 139–140
Opportunity Knocks, 81, 177–178
  applications of, 84–85
  current status, 91
  inference engine, 88–90
  system architecture, 85–88
  usage scenarios, 85
Outcome measures, validated, 256–258
Outdoor navigation
  current status, 91
  inference engine, 88–90
  personal aids, 84–85
  system architecture, 85–88
  usage scenarios, 85
Outpatient care and technology, 11–13,
  281–282
Overload, information, 60

P

Pagers, 25
Particle filtering, 34
Patients
  -centric pervasive healthcare, 140–145
  monitoring, 52–53, 64–65, 291–292
  out-, 11–13, 281–282
  records, electronic, 12, 53–54, 193–195
  as stakeholders in healthcare, 108–110
  view of continuous monitoring,
    138–139
PDAs (personal digital assistants), 22–23,
  25, 106
  -based reminding systems, 84
  heterogeneity between desktop
    computers and, 58
  Opportunity Knocks and, 87
  usability testing and, 199–200
  use by physicians, 56, 61, 203
Peace of mind and pervasive computing, 28
PEAT system, 84
Pens, digital, 65–66
Performance measurements, sports, 114
Personal emergency response systems,
  292–294

Personal navigation system
  current status, 91
  inference engine, 88–90
  public transportation and, 84–85
  system architecture, 85–88
  usage scenarios, 85
Pervasive computing
  acceptance of new technologies in,
    194–195
  acute disease management, 124–127
  analytics and, 279–280
  applications to healthcare, 28–29
  assistive technology and, 29, 100–101
  automated planning using, 37–39
  Bayesian networks (BN) and, 31–32
  chronic disease management, 119–124
  current trends in, 59–69
  defined, 22
  devices, 22–23, 25–26, 218, 221–235,
    257–258, 277–279
  dynamic models, 33–34
  and the elderly, 205–206
  elements of, 25–28, 50–52
  fitness and wellness, 113–115
  goals of, 23–24, 192–193
  independent living, 127–129
  intelligent applications of, 29–39
  as a key driver in healthcare business,
    282
  machine learning and, 34–37
  mobile and, 60–63
  mobile phones as, 24
  networks, 279
  principles of, 24–25
  privacy and security in, 39–42, 200
  probabilistic reasoning and, 30–31
  promoting functional independence,
    210
  quality of care and, 28, 282–283
  representing knowledge, 29–30
  risk management and prevention,
    115–119
  risk management in, 115–119
  safety promotion using, 208–209
  standards and protocols, 26–28
  usability engineering and, 197–202
  usage models, 111–129
Pervasive healthcare
  commercial applications of, 287–294
  continuous monitoring and, 138–140
  enabling technologies, 140–142, 277–280
  future challenges and opportunities,
    156–158, 294–296
  life cycles of unit operations in,
    140–141

market forces behind adoption of, 281–284

new and mixed modalities of interaction in, 210–212

patient-centric approach to, 140–145

system requirements, 142–143

technology trends in, 143–145

wearable biomedical systems in, 145–151

Philips Telemonitoring Services, 121–122

Physical world integration with digital technology, 59

Physicians. *See also* Workers, healthcare collaboration and coordination among, 56–57

nomadic work by, 55–56, 73

use of PDAs by, 56, 61, 203

Picture archiving and communication systems (PACS), 54–55, 61

Polar WM42/WM41 heart rate monitor, 118–119

Population demographics, 4–6

Postmarket surveillance, 230–231

Power-assist manual wheelchairs, 166

Premarket approval (PMA), 228

Preproduction evaluations, 264

Prevention technologies and risk management, 115–119

Privacy and security in pervasive computing, 39–42, 200

Private keys, 40–41

Probabilistic inference, 32

Probabilistic reasoning, 30–31

Probability distribution, 31–32

Prospective memory, 173

Psychological Impact of Assistive Devices Scale (PIADS), 257–258

Public transportation, 84

GPS and, 171

sensing using, 92

Randomized controlled trials (RCTs), 234–235

Rao-Blackwellised particle filters, 89–90

Rapid context switching, 58, 201

Reasoning, probabilistic, 30–31

Regulation of medical devices clinical studies, 232–233, 234–235

conformity and, 228

data issues in, 231–232

definitions in, 222–223

device class and, 225, 226–227

essential principles in, 223–225

ethics and, 233–234

evaluation and, 272–273

postmarket surveillance of, 230–231

standards in, 229–230

Reimbursement issues in healthcare, 220

Reinforcement learning, 36–37

Remote monitoring, 12–13, 121–122, 287–289, 291–292

Research assisted cognition, 81–84

current trends in pervasive computing, 59–69

ethics committees (RECs), 233–234

Return on investment (ROI), 280–281

Risk management and prevention technologies, 115–119

reduction and pervasive computing, 29

Robot technologies, 81–82

Robustness, improving assisted cognition (AC) technologies', 97–100

Q

Quality assurance (QA), 229

Quality of care and pervasive computing, 28, 282–283

Quebec User Evaluation of Satisfaction with Assistive Technology (QUEST 2.0), 257

Questionnaires, user, 256

QuietCare system, 292–294

R

Radio frequency identification (RFID) technology, 25–26, 80, 131

ADL recognition and, 94–95

S

Safety promotion, 208–209

Seamless modal switching, 201

Security and privacy in pervasive computing, 39–42, 200

SenseWear technology, 118

Sensors classification of, 145–146

fabric as, 155–156

networks, 151–152, 267–268

technology, smart, 64–65, 107, 140–142

user evaluations, 267–268

wayfinding, 169–172

SHL Telemedicine, 126–127

Short-distance communication methods, 107
Similarity-based modeling (SBM), 280
Simplicity and pervasive computing,
  24–25
Sleep, poor, 116
Smartcard technology, 204–205
Smart environments, 175–179
Smart sensor technology, 64–65, 107
Smart Shirt, 152–155
Socioeconomic status
  and disease, 7–8
  and Internet access, 9
Speech recognition, 65, 211
Stakeholders in healthcare, 108–110, 183–184,
  280, 283–284
Standards
  medical device, 229–230
  and protocols of pervasive computing
    devices, 26–28
State-space search, 38
Stryker Endosuite system, 65
Sudden infant death syndrome (SIDS),
  155–156
SunTech Medical, 146
Supervised learning, 35
Surveillance, postmarket, 230–231
Surveys, user, 254–256
Suunto, 114
System development life cycles (SDLC),
  195–196

T

TCP/IP (Transmission Control Protocol/
  Internet Protocol), 27
Teams
  formation, 250–251
  recruiting and selecting evaluators for,
    251–254
Telemedicine, 12–13
Telemonitoring, 121–122, 291–292
Testing, usability, 198–200
Textiles
  as sensors, 155–156
  in wearable biomedical systems, 152
Therapeutic Goods Administration (Japan),
  222
Tivoli device management system, 26
To Err Is Human, 15
Trans-European Network Home Care
  Management System (TEN-HMS), 122
Trials, clinical, 232–233, 234–235
Tunstall Lifeline 4000+, 129
Turvy software, 268–270

U

Ubiquitous computing, 22–23.
  See also Pervasive computing
Ultra-Wide-Band (UWB), 279
Unified modeling language (UML), 202
Unobtrusiveness, 200
Unsupervised learning, 35–36
Usability engineering, 197–202, 230
Usability inspection, 200–202
Usability requirements for risk prevention,
  117
Usability testing, 198–200
Usage models
  acute disease management and early
    discharge, 124–127
  chronic disease management, 119–124
  fitness and wellness, 113–115
  independent living, 127–129
  in pervasive healthcare, 111–113
  risk management and prevention,
    115–119
User evaluation. *See also* Users
  adverse incident evaluation and,
    257–258, 272–273
  case studies, 265–273
  conventional engineering design
    method, 247–248
  in design, implementing, 250
  ethics, 253–254
  final, 263–264
  first, 258–260
  informed consent and, 253–254
  initial surveys, 254–256
  intellectual property protection and,
    258
  interim, 263
  by people with cognitive disabilities,
    253–254, 261–262, 265–267
  preproduction, 264
  recruiting and selecting evaluators for,
    251–254
  roles of, 245–246
  supporting users and devices, 262–263
  team formation and, 250–251
  user-centered design and, 248–250
  in user-centered engineering design,
    246–250
  validated outcome measures and,
    256–258
Users. *See also* User evaluation
  authentication and security, 41, 69
  controllability of systems by, 201
  interfaces (UI), 108
  usability testing and, 197–202

V

Validated outcome measures, 256–258
Versus Technology, 68
Veterans Health Administration (VA),
   281–282, 285
Vigilance, 230–231
Virtual Usability Laboratory (VUL), 204
Vocera Communications, 68
Voice output communication aids (VOCAs),
   162–165

W

Walkers, 165–168
Wandering alert systems, 82
Wayfinding technologies, 81, 169–172
Wearable activity recognition systems,
   84, 279
      ambulatory blood pressure, 146–148
      ambulatory cardiac monitoring,
         148–149
      continuous glucose monitoring, 149–151
      integrated multiparameter, 151–156
      Smart Shirt, 152–155
      textiles in, 152
Web sites, health-related, 9–10
WelchAllyn, 146

Wheelchairs, 165–168
White-coat hypertension, 146
WiFi networks, 60, 80, 279
Wireless networks
      advances in, 279
      communication in hospitals, 61, 73
      patient monitoring, 52–53
      sensor networks for patient monitoring,
         64–65
Wizard of Oz technique for evaluations,
      268–270
Workers, healthcare. *See also* Physicians
      collaboration and coordination among,
         56–57, 66–68, 73–74
      nomadic work by, 55–56, 73
      view of continuous monitoring, 139
Workflow modeling, 202
World Health Organization, 8, 116

X

Xerox PARC, 59
XML (eXtensible Markup Language), 28

Z

ZigBee/IEEE 802.15.4 standard, 130, 156, 279